VARIATIONAL PRINCIPLES AND THE NUMERICAL SOLUTION OF SCATTERING PROBLEMS

VARIATIONAL PRINCIPLES AND THE NUMERICAL SOLUTION OF SCATTERING PROBLEMS

SADHAN K. ADHIKARI
Institute of Theoretical Physics
Universidade Estadual Paulista
São Paulo, Brazil

A Wiley-Interscience Publication
JOHN WILEY & SONS, INC.
New York / Chichester / Weinheim / Brisbane / Singapore / Toronto

This text is printed on acid-free paper.

Library of Congress Cataloging in Publication Data:

Adhikari, Sadhan K., 1948–
Variational principles and the numerical solution of scattering problems / by Sadhan K. Adhikari.
p. cm.
Included bibliographical references and index.
ISBN 0-471-18193-5 (cloth: alk. paper)
1. Scattering (Physics) -- Numerical analysis. 2. Variational principles. I. Title.
QC794.6.S3A23 1998
539.7'58 -- dc21 97-24719

Printed in the United States of America

10 9 8 7 6 5 4 3 2 1

To my
father, the late Mr. Nalini Ranjan,
mother, Dr. Mira,
wife, Ratnabali, and
son, Avishek

CONTENTS

FOREWORD

Variational principles were introduced into quantum scattering theory by L. Hulthén, W. Kohn, and J. Schwinger during the 1940s, and it soon became clear that the application of these principles provided an effective method of obtaining approximate numerical solutions of the scattering problems encountered in atomic, molecular, and nuclear physics. In the half century since their inception, variational approximations have been the subject of intense theoretical analysis, important new variants have been discovered, and many successful applications have demonstrated their utility. Despite the importance of the subject, no monograph on variational methods has been published for nearly two decades, and the new advances are contained in research articles scattered over several journals, making it difficult for students to obtain an overview of the present position. The decision of Professor Adhikari to remedy this lacuna in the literature is greatly welcomed.

Professor Adhikari has published many important papers on quantum scattering theory and its applications, and his book, with K. L. Kowalski, titled *Dynamical Collision Theory and Its Applications* is one of the best introductions to modern theoretical developments. His present monograph, written with the authority to be expected from his experience, provides a lucid account of all the main variational methods together with examples of their application. Students wishing to learn the foundations of the subject are well served and, in addition, much is discussed that will interest more senior researchers. This book deserves to be read widely by all those concerned with scattering theory and its applications, and I wish it every success.

B.H. BRANSDEN

Professor of Physics
University of Durham
Durham, UK

PREFACE

In recent years there has been considerable progress in the development of sophisticated numerical methods for treating nuclear, atomic, and molecular scattering problems involving many, but finite, active degrees of freedom. After certain approximations, most of these problems are modeled by a single (or a set of coupled) effective differential or integrodifferential Schrödinger equation(s). Unfortunately, in many problems of physical interest the number of the coupled equations could be very large, which prohibits a direct solution. Hence, further approximations are needed for their numerical solutions.

Imposition of scattering boundary conditions is numerically the most difficult task in obtaining a solution of the differential or integrodifferential Schrödinger equation for scattering problems in the presence of many open channels. In such cases, momentum-space scattering integral equations developed by Lippmann and Schwinger confer a great advantage in incorporating them automatically. For simple problems of scattering, either the differential Schrödinger equation or integral Lippmann–Schwinger equation can be solved numerically.

However, the need for intelligent solution schemes for realistic scattering problems can hardly be overemphasized. Variational principles have proved to be very valuable for the numerical solution of scattering problems. They have frequently been used in the numerical solution of collision models in atomic, molecular, and nuclear physics. The availability of modern supercomputers have facilitated solution of these models. Nevertheless, descriptions of this progress and its logical development are available only in some review articles, which have emphasized primarily the contributions and achievements of the authors of these articles. They do not contain a comprehensive and coherent development of this progress in nuclear, atomic, and molecular scattering problems. There have been many advances since publication of the early monographs on variational methods by Arthurs (1970), Demkov (1963), Mercier (1963), and Michlin (1964). A more recent monograph by Nesbet (1980) deals with a special class of variational principles and their applications to electron–atom scattering. Most of the major benchmark calculations with variational methods were performed following publication of Nesbet's monograph.

Thus, in this rapidly growing field, the present book aims to fill the gap between these older monographs and the present state of affairs. The objective is to present a coherent treatment of most of these variational and some nonvariational methods for solving scattering problems. We also describe the considerable progress made in recent years in the numerical solution of realistic collision problems. The principal aim is to emphasize the role of variational principles in this development. However, general numerical solution procedures to be used in applications of the variational principles are also described for the sake of completeness. This includes, for example, discretization of infinite integrals, development of nonsingular scattering equations for dealing with the troublesome principal-value prescription, and iterative solution of scattering equations. In addition to an account of the formal development of the numerical methods and variational principles, we present a complete description of their applications to model test problems and realistic benchmark calculations in nuclear, atomic, and molecular physics.The formal development presented in this book is based primarily on the momentum-space scattering integral equations of Lippmann–Schwinger type. However, most numerical applications can be made in either configuration or momentum space.

Most of the book is at a level accessible to graduate students and nonspecialists and could be used as a graduate text for students who have had a two-semester quantum mechanics course including a couple of chapters on scattering theory. A formal course in scattering theory is not a prerequisite. A self-contained discussion of scattering theory appropriate for the present development is given in Chapter 1. The present discussion of scattering theory and variational principles is directed toward numerical applications. Hence, many formal and mathematical aspects of the subject are omitted. The book would also be useful to research workers in nuclear, atomic, and molecular physics interested in performing numerical scattering calculations. Chapters 1 to 4 and Sections 6.1 to 6.4 are written in a pedagogic style appropriate for second-year graduate students. The remaining parts of the book are intended for research workers. In Chapter 5 we present an extension of variational principles to various realistic multichannel problems. In Sections 6.5 to 6.11 we present an account of numerical applications to realistic scattering problems of the methods presented in this book with up-to-date references. However, the book should not be considered as a comprehensive review of all the topics discussed, such as electron–atom, electron–molecule, atom–diatom scattering, and molecular photodissociation. Our aim is to cover only the underlying numerical methods.

The author is pleased to extend thanks and gratitude to his friends and collaborators R. D. Amado, A. C. Fonseca, E. Gerjuoy, W. Glöckle, M. A. F. Gomes, M. S. Hussein, K. L. Kowalski, G. Krein, L. Tomio, and D. G. Truhlar for encouragement, support, and assistance.

Most of the research reported in this book involving the author was funded by Conselho Nacional de Desenvolvimento–Cientifico e Tecnologico of Brazil over the last two decades. The author gratefully acknowledges a John Simon Guggenheim Memorial Foundation Fellowship during the course of writing this book.

SADHAN K. ADHIKARI

São Paulo, 1998

VARIATIONAL PRINCIPLES AND THE NUMERICAL SOLUTION OF SCATTERING PROBLEMS

CHAPTER 1

SCATTERING THEORY

1.1 INTRODUCTION

Scattering processes are very common in many areas of physical science, such as in particle, nuclear, atomic, molecular, and chemical, physics, and in chemistry just to mention a few. Usually, these processes are formulated by use of the nonrelativistic quantum-mechanical Schrödinger equation for a fixed number of constituents as its dynamical basis. The fundamental interactions between the constituents are often represented by simple potentials. The potentials are in many cases taken to be model ones that can be handled easily. There have also been attempts to derive these potentials from fundamental many-body theory after making approximations. In this second category are the meson-exchange field-theoretic potentials among hadrons and energy-dependent cluster–cluster potentials of nuclear, atomic, and molecular physics derived from multiparticle Schrödinger equations. Once the potentials among the constituents are decided upon, the Schrödinger equation for the constituents is solved consistent with the boundary conditions of the problem [1].

This last task of the numerical solution of the problem is a nontrivial but nevertheless important one. Only after numerical solution of the scattering model can physical observables be determined. However, there are many ways of finding the solution. The purpose of this book is to make some of the solution procedures understandable to nonspecialists. The emphasis is on the use of momentum-space methods and variational principles.

Scattering refers to an interaction between two classes of objects, frequently referred to as *particles*. They are not necessarily structureless fundamental particles such as photons or electrons, but could be composite quantum objects such as nucleons, nuclei, atoms, or molecules. Usually, in a typical scattering experiment in the laboratory, a target composed of essentially noninteracting objects is bombarded with a (collimated and/or monoenergetic) beam of projectiles. In high-energy particle physics, there could be a colliding beam experiment where two (collimated and monoenergetic) beams of objects are allowed to interact in a small region in space. The colliding beam experiment has the advantage of increasing the center of mass energy without increasing the energy of particles in each beam. In both cases, after interaction of the initial objects, a collection of final objects may emerge, and some of these noninteracting objects are detected in a region far away from the region of interaction. In a scattering experiment, a careful study is made of the angular distribution or of the spectrum of final-state particles. The experimental results are then explained using a nonrelativistic quantum-mechanical model with an interaction potential. If the interaction potential among the two types of objects is not precisely known, a scattering study gives us the opportunity to extract information about the interaction potential. Although the Coulomb interaction among charged particles is well known, the interaction potential between two atoms or between an electron and an atom is not precisely known.

Interaction between the noninteracting fragments in the initial and final states is localized in space and time (e.g., it operates during a short interval of time and in a small region in space when the noninteracting fragments are close to each other). (Even a long-range Coulomb interaction may be shielded at large distances in a scattering experiment and hence is operative throughout a small region in space.) In nature, one could have scattering involving more than two objects in the initial state, which is difficult to study in the laboratory. For example, in a chemical or stellar process, three free objects in the initial state may undergo a chemical or nuclear reaction. In the present book we confine the discussion to two objects in the initial state. Each such object is assumed to have a very large lifetime compared to the interaction time and is usually termed a particle with *internal structure*. The complete process of interaction among these particles is termed a *collision* [2].

Scattering may involve redistribution or rearrangement of the constituents of the two initial objects. In atomic physics the constituents are nuclei and electrons, in molecular physics they are atoms, and in nuclear physics they are nucleons. The freely moving asymptotic components in the final state could be either the constituents or composites formed from them. Evidently, scattering encompasses the entire history of a process, from the infinite past to the infinite future. The physical process is explicitly time dependent and calls for treatment in terms of wave packets satisfying a time-dependent Schrödinger equation. However, a steady-state description of the problem based on the nonrelativistic time-independent Schrödinger equation with appropriate asymptotic boundary conditions is possible and is usually employed in a theoretical analysis of scattering [1].

The simplest example of scattering is that of a single particle that interacts with a finite-range fixed center of force. Prior to the interaction or the collision, as well as following it, the particle generally moves with a constant velocity in different directions. This is usually known as *potential scattering* and the interaction can be represented by a one-body potential. Potential scattering can also describe the simplest case of scattering between two objects in the center-of-mass frame, where there is no change in the structures of the objects as a result of collision. This means that the objects are the same in the infinite past and the future. Such a process is usually called *elastic scattering*. The initial and final configurations in elastic scattering differ at most in the velocities and spin orientations of each particle [2].

In more complicated situations, the noninteracting objects before and after collision need not be the same. For example, two molecules might scatter into a final configuration consisting of their dissociated atoms or excited states. They could even form two new molecules. Usually, each such process is known by a specific name. When the final objects are the excited states of the initial objects, the process is termed *inelastic scattering*. When more than two objects are formed in the final state, the process is termed *reaction* or *breakup scattering*. When two new final objects, with internal structures distinct from those of the initial objects, are formed, the process is termed *rearrangement scattering*.

All the situations described above can be illustrated by a system comprised of three particles. If we label the particles A, B, and C, the various scattering possibilities are

$$A + BC \to A + BC, \quad \text{elastic scattering,} \tag{1.1}$$

$$\to A + (BC)^*, \quad \text{inelastic scattering,} \tag{1.2}$$

$$\to B + CA, \quad \text{rearrangement scattering,} \tag{1.3}$$

$$\to A + B + C, \quad \text{breakup.} \tag{1.4}$$

Here BC denotes the ground state of particles B and C and $(BC)^*$ denotes an excited state. The time-reversed processes corresponding to (1.1), (1.2), and (1.3) are also termed elastic, inelastic, and rearrangement scattering, respectively. The time reversed breakup process is called *recombination*. In chemistry and chemical physics, the term *nonreactive scattering* is often used to specify inelastic scattering of the type (1.2), and the term *reactive scattering* is then used to specify rearrangement scattering of the type (1.3). In reactive or rearrangement scattering, the cluster CA may appear in either a ground or excited state. In chemical, atomic, and molecular physics a special type of breakup, where a bound electron is removed from an atom or molecule, is of interest. Such a process is often called *ionization*.

Elastic scattering (1.1) takes place at any positive center-of-mass (CM) energy. Inelastic (1.2) and breakup scattering (1.3) are allowed if the incident CM energies are greater than certain positive values called *threshold* of inelastic or breakup scattering, respectively. For CM energies above (below) these thresholds, these channels are called *open* (*closed*). The threshold for elastic scattering is at zero energy in both the laboratory and CM frames. The thresholds for different inelastic and breakup reactions are at different positive energies. The threshold for rearrangement scattering could be above or below the threshold for elastic scattering depending on the binding-energy difference between clusters BC and CA, which could be either positive or negative.

In process (1.3), the convenient coordinates for studying the fragments in the initial (final) state are the relative coordinates between particles B (C) and C (A) and between particle A (B) and the bound state BC (CA). Thus it is not possible to use a single set of these coordinates for both boundary conditions. This change in coordinates from the initial to the final state makes rearrangement scattering extremely difficult to handle by conventional means, and special care must be needed.

There are several ways to circumvent this problem [3]. One can use a hyperspherical system that treats different coordinates more democratically than the usual Euclidean system. Then the boundary condition becomes quite complicated. Another possibility is to expand the wave function in basis functions defined in more than one arrangement, but this leads to coupled integrodifferential equations in configuration space for the radial wave function with nonlocal exchange potential. Compared to the usual coupled differential equations in the case of local potentials, these equations with nonlocal potentials are difficult to treat numerically. However, using the Green's function technique, they can be transformed into coupled integral equations in momentum space for scattering amplitudes which are no more difficult to treat numerically than similar equations for local potentials. It is these momentum-space equations that are often solved numerically for realistic scattering processes. Several variational principles are found to exist for these amplitudes. These amplitudes can then be expanded in terms of square-integrable or $\mathcal{L}^2$ basis functions, which reduces these variational principles to a set of algebraic equations and minimizes the computational effort by a large factor. A large part of the book is dedicated to studying these numerical methods.

Each of the initial and final configurations of the processes described above corresponds to what we call a *physical channel* or simply a *channel*. Channels are associated with the partitions of the particles into clusters, but there could be more than one channel for each partition. For example, in the example above, let the breakup of ABC into clusters A and BC be termed partition a. Let the asymptotic states composed of A and BC define channel α_1, and of A and $(BC)^*$ define channel α_2. Then channels α_1 and α_2 both correspond to partition a. Similar definitions of channels and partitions apply to more complex situations involving four-, five-, and N-particle systems.

It is often convenient to work in the CM frame and break up the full Hamiltonian H, defined by

$$H = \sum_i H_i + \sum_{i<j} V_{ij}, \tag{1.5}$$

into parts corresponding to different partitions. Here H_i is the kinetic energy of particle i and V_{ij} is the potential among particles i and j. Corresponding to partition a, the Hamiltonian can conveniently be broken into a part H_a internal to partition a and a part V^a external to partition a, so that

$$H = H_a + V^a, \tag{1.6}$$

where H_a includes all the kinetic energy and interaction potentials internal to partition a, and V^a only includes the interaction potentials between clusters of partition a. For example, for partition $a \equiv [ij, klm]$, $V^a = V_{ik} + V_{il} + V_{im} + V_{jk} + V_{jl} + V_{jm}$ and $H_a = H_0 + V_{ij} + V_{kl} + V_{km} + V_{lm}$, with H_0 the total kinetic energy. Asymptotically, free movement of clusters ij and klm is governed by the Hamiltonian H_a, and the short-range interaction potential V^a is effective only during collisions. Often, channel indices (e.g., α) are used, and will be used in the following, to label partitions, such as in Eq. (1.6). This is correct if there is one channel in each partition, but when there is more than one channel (e.g., α, β...) corresponding to a single partition, quantities such as H_α or V^α could be the same for two different channels.

Two-cluster channels play a distinct role in scattering theory because of the space-time localization of two-cluster collisions. The two-cluster channels are often denoted by lowercase Greek indices: $\alpha, \beta, \gamma, \ldots$. We use, with some exceptions, lower case Latin letters beginning with i to refer to individual particles or to expansion functions. The Lowercase Latin letters beginning with a refer to distinct partitions of N particles into clusters.

Once the interaction potential among the interacting objects is known, use of quantum mechanics allows one to calculate the physical observables. Numerically, there are various ways to find a solution. The Schrödinger equation together with asymptotic boundary conditions is enough to implement a solution algorithm. For two-body problems this is achieved by direct numerical integration of the time-independent Schrödinger differential equation in configuration space with the interactions expressed in terms of local potentials. This elementary approach is not convenient when the potentials are more complicated or when the scattering process involves more than one channel, which is often the case in realistic situations. Direct numerical integration of differential equations is a delicate task in realistic situations, and constant attention must be paid to step size, error propagation, and similar factors.

Imposition of scattering boundary conditions is numerically the most difficult task in obtaining a solution to the differential Schrödinger equation for scattering problems in the presence of many open channels. In such cases, momentum-space scattering integral equations developed by Lippmann and Schwinger provide a great advantage. Such equations incorporate scattering boundary conditions automatically [4]. Lippmann and Schwinger expressed the scattering problem in momentum space in terms of a Fredholm integral equation of the second kind [1]. The original development of this strategy was strictly correct only for single-channel potential scattering. Later, Lippmann extended this work to include multichannel scattering [4]. The majority of treatments presented in this book are based on momentum-space scattering integral equations of the Lippmann–Schwinger type.

For the single-channel two-body problem, after discretization of the momentum integral, the original Lippmann–Schwinger equation is transformed into a set of linear algebraic equations that can be solved by the technique of matrix inversion or another method. In the process of discretization, one must handle a principal-value integral, which requires a certain amount of care. Many ways of handling such integrals are currently known. The matrix-inversion procedure that one encounters in solution of the Lippmann–Schwinger equations may involve an accumulation of numerical errors, which may make it difficult to obtain accurate results.

The need for intelligent solution schemes for realistic scattering problems can hardly be overemphasized. Among these schemes are the variational and iterative Neumann series approaches [1,2]. Variational principles have proved to be very valuable for the numerical solution of scattering problems. Specific solution algorithms based on these general schemes have oftent been used for the numerical solution of collision models in atomic, molecular, and nuclear physics.

There are broadly two types of variational approaches for the solution of scattering problems. The first type, discussed in Chapter 3, is based on the Schrödinger equation and exploits stationary expressions for some observables, such as scattering phase shifts. The second type, described in Chapter 4, is based on the operator nature of the transition matrix and exploits stationary expressions for this operator. The observables are finally related to a matrix element of this operator. In both cases the solution is expanded in terms of certain known basis functions, and the expansion coefficients are then determined variationally by solving a set of linear algebraic equations whose dimensions are much smaller than those of the set of algebraic equations encountered in the direct solution. For rigorous application of the variational methods, the basis functions are taken as a complete set of orthonormal functions. But in practical implementation it is seldom clear how such a basis set is to be formulated, and often, linearly dependent basis functions are used. In these methods the numerical task involves the calculation of some integrals in either configuration or momentum space and subsequent solution of algebraic equations.

The variational approach also poses difficulties, however. The success of any variational calculation depends on a correct initial guess as to the solution. Unless the initial guess is appropriate and one finds an accurate result with a small number of basis functions, the convergence could be slow. As the number of basis functions increases, there is competition between increased accuracy and the accumulation of numerical error. Under many situations, the latter dominates and makes it very difficult to arrive at an accurate result. In practice, a linear set of equations may become nearly singular when the number of (linearly dependent) expansion functions is increased, and this also puts a limit on the final accuracy. Also, since continuum wave functions of scattering are not square integrable, integrals may be encountered in the variational approach with non-$\mathcal{L}^2$ functions, which are more complicated to evaluate than those encountered in bound-state calculations. Despite these difficulties, the variational approach has emerged in the last two decades as one of the most powerful tools for solving scattering problems, and most of the present book is dedicated to this topic.

The advantage of using the iterative Neumann series solution (Section 2.5) of the Lippmann–Schwinger equation [1,2] is that it involves only successive matrix vector multiplications, where the accumulation of numerical error is small. Hence, if the Neumann series converges satisfactorily, it may lead to high-precision results. Unfortunately, in most problems of scattering, the Neumann series diverges except at very high energies. In these cases, by introducing a subtraction term in the original Lippmann–Schwinger equation, one can write an auxiliary nonsingular scattering integral equation, called the Γ-matrix equation (Section 2.4), which permits a rapidly convergent iterative solution for the scattering problem at low energies. If convergence of the iterative solution of the Γ-matrix equation is not satisfactory, a solution can be obtained by other means. For example, one can use variational or similar methods for its solution. One advantage of dealing with the real Γ-matrix equation is that there is no principal-value integral in this equation as in the original Lippmann–Schwinger equation. However, after obtaining the Γ matrix, one needs to evaluate a single principal-value integral over the elements of the Γ matrix, which can easily be done. Solution of the scattering problem employing an auxiliary nonsingular equation such as the Γ-matrix equation, is another powerful approach to obtaining accurate results.

The situation is far more complicated in the many-particle multichannel case, and most problems of physical interest fall into this category. In this case there are a host of Lippmann–Schwinger equations, and the solution of anything less than a minimum number of Lippmann–Schwinger equations is nonunique [2, 5]. Although this set of Lippmann–Schwinger equations handles the scattering solution, it is impossible to extract a numerical solution because of the decoupled and ill-defined nature of this set of equations. Rigorously, for short-range potentials, these multichannel scattering problems can be reformulated in terms of Faddeev–Yakubovskii scattering integral equations [6] with unique solutions. The three- and four-particle versions of these equations have been solved for several nuclear scattering problems. When the number of particles is greater than four, these equations become too complicated to be of practical numerical use, and for coulombic long-range potentials (of atomic or molecular physics), these equations become ill defined. In the latter case, approximate multichannel scattering models with unique solutions have been proposed and used with great success in nuclear, atomic, and molecular physics. They are the coupled reactions channel [7,8] and resonating group equations [9] of nuclear physics and close-coupling equations [7,8] of atomic and molecular physics. All these approximate equations can be cast into the form of a set of coupled integral equations with unique solutions. This coupled set of equations can be solved numerically as in the case of a single-channel Lippmann–Schwinger equation. In Chapter 5 we deal with several realistic scattering problems of the multichannel type and show how the variational principles developed in Chapters 3 and 4 can be applied to these problems. Finally, in Chapter 6 we present numerical results obtained by various methods for model few-channel and realistic multichannel problems.

In the remainder of this chapter we give a brief summary of quantum-mechanical scattering theory relevant for future development. No attempt is made to present a mathematically rigorous theory of scattering. We present a plausible description of scattering, that will be useful for future development. For a more complete account the reader is referred to a number of excellent texts in scattering theory

[10]. In Section 1.2 we discuss how time-independent treatment is possible for apparently time-dependent scattering process. In Sections 1.3, 1.4, and 1.5 we describe the usual wave-function, t-matrix, and K-matrix formulations of scattering, respectively. Finally, in Section 1.6 we describe a multichannel formulation of scattering.

1.2 TIME-INDEPENDENT DESCRIPTION OF SCATTERING

The scattering process is explicitly time dependent. However, in numerical treatment of scattering and bound states, a time-independent formulation is generally used. We would like to demonstrate how a time-independent description of scattering is possible. Both bound-state and scattering problems are governed by the time-dependent Schrödinger equation

$$i\hbar\frac{\partial}{\partial t}|\psi(t)\rangle = \mathcal{H}|\psi(t)\rangle, \tag{1.7}$$

where $\mathcal{H}$ is the total Hamiltonian of the system: $\mathcal{H} = \mathcal{H}_0 + U$. Here $\mathcal{H}_0$ is the total kinetic energy, also known as the free Hamiltonian, and U is the interaction potential. For a time-independent potential, the time dependence of the bound-state problem is given by the trivial phase factor $\exp(-i\mathcal{E}_b t/\hbar)$, where $\mathcal{E}_b$ is the bound-state energy. Hence, all expectation values and probabilities involve $|\psi(t)|^2$, are time-independent and do not change with time. Such a bound state is absolutely stable and does not change with time.

Although the scattering process is explicitly time dependent, a time-independent description is plausible [1,2]. Suppose that the scattering center is at the origin and the scattering takes place at $t = 0$. The incident particle can be described by a wave packet $|\phi(t)\rangle$, satisfying the Schrödinger equation

$$i\hbar\frac{\partial}{\partial t}|\phi(t)\rangle = \mathcal{H}_0|\phi(t)\rangle, \tag{1.8}$$

with the normalization condition

$$\langle\phi(t)|\phi(t)\rangle = \int d^3x\ \phi^*(\mathbf{x}, t)\phi(\mathbf{x}, t) = 1. \tag{1.9}$$

Here $\phi(\mathbf{x}, t) \equiv \langle\mathbf{x}|\phi(t)\rangle$ is the configuration-space representation of the wave function. The motion of the incident particle is described by the free-particle state $\phi(t)$ in the infinite past, far away from the scattering center. This state evolves with time in a complicated fashion as the particle interacts with the force center for $t \approx 0$, and finally, the particle is detected in the infinite future away from the scattering center.

Any solution to Eq. (1.8) can be written as

$$\phi(\mathbf{x}, t) = (2\pi)^{-3/2} \int d^3k\, a(\mathbf{k}) \exp\left[i\left(\mathbf{k} \cdot \mathbf{x} - \frac{\mathcal{E}_k t}{\hbar}\right)\right], \qquad (1.10)$$

where $\mathcal{E}_k = \hbar^2 k^2/2m$, with m the reduced mass. Equation (1.10) represents a superposition of plane waves of wave vector $\mathbf{k}$ and energy $\mathcal{E}_k$. The normalization condition (1.9) imposes the following constraint on the weight function $a(\mathbf{k})$:

$$\int d^3k\, |a(\mathbf{k})|^2 = 1, \qquad (1.11)$$

where $|a(\mathbf{k})|^2$ is the probability that the free particle has momentum $\mathbf{k}$. If the free-particle state closely approximates particle of definite momentum $\mathbf{k_0}$, $|a(\mathbf{k})|$ should be sharply peaked about $\mathbf{k} = \mathbf{k_0}$. The ideal case of a wave-packet representation of a particle of fixed momentum $\mathbf{k_0}$ is given by $a(\mathbf{k}) = \delta(\mathbf{k} - \mathbf{k_0})$. However, as the wave packets are solutions of the Schrödinger equation, they spread in space with time. Hence, for a realistic treatment of scattering, it is useful to know if the wave-packet description of scattering is meaningful.

The dynamical scattering problem is governed by the Schrödinger equation (1.7). As in the infinite past, the scattering state is essentially the free-particle state, we have the asymptotic boundary condition

$$\lim_{t \to -\infty} \left[|\psi^{(+)}(t)\rangle - |\phi(t)\rangle \right] = 0. \qquad (1.12)$$

The $+$ superscript in this equation corresponds to this specific boundary condition. The time limits in Eq. (1.12) and below are strong limits, as discussed by Newton [1].

As only the asymptotic limit (1.12) is used to define a mathematical formulation of the scattering process, it is reasonable that for time-independent potentials it should be possible to develop a time-independent scattering theory. For such potentials, time evolution of the scattering state is given by

$$|\psi^{(+)}(t)\rangle = \exp\left(-\frac{i\mathcal{H}t}{\hbar}\right)|\psi^{(+)}(0)\rangle. \tag{1.13}$$

The time-evolution of the free particle state is given by

$$|\phi^{(+)}(t)\rangle = \exp\left(-\frac{i\mathcal{H}_0 t}{\hbar}\right)|\phi^{(+)}(0)\rangle. \tag{1.14}$$

From Eqs. (1.12), (1.13), and (1.14), we find the mapping

$$|\psi^{(+)}(t)\rangle = \Omega^{(+)}|\phi(t)\rangle \equiv \left[\lim_{t\to-\infty}\left\{\exp\left(\frac{i\mathcal{H}t}{\hbar}\right)\exp\left(-\frac{i\mathcal{H}_0 t}{\hbar}\right)\right\}\right]|\phi(t)\rangle, \tag{1.15}$$

where $\Omega^{(+)}$ is the time-independent Möller operator, which maps a free state at time t to the scattering state at time t. A sufficient condition for the existence of the limit in Eq. (1.15) is that the potential should have a finite norm [1]. For a local potential this condition is expressed by Eq. (1.72) in Section 1.3, which excludes potentials with a Coulomb tail and other singular behaviors. The full evolution of scattering can be determined by the Möller operator, $\Omega^{(+)}$, of Eq. (1.15). This operator does not depend on details of the underlying wave packets. As this operator is time independent, it seems plausible to develop a time-independent formulation of scattering for usual time-independent nonsingular short-range potentials. One can formulate the scattering theory in terms of the Möller operator, $\Omega^{(+)}$. However, in this book we consider an alternative approach in close analogy with the bound-state problem.

A time-independent formulation of scattering is possible if in the wave-packet description one considers idealized states of well-defined energy and momentum. Corresponding to the wave packet (1.10), we assume the representation-independent form

$$|\phi(t)\rangle = \int d^3k a(\mathbf{k}) \exp\left(-\frac{i\mathcal{E}_k t}{\hbar}\right)|\phi_{\mathbf{k}}\rangle, \tag{1.16}$$

where $|\phi_{\mathbf{k}}\rangle$ is an idealized (plane-wave) state of well-defined momentum $\mathbf{k}$ and satisfies the free-particle Schrödinger equation

$$(\mathcal{E}_k - \mathcal{H}_0)|\phi_{\mathbf{k}}\rangle = 0. \tag{1.17}$$

The configuration-space representation of $\phi_{\mathbf{k}}$ is defined by

$$\phi_{\mathbf{k}}(\mathbf{r}) \equiv \langle \mathbf{r}|\phi_{\mathbf{k}}\rangle = (2\pi)^{-3/2} \exp(i\mathbf{k}\cdot\mathbf{r}). \tag{1.18}$$

If the limit of Eq. (1.12) holds and if a time-independent Möller operator exists, it is reasonable to take the scattering wave function to be

$$|\psi^{(+)}(t)\rangle = \int d^3k\, a(\mathbf{k}) \exp\left(-\frac{i\mathcal{E}_k t}{\hbar}\right)|\psi_{\mathbf{k}}^{(+)}\rangle, \tag{1.19}$$

where $|\psi_{\mathbf{k}}^{(+)}\rangle$ is an idealized scattering state of well-defined momentum $\mathbf{k}$ and satisfies the Schrödinger equation

$$(\mathcal{E}_k - \mathcal{H})|\psi_{\mathbf{k}}^{(+)}\rangle = 0. \tag{1.20}$$

Time-independent mapping (1.15) between the free and scattering states implies that for time-independent potentials, the time dependence of both the free and scattering states are governed by the trivial phase factor $\exp(-i\mathcal{E}_k t/\hbar)$, as in Eqs. (1.16) and (1.19). Hence, a determination of the time-independent state vector $|\psi_{\mathbf{k}}^{(+)}\rangle$ consistent with scattering boundary conditions is enough for a time-dependent description of scattering. The interesting aspect of the time-independent realization of scattering implicit in Eqs. (1.17) and (1.20) is that in the scattering process, the direction of the momentum vector $\mathbf{k}$ can change, maintaining $\mathcal{E}_k$ constant. Hence, $|\psi_{\mathbf{k}}^{(+)}\rangle$ can represent all of the scattering possibilities from an initial state $|\phi_{\mathbf{k}}\rangle$. In the next section we formulate scattering theory in terms of this state vector.

1.3 WAVE-FUNCTION DESCRIPTION

Let us first consider the simplest case of single-channel scattering of a particle interacting with a center of force via a local short-range potential $U(r)$. This treatment is also valid for the scattering of two particles in the CM frame. This is possible because one can separate the uninteresting motion of the CM from the relative motion of the particles. This relative motion with a reduced mass is responsible for scattering and can be treated in the same fashion as the movement of a particle in a central field. The wave function $\psi_{\mathbf{k}}^{(+)}(\mathbf{r}) \equiv \langle \mathbf{r}|\psi_{\mathbf{k}}^{(+)}\rangle$ satisfies the configuration-space Schrödinger equation

$$(\mathcal{E}_k - \mathcal{H})\psi_{\mathbf{k}}^{(+)}(\mathbf{r}) = 0. \tag{1.21}$$

Here the interaction potential U is assumed to decrease faster than $1/r$ at ∞ and be nonsingular elsewhere, $\mathbf{k}$ is the incident wave vector, and $\mathcal{E}_k$ is the total energy of the system in the CM frame, so that $k^2\hbar^2/2m = \mathcal{E}_k$, with m the (reduced) mass. The subscript $^{(+)}$ on ψ denotes that we are looking for a scattering solution with in outgoing-wave boundary condition (1.23).

For bound-state problems, Eq. (1.21) permits solution only at specific discrete negative energies consistent with the exponentially decaying asymptotic boundary condition. These energies are bound-state energies. The correct physical boundary condition excludes solution of Eq. (1.21) at all other negative energies.

For the scattering problem, the total energy $\mathcal{E}_k$ is given and the solution of Eq. (1.21) gives the asymptotic boundary condition for the wave function. Usually, it is convenient to rewrite Eq. (1.21) as

$$(E - H)|\psi_{\mathbf{k}}^{(+)}\rangle \equiv [\nabla_{\mathbf{r}}^2 + k^2 - V(\mathbf{r})]\psi_{\mathbf{k}}^{(+)}(\mathbf{r}) = 0, \tag{1.22}$$

where $V(\mathbf{r}) = 2mU(\mathbf{r})/\hbar^2$ and $E = k^2 = 2m\mathcal{E}_k/\hbar^2$. Unless explicitly noted to the contrary, in the remainder of the book, V (E, H, H_0) denotes potential (CM energy, full Hamiltonian, free Hamiltonian) in units of $\hbar^2/2m$. The asymptotic boundary condition satisfied by this wave function is

$$\lim_{r\to\infty} \psi_{\mathbf{k}}^{(+)}(\mathbf{r}) \to \frac{1}{(2\pi)^{3/2}}\left[\exp(ikz) + f(\theta)\frac{\exp(ikr)}{r}\right], \tag{1.23}$$

corresponding to the incident plane wave in the z direction and a scattered spherical outgoing wave. Here θ is the angle between the direction of the scattered wave ($\mathbf{k}'$) and the original direction ($\mathbf{k}$), and $f(\theta)$ is the amplitude of the spherical outgoing wave. The quantity $f(\theta)$ is called the *scattering amplitude*. The azimuthal angle ϕ does not appear because of the spherical symmetry of the potential. Equation (1.23) represents the steady-state boundary condition. This means that one has a monoenergetic collimated parallel beam of incident particles that can be approximated by the incident plane wave. The target is composed of well-separated isolated scattering centers. Most of the incident particles of the monoenergetic beam are not scattered and move in the forward direction. Only a small number of particles from the incident beam are scattered and form the outgoing spherical wave of Eq. (1.23). The amplitude $f(\theta)$ of this spherical wave is a measure of scattering and hence is called the *scattering amplitude*.

The probability current density for the scattered spherical wave is given by $\mathcal{J} = v|f(\theta)|^2/r^2$, where $v = \hbar k/m$, and the number of particles detected per unit time in a solid angle $d\Omega \equiv \sin\theta\, d\theta\, d\phi$ is $N = \mathcal{J} r^2 d\Omega$. The differential cross section is the number of particles scattered into unit solid angle per a unit time per unit incident flux and is given by

$$\frac{d\sigma}{d\Omega} = |f(\theta)|^2, \tag{1.24}$$

and the total cross section by

$$\sigma(E) = \int |f(\theta)|^2 \, d\Omega. \tag{1.25}$$

Total cross section is the total number of particles scattered per unit time per unit incident flux and gives the total probability of scattering taking place at energy $E = k^2$.

To find a numerical solution of the scattering problem, a partial-wave expansion is useful. The plane wave is expanded by

$$\exp(ikz) = \sum_{L=0}^{\infty} i^L (2L+1) j_L(kr) P_L(\cos\theta), \tag{1.26}$$

where $P_L(\cos\theta)$ is the usual Legendre polynomial, θ the azimuthal angle of $\mathbf{k}$, and $j_L(kr)$ the spherical Bessel function, which satisfies the partial-wave free-particle Schrödinger equation

$$\left[-\frac{d^2}{dr^2} + \frac{L(L+1)}{r^2} - k^2 \right] \{ kr j_L(kr) \} = 0. \tag{1.27}$$

and the boundary conditions

$$\lim_{r \to 0} kr j_L(kr) \to \frac{(kr)^{L+1}}{(2L+1)!!}, \tag{1.28}$$

$$\lim_{r \to \infty} kr j_L(kr) \to \sin\left(kr - \frac{L\pi}{2} \right), \tag{1.29}$$

where $(2L+1)!! = (2L+1)(2L-1)\cdots(5)(3)(1)$. As $r \to 0$, the k^2 term in Eq. (1.27) can be neglected and Eq. (1.28) is obtained. The free-particle Schrödinger equation (1.27) is also satisfied by the spherical Neumann and Hankel functions, $n_L(kr)$ and $h_L^{(1)}(kr) = j_L(kr) + in_L(kr)$, respectively. These functions satisfy the boundary conditions

$$\lim_{r\to 0} krn_L(kr) \to -(kr)^{-L}(2L-1)!!, \tag{1.30}$$

$$\lim_{r\to\infty} krn_L(kr) \to -\cos\left(kr - \frac{L\pi}{2}\right), \tag{1.31}$$

$$\lim_{r\to\infty} krh_L^{(1)}(kr) \to (-i)^{L+1}\exp(ikr). \tag{1.32}$$

The scattering wave function is also expanded in partial waves:

$$\psi_{\mathbf{k}}^{(+)}(\mathbf{r}) = \frac{1}{(2\pi)^{3/2}} \sum_{L=0}^{\infty} i^L \frac{\psi_L^{(+)}(r)}{kr}(2L+1)P_L(\cos\theta), \tag{1.33}$$

where the k dependence of the wave-function components, $\psi_L^{(+)}(r)$, is suppressed. These components satisfy

$$\left[-\frac{d^2}{dr^2} + \frac{L(L+1)}{r^2} + V_L(r) - k^2\right]\psi_L^{(+)}(r) = 0, \tag{1.34}$$

and the boundary conditions

$$\lim_{r\to 0} \psi_L^{(+)}(r) \sim (kr)^{L+1}, \tag{1.35}$$

$$\lim_{r\to\infty} \psi_L^{(+)}(r) \to \exp(i\delta_L)\sin\left(kr - \frac{L\pi}{2} + \delta_L\right), \tag{1.36}$$

$$\equiv \sin\left(\frac{kr - L\pi}{2}\right) + [\exp(i\delta_L)\sin\delta_L]i^{-L}\exp(ikr), \tag{1.37}$$

$$\equiv \left(\frac{i}{2}\right)\left[\exp\left\{-i\left(kr - \frac{L\pi}{2}\right)\right\}\right.$$

$$\left. - \exp(2i\delta_L)\exp\left\{i\left(kr - \frac{L\pi}{2}\right)\right\}\right], \tag{1.37a}$$

where δ_L is the partial-wave phase shift and is the change in the phase of the solution in the asymptotic region in relation to the incident plane wave of Eq. (1.29). In approximate solution of the scattering problem, condition (1.35) is often not taken into consideration and is replaced by the simple condition

$$\psi_L^{(+)}(0) = 0. \tag{1.38}$$

Equation (1.34), satisfied by components $\psi_L^{(+)}(r)$, is essentially the same as a one-dimensional Schrödinger equation, and one can develop scattering theory as in one dimension. However, we shall maintain the three-dimensional phase space and work with components $\psi_L^{(+)}(r)/(kr)$ and $j_L(kr)$ of scattered and plane waves, respectively.

Equation (1.34) is also satisfied by the bound-state wave function $\psi_{LB}(r)$ of binding energy α^2, $k^2 = -\alpha^2$, with boundary conditions (1.35) and

$$\lim_{r\to\infty} \psi_{LB}(r) \to \exp(-\alpha r). \tag{1.39}$$

In the asymptotic region, $r \to \infty$, the potential term can be neglected in Eq. (1.34), which reduces to the free-particle equation (1.27), and Eq. (1.39) is obtained.

The general solution of Eq. (1.34), at positive energies, can be written as a linear combination of spherical Bessel and Neumann functions, $j_L(kr)$ and $n_L(kr)$, respectively. The specific combination (1.36) leads to a simple and useful parametrization for the scattering amplitude $f(\theta)$ and is also consistent with the partial-wave form of the asymptotic boundary condition (1.23). The first term on the right-hand side of Eq. (1.37) is the partial-wave asymptotic form of the plane wave in Eq. (1.23). The last term in Eq. (1.37) is the outgoing spherical wave, provided that the partial-wave scattering amplitude is $\exp(i\delta_L)\sin\delta_L/k$.

In Eq. (1.34), for large L and at low energies, only the centrifugal barrier term $L(L+1)/r^2$ should dominate, and this equation essentially reduces to a free-particle equation. This means that in higher partial waves, there is no scattering at low energies, and for all practical purpose, infinite expansions (1.26) and (1.33) can be truncated at a small value of L. This is a great advantage when dealing with a few one-dimensional scattering equations (1.34) rather than the single three-dimensional scattering equation (1.22).

If we use the partial-wave projections given by Eqs. (1.26) and (1.33) in Eq. (1.23) and the asymptotic properties (1.29) and (1.36), we obtain the following partial-wave expansion for the scattering amplitude:

$$f(\theta) = \frac{1}{2ik} \sum_{L=0}^{\infty} (2L+1)[\exp(2i\delta_L) - 1] P_L(\cos\theta). \tag{1.40}$$

Further use of Eq. (1.40) in Eqs. (1.24) and (1.25) leads to

$$\sigma(E) = \sum_{L=0}^{\infty} (2L+1)\sigma_L \equiv \frac{4\pi}{k^2} \sum_{L=0}^{\infty} (2L+1) \sin^2 \delta_L, \tag{1.41}$$

where σ_L are the partial-wave cross sections. The following optical theorem provides a measure of flux conservation and relates the imaginary part of the forward scattering amplitude to the total cross section, and is derived from the foregoing expressions for $f(\theta)$ and σ:

$$\Im[f(0)] = \frac{k}{4\pi}\sigma, \tag{1.42}$$

where $\Im$ denotes the imaginary part.

The scattering amplitude $f(\theta)$ and the total cross section $\sigma(E)$ can conveniently be written in terms of the unitary S-matrix elements defined by

$$S_L(k) = \exp(2i\delta_L). \tag{1.43}$$

In terms of this S matrix we have

$$f(\theta) = \frac{1}{2ik} \sum_{L=0}^{\infty} (2L+1)[S_L(k) - 1] P_L(\cos\theta), \tag{1.44}$$

$$\sigma(E) = \frac{\pi}{k^2} \sum_{L=0}^{\infty} (2L+1)|1 - S_L(k)|^2. \tag{1.45}$$

For a real Hermitian potential, the phase shifts are real and one has

$$|S_L(k)|^2 = 1, \tag{1.46}$$

which is the condition of unitarity or probability conservation.

It is often useful to define the t-matrix elements by

$$t_L(k) = -\frac{\exp(i\delta_L)\sin\delta_L}{k}. \tag{1.47}$$

In terms of the t matrix the scattering amplitude can be written as

$$f(\theta) = -\sum_{L=0}^{\infty}(2L+1)t_L(k)P_L(\cos\theta). \tag{1.48}$$

For potential scattering with real Hermitian potential, the phase shifts have to be real. From Eq. (1.36) we find that an imaginary part in δ_L leads to an exponentially growing or decaying part in the wave function or to probability or flux nonconservation in the scattering process. An imaginary part of δ_L leading to an exponentially decaying part in the wave function corresponds to absorption. For example, if inelastic channels are energetically open, incident flux of particles will not be conserved unless these channels are explicitly taken into account. Isolated elastic scattering in such a case can be modeled by a complex potential and complex phase shifts leading to absorption.

For a complex phase shift in Eq. (1.43), expressions (1.44) and (1.45) for the scattering amplitude and elastic scattering cross section, respectively, remain valid. Now the S matrix is not unitary and Eq. (1.46) does not hold, as some of the incident flux is lost to inelastic channels. The difference in scattering by a real and a complex potential is explicit in Eq. (1.37a), where the asymptotic form is written as a linear combination of spherically incoming and outgoing waves. For a real potential the difference in intensities of these two waves is zero. For a complex potential some incident flux is transferred to inelastic channels. From Eq. (1.37a) the loss of flux is proportional to $(1 - |S_L(k))|^2)$. For a real potential Eq. (1.46) holds and the loss of flux is zero. For a complex potential this part of scattering is expressed by the reaction cross section, defined by

$$\sigma_{\text{re}}(E) = \frac{\pi}{k^2}\sum_{L=0}^{\infty}(2L+1)(1 - |S_L(k)|^2). \tag{1.49}$$

For a real phase shift, $\sigma_{\text{re}} = 0$. The total cross section is now the sum of elastic and reaction cross sections given by Eqs. (1.45) and (1.49) and is given by

$$\sigma_{\text{tot}}(E) = \frac{\pi}{k^2}\sum_{L=0}^{\infty}(2L+1)(1 - |S_L(k)|^2 + |1 - S_L(k)|^2) \tag{1.50}$$

$$= \frac{2\pi}{k^2}\sum_{L=0}^{\infty}(2L+1)(1 - \Re[S_L(k)]), \tag{1.51}$$

where $\Re$ denotes the real part. From Eqs. (1.44) and (1.51) we find that the optical theorem (1.42) holds with σ replaced by σ_{tot} of (1.51).

In a straightforward approach to a numerical solution of the scattering problem, one integrates Eq. (1.34) consistent with boundary conditions (1.35) and (1.36). This procedure results in the scattering phase shifts δ_L in each partial wave. The phase shifts can then be used to calculate all scattering observables. For a simple local potential, this is a relatively easy task. The task may become far from routine when either the potential becomes nonlocal or the problem involves many physical channels or both. In the presence of a nonlocal potential, the ordinary differential equation (1.34) becomes the integrodifferential equation

$$\left[-\frac{d^2}{dr^2} + \frac{L(L+1)}{r^2} - k^2\right]\psi_L^{(+)}(r) + \int d^3r' V_L(r, r')\psi_L^{(+)}(r') = 0, \tag{1.52}$$

In the presence of many physical channels, such a model equation becomes a set of coupled integrodifferential equations. Under such situations, use of momentum-space integral equations and intelligent solution methods provides an advantage.

Instead of solving the differential equation (1.34) or (1.52), one could formulate the scattering theory in terms of integral equations. In solving the differential equations, we have to impose boundary conditions (1.35) and (1.36) on the solution. Any solution of Eq. (1.34) does not automatically satisfy these physical scattering boundary conditions. The scattering integral equations automatically incorporate these boundary conditions. They are equivalent to the differential equations of scattering and the correct boundary condition. To derive the integral equations of scattering, Eq. (1.22) is rewritten as

$$(E - H_0)\psi_{\mathbf{k}}^{(+)}(\mathbf{r}) = V(\mathbf{r})\psi_{\mathbf{k}}^{(+)}(\mathbf{r}). \tag{1.53}$$

To transform this equation into an integral equation, the operator $(E - H_0)$ needs to be inverted. The inverse of this operator is written as

$$G_0(k^2) \equiv \frac{1}{k^2 - H_0} = \int d^3q \frac{|\phi_{\mathbf{q}}\rangle\langle\phi_{\mathbf{q}}|}{k^2 - q^2}, \tag{1.54}$$

where the momentum eigenstate $|\phi_{\mathbf{q}}\rangle$ satisfies

$$(E - H_0)|\phi_{\mathbf{q}}\rangle \equiv (k^2 + \nabla_{\mathbf{r}}^2)\phi_{\mathbf{q}}(\mathbf{r}) = 0 \tag{1.55}$$

and is defined by Eq. (1.18). In writing Eq. (1.54) we have used $E = k^2$ and the completeness relation of momentum eigenstates,

$$\int d^3q\, |\phi_{\mathbf{q}}\rangle\langle\phi_{\mathbf{q}}| = I, \tag{1.56}$$

where I is the identity operator. In the scattering region, the kinetic energy operator is positive definite. Hence, operator (1.54) is singular. The singular integral on the right-hand side of this equation is meaningful only if a definite integration prescription is provided at the singular point $q = k$. From the theory of complex variables, one knows that the inverse operator in Eq. (1.54) becomes meaningful if the principal-value part of the integral in Eq. (1.54) is considered or if an infinitesimally small imaginary part is included in the denominator. This is a known trick in mathematical physics. Lippmann and Schwinger [4] were the first to use it in scattering theory to determine the appropriate inverse consistent with the asymptotic boundary condition (1.23). Two possibilities for this inverse operator, known as the Green's function, $G_0^{(\pm)}(k^2)$, are obtained if we consider the configuration-space matrix elements of Eq. (1.54):

$$G_0^{(\pm)}(\mathbf{r}, \mathbf{r}', k^2) \equiv \langle\mathbf{r}|G_0^{(\pm)}(k^2)|\mathbf{r}'\rangle = \int \frac{d^3q}{(2\pi)^3} \frac{\exp[i\mathbf{q}\cdot(\mathbf{r}' - \mathbf{r})]}{k^2 - q^2 \pm i0} \tag{1.57}$$

$$= -\frac{1}{4\pi} \frac{\exp(\pm ik|\mathbf{r}' - \mathbf{r}|)}{|\mathbf{r}' - \mathbf{r}|}. \tag{1.58}$$

The quantity $\pm i0$ in the denominator of Eq. (1.57) means a limiting $\lim_{\epsilon\to 0}$ procedure with a infinitely small imaginary part $\pm i\epsilon$ included in the denominator. A Green's function with a $+$ $(-)$ superscript is called an *outgoing-wave* (*incoming-wave*) *Green's function*. The third possibility of taking only the real principal-value part of the integral in Eq. (1.54), is considered in Section 1.5 in relation to the K-matrix description of scattering. To derive (1.58), first the angular integrals in Eq. (1.57) are evaluated, leading to

$$G_0^{(\pm)}(\mathbf{r}, \mathbf{r}', k^2) = \frac{1}{4\pi^2 |\mathbf{r}' - \mathbf{r}|} \int_{-\infty}^{\infty} dq \frac{q \sin(q|\mathbf{r}' - \mathbf{r}|)}{k^2 - q^2 \pm i0}. \tag{1.59}$$

The remaining integral is then performed by Cauchy's theorem to obtain Eq. (1.58). The Green's function (1.57) is complex above the scattering threshold at $k^2 = 0$ and possesses a branch cut along the entire real-energy axis known as the *unitarity cut*.

The Green's function is symmetric – $G_0^{(\pm)}(\mathbf{r}', \mathbf{r}, k^2) = G_0^{(\pm)}(\mathbf{r}, \mathbf{r}', k^2)$ – and satisfies the differential equation

$$(\nabla_{\mathbf{r}}^2 + k^2) G_0^{(\pm)}(\mathbf{r}, \mathbf{r}', k^2) = \delta^3(\mathbf{r} - \mathbf{r}'). \tag{1.60}$$

Equation (1.60) is obtained by operating $E - H_0 \equiv k^2 + \nabla_{\mathbf{r}}^2$ on Eqs. (1.54) and (1.57). The asymptotic behavior of the Green's function (1.58) as $r \to \infty$ is given by

$$\lim_{r \to \infty} G_0^{(\pm)}(\mathbf{r}, \mathbf{r}', k^2) \to -\frac{1}{4\pi} \frac{\exp(\pm ikr)}{r} \exp(\mp i\mathbf{k}' \cdot \mathbf{r}'), \tag{1.61}$$

where the limit is taken in the direction of the outgoing wave vector,

$$\mathbf{k}' = k \frac{\mathbf{r}}{r}. \tag{1.62}$$

If we impose boundary condition (1.61) on the solution of the differential equation (1.60), the Green's function (1.58) is obtained [1]. If we formulate the scattering theory in terms of well-defined Green's functions, the scattering boundary conditions will be incorporated automatically.

The Green's function $G^{(+)}(k^2)$ has a spherical outgoing wave in the asymptotic region as can be seen in Eq. (1.61), and this Green's function can be used to construct the physical scattering wave function. The other Green's function, $G^{(-)}(k^2)$, has, asymptotically, an incoming spherical wave. Applying the outgoing-wave Green's function $G^{(+)}(k^2)$ to Eq. (1.53), one obtains the following Lippmann–Schwinger scattering equation for the outgoing-wave physical scattering wave function, $\psi_{\mathbf{k}}^{(+)}(\mathbf{r})$,

$$\psi_{\mathbf{k}}^{(+)}(\mathbf{r}) = \phi_{\mathbf{k}}(\mathbf{r}) + \int d^3r' G_0^{(+)}(\mathbf{r}, \mathbf{r}', k^2) V(\mathbf{r}') \psi_{\mathbf{k}}^{(+)}(\mathbf{r}'). \tag{1.63}$$

If we operate on the left of Eq. (1.63) by $(E - H_0)$, equivalence of this equation with the Schrödinger equation is established. An integral equation of type (1.63) is often written in formal operator form,

$$|\psi_{\mathbf{k}}^{(+)}\rangle = |\phi_{\mathbf{k}}\rangle + G_0^{(+)}(k^2) V |\psi_{\mathbf{k}}^{(+)}\rangle, \tag{1.64}$$

with

$$G_0^{(\pm)}(k^2) = (k^2 - H_0 \pm i0)^{-1}, \tag{1.65}$$

where both the explicit configuration-space representation and the intermediate-state integration have been suppressed. The quantity $\mathcal{K} \equiv G_0^{(+)}(k^2) V$ is called the *kernel* of the integral equation (1.64). Similar formal operator forms for scattering equations are often written. The same operator equation can be expressed in either configuration or momentum representation. The observables are related to certain (configuration or momentum-space) matrix elements of some of the operators. If the incoming-wave Green's function $G_0^{(-)}(k^2)$ is used, rather than the outgoing-wave Green's function, one obtains the incoming-wave wave function, $|\psi_{\mathbf{k}}^{(-)}\rangle$, which satisfies the following Lippmann-Schwinger equation:

$$|\psi_{\mathbf{k}}^{(-)}\rangle = |\phi_{\mathbf{k}}\rangle + G_0^{(-)}(k^2) V |\psi_{\mathbf{k}}^{(-)}\rangle. \tag{1.66}$$

Asymptotically, the wave function $\psi_{\mathbf{k}}^{(-)}(\mathbf{r})$ has a plane wave and a spherical incoming wave. The conjugate wave functions satisfy

$$\langle\psi_{\mathbf{k}}^{(-)}| = \langle\phi_{\mathbf{k}}| + \langle\psi_{\mathbf{k}}^{(-)}| V G_0^{(+)}(k^2), \tag{1.67}$$

$$\langle\psi_{\mathbf{k}}^{(+)}| = \langle\phi_{\mathbf{k}}| + \langle\psi_{\mathbf{k}}^{(+)}| V G_0^{(-)}(k^2). \tag{1.68}$$

We have the normalization conditions

$$\langle \psi_{\mathbf{p}}^{(\pm)} | \psi_{\mathbf{p}'}^{(\pm)} \rangle = \langle \phi_{\mathbf{p}} | \phi_{\mathbf{p}'} \rangle = \delta(\mathbf{p} - \mathbf{p}'). \tag{1.69}$$

For short-range nonsingular potentials, the Lippmann–Schwinger equations (1.63) and (1.66) have compact or Fredholm kernels [1]. A sufficient condition for compactness of the kernel $\mathcal{K}$ is to possess a finite Hilbert–Schmidt norm:

$$|\mathcal{K}|_{\mathrm{HS}}^2 \equiv \mathrm{Tr}[\mathcal{K}\mathcal{K}^\dagger] < \infty. \tag{1.70}$$

This norm can be calculated [1,6] for $\mathcal{K} = G_0^{(+)}(k^2)V$:

$$\begin{aligned}
|\mathcal{K}|_{\mathrm{HS}}^2 &= \int d^3r \int d^3r' \, |G_0^{(+)}(\mathbf{r}, \mathbf{r}', k^2)|^2 V^2(r') \\
&= \lim_{\epsilon \to 0} \int d^3r \int d^3r' \left| \frac{1}{(2\pi)^3} \int d^3q \frac{\exp(i\mathbf{q}.(\mathbf{r}' - \mathbf{r}))}{k^2 - q^2 + i\epsilon} \right|^2 V^2(r') \\
&= \lim_{\epsilon \to 0} \frac{1}{8\pi\sqrt{\epsilon}} \int d^3r V^2(r).
\end{aligned} \tag{1.71}$$

The condition of compactness is given by the following condition on the potential:

$$|V|_{\mathrm{HS}}^2 \equiv \int d^3r \, V^2(r) < \infty, \tag{1.72}$$

as long as $\epsilon \neq 0$. When the condition of compactness is satisfied, the scattering integral equation is called *Fredholm* or *compact* [1]. This condition requires that the potential decay faster than the Coulomb potential at infinity and not possess other singularities at short distances, which could make the Lippmann–Schwinger equation non-Fredholm. We shall assume that this condition of compactness is always satisfied. An integral equation with a Fredholm kernel has a unique and well-defined numerical solution [1]. For compact potentials, the homogeneous version of Eqs. (1.63) and (1.66) does not have a solution in the scattering region, and the solutions to Eqs. (1.64) and (1.66) are unique. For local Hermitian potentials, the solution of the homogeneous equation

$$|\psi_B\rangle = G_0^{(+)}(k^2)V|\psi_B\rangle \tag{1.73}$$

corresponds to a bound-state wave function at negative energies.

Using the asymptotic behavior (1.61) of the outgoing-wave Green's function, it can be verified that the asymptotic behavior of the outgoing-wave, or physical, solution of the Lippmann–Schwinger equation (1.63) is the same as that given by Eq. (1.23). If we take the $r \to \infty$ limit in the Lippmann-Schwinger equation (1.63) and use Eq. (1.61), we obtain the desired asymptotic behavior (1.23), provided that

$$f(\theta) = -2\pi^2 \int d^3r\phi_{\mathbf{k}'}(\mathbf{r})V(\mathbf{r})\psi_{\mathbf{k}}^{(+)}(\mathbf{r}) \equiv -2\pi^2\langle\phi_{\mathbf{k}'}|V|\psi_{\mathbf{k}}^{(+)}\rangle. \tag{1.74}$$

This is a consistency condition relating the scattering wave function and the scattering amplitude. Equation (1.74) implies that unlike the case of solution of the Schrödinger equation, where the asymptotic boundary condition has to be imposed in order to extract the scattering solution, solution of the Lippmann–Schwinger equation (1.63) automatically includes the scattering boundary condition. For usual short-range nonsingular potentials, Eq. (1.63) is Fredholm and hence is amenable to standard numerical techniques.

The asymptotic behavior of the solution to Eq. (1.73) at positive energies will not have an incoming plane wave, hence does not lead to an acceptable scattering solution. However, at negative energy $k^2 = -\alpha^2$, by using Eq. (1.61) we find that the asymptotic behavior of the solution to Eq. (1.73) is consistent with that for the bound state given by Eq. (1.39). Hence, Eq. (1.73) has only bound-state solutions.

For numerical treatment, one often employs the partial-wave projection technique. Then the partial-wave scattering integral equation for the wave function also satisfies the required boundary condition (1.37). The partial-wave expansion for the Green's function in configuration space is given by

$$G_0^{(+)}(\mathbf{r},\mathbf{r}',k^2) = \frac{1}{4\pi}\sum_{L=0}^{\infty}(2L+1)G_L^{(+)}(r,r',k^2)P_L(\cos\theta), \tag{1.75}$$

where θ is the angle between vectors $\mathbf{r}$ and $\mathbf{r}'$. The partial-wave components of the Green's function are given by

$$G_L^{(+)}(r,r',k^2) = 2\pi\int_{-1}^{1} dxP_L(x)G_0^{(+)}(\mathbf{r},\mathbf{r}',k^2), \tag{1.76}$$

where $x = \cos\theta$. Using the Green's function (1.58), integration of Eq. (1.76) leads to

$$G_L^{(+)}(r, r', k^2) = -ikj_L(kr_<)h_L^{(1)}(kr_>), \tag{1.77}$$

for $k^2 > 0$ and

$$G_L^{(+)}(r, r', k^2) = kj_L(ikr_<)h_L^{(1)}(ikr_>), \tag{1.78}$$

for $k^2 < 0$. Here $<$ ($>$) denotes the smaller (larger) of the variables r and r', and $j_L, h_L^{(1)}$ are the usual spherical Bessel and Hankel functions, respectively. Using Eqs. (1.26), (1.33), and (1.75), the partial-wave projection of Eq. (1.63) is given by

$$\frac{\psi_L^{(+)}(r)}{kr} = j_L(kr) + \int_0^\infty dr'\, r'^2 G_L^{(+)}(r, r', k^2) V(r') \frac{\psi_L^{(+)}(r')}{kr'}. \tag{1.79}$$

As $r \to \infty$, the outgoing-wave Green's function (1.77) has the asymptotic behavior

$$\lim_{r\to\infty} G_L^{(+)}(r, r', k^2) \to -i^{-L} k \frac{\exp(ikr)}{kr} j_L(kr'). \tag{1.80}$$

In deriving this limit, we have used Eq. (1.32). In the asymptotic region, using the limiting expression (1.80) for the Green's function, Eq. (1.79) can be written as

$$\frac{\psi_L^{(+)}(r)}{kr} = j_L(kr) - i^{-L} k \frac{\exp(ikr)}{kr} \left[\int_0^\infty dr' r'^2 j_L(kr') V(r') \frac{\psi_L^{(+)}(r')}{kr'} \right]. \tag{1.81}$$

If the quantity in brackets in Eq. (1.81) is identified as the partial-wave t matrix $t_L(k)$ of Eq. (1.47).

$$\frac{1}{k} \int_0^\infty dr' r' j_L(kr') V(r') \psi_L^{(+)}(r') = -\frac{\exp(i\delta_L) \sin\delta_L}{k}, \tag{1.82}$$

then using the asymptotic form (1.29) of the Bessel function, the large r limit of Eq. (1.81) becomes

$$\lim_{r\to\infty} \frac{\psi_L^{(+)}(r)}{kr} \to \frac{\sin(kr - L\pi/2)}{kr} + i^{-L} \frac{\exp(ikr)}{kr} \exp(i\delta_L) \sin\delta_L. \tag{1.83}$$

Equation (1.82) is the partial-wave form of Eq. (1.74), as we shall see explicitly in Section 1.4. Equation (1.83) is identical to the initial boundary condition (1.37). Hence, solution of the partial-wave Lippmann–Schwinger equation (1.79) satisfies the asymptotic boundary condition (1.37). Also, as $r \to 0$, one can see using Eq. (1.77) in Eq. (1.79) that $\psi_L^{(+)}(r)$ satisfies Eq. (1.35).

The partial-wave integral equation for the wave-function component (1.79) is one with a compact kernel and can be solved numerically. The scattering cross section or phase shifts are related to its asymptotic behavior via Eqs. (1.36), (1.40), and (1.41). However, for numerical computation it is often convenient to use the transition-matrix description of scattering, which we describe in the following section.

1.4 TRANSITION-MATRIX DESCRIPTION

The transition matrix is the momentum-space matrix element of the operator $t(k^2)$, called the *transition operator*, often called the *t matrix*. Unlike in the wave-function description of scattering, where the scattering observables are extracted from the asymptotic behavior of the numerically computed wave function, the t matrix is directly related to scattering observables: phase shifts and cross sections. In realistic scattering problems, especially, the transition-matrix approach offers a considerable advantage over the wave-function description of scattering.

The transition matrix is defined by

$$\langle \phi_{\mathbf{k}'} | t(k^2) | \phi_{\mathbf{k}} \rangle \equiv \langle \phi_{\mathbf{k}'} | V | \psi_{\mathbf{k}}^{(+)} \rangle = \langle \psi_{\mathbf{k}'}^{(-)} | V | \phi_{\mathbf{k}} \rangle. \tag{1.84}$$

From Eqs. (1.74) and (1.84) we have

$$f(\theta) = -2\pi^2 \langle \phi_{\mathbf{k}'} | t(k^2) | \phi_{\mathbf{k}} \rangle. \tag{1.85}$$

In terms of this t matrix, the differential cross section (1.24) becomes

$$\frac{d\sigma}{d\Omega} = (2\pi^2)^2 \frac{(2m)^2}{(\hbar^2)^2} |\langle \phi_{\mathbf{k}'} | t(k^2) | \phi_{\mathbf{k}} \rangle|^2 \tag{1.86}$$

$$= (2\pi)^4 \frac{m^2}{\hbar^4} |\langle \phi_{\mathbf{k}'} | t(k^2) | \phi_{\mathbf{k}} \rangle|^2. \tag{1.87}$$

In Eqs. (1.86) and (1.87) all the factors of reduced mass m and $\hbar$ have been restored. Hence, the t matrix is closely related to the physical scattering amplitude $f(\theta)$ and the cross section $d\sigma/d\Omega$. Formal solutions of the Lippmann–Schwinger equations (1.64) and (1.66) can be written as

$$|\psi_{\mathbf{k}}^{(\pm)}\rangle = (I - G_0^{(\pm)}V)^{-1}|\phi_{\mathbf{k}}\rangle = |\phi_{\mathbf{k}}\rangle + G^{(\pm)}(k^2)V|\phi_{\mathbf{k}}\rangle, \tag{1.88}$$

where

$$G^{(\pm)}(k^2) = (k^2 - H \pm i0)^{-1} \tag{1.89}$$

is the full Green's function and satisfies the outgoing-wave (+) and incoming-wave (−) boundary conditions. Equation (1.88) can be established by using the following identity satisfied by the Green's functions:

$$\begin{aligned} G^{(\pm)}(k^2) &= G_0^{(\pm)}(k^2) + G^{(\pm)}(k^2)VG_0^{(\pm)}(k^2) \\ &= G_0^{(\pm)}(k^2) + G_0^{(\pm)}(k^2)VG^{(\pm)}(k^2). \end{aligned} \tag{1.90}$$

It can be seen from definition (1.84) and Eq. (1.88) that the transition operator is defined by

$$t(k^2) = V + VG^{(+)}(k^2)V. \tag{1.91}$$

The t-matrix elements of Eq. (1.84) are the momentum-space matrix elements of the transition operator $t(k^2)$. One has the useful identities

$$G^{(+)}(k^2)V = G_0^{(+)}(k^2)t(k^2), \qquad VG^{(+)}(k^2) = t(k^2)G_0^{(+)}(k^2). \tag{1.92}$$

From Eqs. (1.91) and (1.92) we find that the t matrix satisfies the following formal Lippmann–Schwinger integral equations:

$$t(k^2) = V + VG_0^{(+)}(k^2)t(k^2), \tag{1.93}$$

$$t(k^2) = V + t(k^2)G_0^{(+)}(k^2)V. \tag{1.94}$$

If we use the completeness relation for the bound and scattering eigenstates $|\psi_B\rangle$ and $|\psi_{\mathbf{q}}^{(\pm)}\rangle$, respectively, of the Hamiltonian H,

$$\sum_B |\psi_B\rangle\langle\psi_B| + \int d^3q\, |\psi_{\mathbf{q}}^{(\pm)}\rangle\langle\psi_{\mathbf{q}}^{(\pm)}| = I, \tag{1.95}$$

the spectral representation of the full Green's function $G^{(+)}(k^2)$ of Eq. (1.89) can be written as

$$G^{(+)}(k^2) = \sum_B \frac{|\psi_B\rangle\langle\psi_B|}{k^2 - E_B} + \int d^3q \frac{|\psi_{\mathbf{q}}^{(+)}\rangle\langle\psi_{\mathbf{q}}^{(+)}|}{k^2 - q^2 + i0}. \tag{1.96}$$

The sum is over the discrete bound states $|\psi_B\rangle$ and the integral is over the continuum scattering states $|\psi_{\mathbf{q}}^{(+)}\rangle$, both satisfying the Schrödinger equation.

From (1.96) we see that the matrix elements of the Green's function are analytic functions of energy in the entire complex energy plane except for some simple poles at real energies E_B and a cut across the positive real energy axis. Using Eq. (1.96) in Eq. (1.91) for the t matrix, the following expression is obtained:

$$\langle\phi_{\mathbf{p}}|t(k^2)|\phi_{\mathbf{p}'}\rangle = \langle\phi_{\mathbf{p}}|V|\phi_{\mathbf{p}'}\rangle + \sum_B \frac{f_B(\mathbf{p})^* f_B(\mathbf{p}')}{k^2 - E_B}$$

$$+ \int d^3q \frac{\langle\phi_{\mathbf{p}}|t(q^2)|\phi_{\mathbf{q}}\rangle\langle\phi_{\mathbf{q}}|t(q^2)|\phi_{\mathbf{p}'}\rangle^*}{k^2 - q^2 + i0}. \tag{1.97}$$

In Eq. (1.97) the momentum-space matrix element of a local Hermitian potential $V(\mathbf{r})$ is given by

$$\langle\phi_{\mathbf{p}}|V|\phi_{\mathbf{p}'}\rangle = \int d^3r\, \langle\phi_{\mathbf{p}}|\mathbf{r}\rangle V(\mathbf{r})\langle\mathbf{r}|\phi_{\mathbf{p}}'\rangle \tag{1.98}$$

$$= \frac{1}{(2\pi)^3} \int d^3r \exp(i\mathbf{Q}\cdot\mathbf{r}) V(r) \tag{1.99}$$

$$= \frac{1}{2\pi^2 Q} \int_0^\infty r\, dr \sin(Qr) V(r), \tag{1.100}$$

where $\mathbf{Q} = \mathbf{p}' - \mathbf{p}$. When the potential term of Eq. (1.98) is substituted for the t matrix in Eq. (1.85), we obtain the *Born approximation* or *first*

Born approximation for the scattering amplitude, given by

$$f_B(\theta) = -\frac{1}{4\pi}\frac{2m}{\hbar^2}\int d^3r \exp(i\mathbf{Q}\cdot\mathbf{r})U(\mathbf{r}), \qquad (1.101)$$

where we have restored the factors of reduced mass. Here $\mathbf{Q} = \mathbf{k} - \mathbf{k}'$ and θ is the angle between $\mathbf{k}$ and $\mathbf{k}'$.

In the complex energy plane, the t-matrix elements also have simple poles at the bound states and a branch cut along the real energy axis. Consequently, the t-matrix elements are complex for all real energies above the scattering threshold. This branch cut is a consequence of the branch cut in the free Green's function (1.57) at positive energies and is called the unitarity cut. The quantity

$$f_B(\mathbf{p}) \equiv \langle\phi_{\mathbf{p}}|V|\psi_B\rangle \qquad (1.102)$$

is the bound-state form factor or vertex function. An important feature of Eq. (1.97) is that the residues of the bound-state poles of the t matrix factorize in $\mathbf{p}'$ and $\mathbf{p}$. For real Hermitian potentials and at real positive energies the branch cut implied by Eq. (1.97) can be shown more explicitly as

$$\langle\phi_{\mathbf{p}}|[t(k^2) - t^\dagger(k^2)]|\phi_{\mathbf{p}'}\rangle = \\ -2\pi i\int d^3q\, \langle\phi_{\mathbf{p}}|t(q^2)|\phi_{\mathbf{q}}\rangle\delta(k^2 - q^2)\langle\phi_{\mathbf{q}}|t^\dagger(q^2)|\phi_{\mathbf{p}'}\rangle. \qquad (1.103)$$

In deriving Eq. (1.103), use has been made of the following identity:

$$\frac{1}{k^2 - q^2 \pm i0} = \frac{\mathcal{P}}{k^2 - q^2} \mp i\pi\delta(k^2 - q^2), \qquad (1.104)$$

where the symbol $\mathcal{P}$ denotes that a principal-value prescription is to be used in evaluating the singular integral over the momentum-space Green's function. Equation (1.104) separates the free Green's function in its real and imaginary parts. The t matrix is neither Hermitian nor unitary, but satisfies the Hermitian analyticity property

$$t^\dagger(k^2 + i0) = t(k^2 - i0) \qquad (1.105)$$

with $t(k^2) \equiv t(k^2 + i0)$. The quantity $t(k^2 - i0)$ satisfies Eqs. (1.93) and (1.94), with the outgoing-wave Green's function $G_0^{(+)}(k^2)$ replaced by the incoming-wave Green's function $G_0^{(-)}(k^2)$. Equation (1.103) is the unitarity relation for the full t matrix.

In explicit notation, using the completeness relation (1.56) for the momentum eigenstates, the momentum-space representation of the t-matrix Lippmann–Schwinger equation (1.93) is given by

$$\langle \phi_{\mathbf{p}} | t(k^2) | \phi_{\mathbf{p}'} \rangle = \langle \phi_{\mathbf{p}} | V | \phi_{\mathbf{p}'} \rangle + \int d^3 q \frac{\langle \phi_{\mathbf{p}} | V | \phi_{\mathbf{q}} \rangle \langle \phi_{\mathbf{q}} | t(k^2) | \phi_{\mathbf{p}'} \rangle}{(k^2 - q^2 + i0)}, \tag{1.106}$$

where we have inserted a complete set of momentum eigenstates in the intermediate state.

Solutions of the Lippmann–Schwinger equations (1.63), or equivalently, (1.106), satisfy the Schrödinger equation and the correct asymptotic boundary conditions. So these equations generate unique scattering solutions. The uniqueness of the solution is guaranteed by the Fredholm nature of the kernel, or equivalently, by the absence of the solution of the homogeneous Lippmann–Schwinger equation in the scattering region. The homogeneous version of the t-matrix equation is given by

$$t(k^2) | \phi_{\mathbf{k}} \rangle = V G_0^{(+)}(k^2) t(k^2) | \phi_{\mathbf{k}} \rangle. \tag{1.107}$$

Comparing Eqs. (1.73) and (1.107), we see that these two equations are equivalent with the identification $|\psi_B\rangle = G_0^{(+)}(k^2) t(k^2) | \phi_{\mathbf{k}} \rangle$. Hence, both equations have solutions at the same discrete energies, the bound-state energies. For the usual short-range local Hermitian potentials, these equations do not have a solution in the scattering region.

Often, it is convenient to define an operator, called the S *matrix*, closely related to the t matrix, by

$$\langle \phi_{\mathbf{p}} | S | \phi_{\mathbf{p}'} \rangle = \langle \psi_{\mathbf{p}}^{(-)} | \psi_{\mathbf{p}'}^{(+)} \rangle. \tag{1.108}$$

Using definitions (1.88) of the outgoing- and incoming-wave wave functions, the S-matrix element can be rewritten as

$$\langle\phi_{\mathbf{p}}|S|\phi_{\mathbf{p}'}\rangle = \langle\phi_{\mathbf{p}}|\psi_{\mathbf{p}'}^{(+)}\rangle + \langle\phi_{\mathbf{p}}|V(p^2 - H + i0)^{-1}|\psi_{\mathbf{p}'}^{(+)}\rangle \tag{1.109}$$

$$= \langle\phi_{\mathbf{p}}|\phi_{\mathbf{p}'}\rangle + (p^2 - p'^2 + i0)^{-1}\langle\phi_{\mathbf{p}}|V|\psi_{\mathbf{p}'}^{(+)}\rangle$$
$$+ (p'^2 - p^2 + i0)^{-1}\langle\phi_{\mathbf{p}}|V|\psi_{\mathbf{p}'}^{(+)}\rangle \tag{1.110}$$

$$= \delta^3(\mathbf{p} - \mathbf{p}') - 2i\pi\delta(p^2 - p'^2)\langle\phi_{\mathbf{p}}|t(p^2)|\phi_{\mathbf{p}'}\rangle. \tag{1.111}$$

From Eqs. (1.69) and (1.108), the completeness relation (1.95) of the orthonormal functions $|\psi_B\rangle$ and $|\psi_{\mathbf{q}}^{(\pm)}\rangle$, and the property

$$\langle\phi_{\mathbf{p}}|S^\dagger|\phi_{\mathbf{q}}\rangle = \langle\psi_{\mathbf{q}}^{(-)}|\psi_{\mathbf{p}}^{(+)}\rangle^* = \langle\psi_{\mathbf{p}}^{(+)}|\psi_{\mathbf{q}}^{(-)}\rangle, \tag{1.112}$$

it follows that the S matrix is unitary:

$$\int d^3q\,\langle\phi_{\mathbf{p}}|S^\dagger|\phi_{\mathbf{q}}\rangle\langle\phi_{\mathbf{q}}|S|\phi_{\mathbf{p}'}\rangle =$$
$$\int d^3q\,\langle\psi_{\mathbf{p}}^{(+)}|\psi_{\mathbf{q}}^{(-)}\rangle\langle\psi_{\mathbf{q}}^{(-)}|\psi_{\mathbf{p}'}^{(+)}\rangle = \delta^3(\mathbf{p} - \mathbf{p}'), \tag{1.113}$$

or

$$SS^\dagger = S^\dagger S = I. \tag{1.114}$$

The partial-wave projection for the t matrix is defined by

$$\langle\phi_{\mathbf{p}}|t(k^2)|\phi_{\mathbf{p}'}\rangle = \frac{1}{2\pi^2}\sum_{L=0}^{\infty}(2L+1)t_L(p,p',k^2)P_L(\cos\theta) \tag{1.115}$$

$$= \frac{2}{\pi}\sum_{L=0}^{\infty}\sum_{M=-L}^{L} t_L(p,p',k^2)Y_{LM}(\hat{\mathbf{p}})Y_{LM}^*(\hat{\mathbf{p}}'), \tag{1.116}$$

where θ is the angle between vectors $\mathbf{p}'$ and $\mathbf{p}$, Y_{LM} are the spherical harmonics, and $\hat{\mathbf{p}} = \mathbf{p}/p$. A similar partial-wave projection exists for the potential

$$\langle \phi_{\mathbf{p}} | V | \phi_{\mathbf{p}'} \rangle = \frac{1}{2\pi^2} \sum_{L=0}^{\infty} (2L+1) V_L(p, p') P_L(\cos\theta) \tag{1.117}$$

$$= \frac{2}{\pi} \sum_{L=0}^{\infty} \sum_{M=-L}^{L} V_L(p, p') Y_{LM}(\hat{\mathbf{p}}) Y_{LM}^*(\hat{\mathbf{p}}'), \tag{1.118}$$

so that the partial-wave components are defined by

$$V_L(p, p') = \pi^2 \int_{-1}^{1} dx\, P_L(x) \langle \phi_{\mathbf{p}} | V | \phi_{\mathbf{p}'} \rangle, \tag{1.119}$$

where $x = \cos\theta$. The partial-wave momentum-space expression of a local potential $V(r)$ is given by

$$V_L(p, q) = \int_0^{\infty} j_L(pr) V(r) j_L(qr) r^2\, dr. \tag{1.120}$$

Equation (1.120) is the partial-wave projection of Eq. (1.98).

Using partial-wave projections (1.116) and (1.118) in the t-matrix equation (1.106), the following one-dimensional momentum-space partial-wave Lippmann–Schwinger t-matrix equation is obtained:

$$t_L(p, p', k^2) = V_L(p, p') + \frac{2}{\pi} \int_0^{\infty} q^2 dq \frac{V_L(p, q) t_L(q, p', k^2)}{k^2 - q^2 + i0}. \tag{1.121}$$

The t matrix $t_L(k)$ of Eq. (1.47) is a special element of the general partial-wave t-matrix elements of Eq. (1.116): $t_L(k) = t_L(k, k, k^2)$. The quantity $t_L(k)$ is the on-shell t-matrix element. The general element $t_L(p, p', k^2)$, with $p' \neq k \neq p$, is off-shell and the elements $t_L(k, p', k^2)$ and $t_L(p, k, k^2)$ are referred to as either half-on-shell or half-off-shell. All physical observables are given by the on-shell t-matrix elements. The solution of the momentum-space partial-wave equation (1.121) for the physical scattering observables given by the on-shell t-matrix elements yields as a by-product the off-shell t-matrix elements.

The partial-wave Lippmann–Schwinger t-matrix equation couples the on-shell and off-shell t-matrix elements. The partial-wave projection of Eq. (1.84), with $|\mathbf{k}'| = p, |\mathbf{k}| = k$, leads to the following expression relating half-shell t-matrix elements and the wave function in partial waves :

$$t_L(p,k,k^2) = \int_0^\infty r^2\,dr\,j_L(pr)V(r)\frac{\psi_L^{(+)}(r)}{kr}. \tag{1.122}$$

If we take $p = k$ in this equation, the on-shell partial-wave t matrix, $t_L(k) = t_L(k,k,k^2)$, is given by

$$t_L(k) = -\frac{\exp(i\delta_L)\sin\delta_L}{k} = \int_0^\infty r^2\,dr\,j_L(kr)V(r)\frac{\psi_L^{(+)}(kr)}{kr}, \tag{1.123}$$

where we have used Eq. (1.147). The on-shell partial-wave t matrix of Eq. (1.123) is identical with previous identification (1.82). The Born–Neumann series (Section 2.5) for t_L can be obtained by using the following iterative solution of Eq. (1.79) in Eq. (1.123) [1]:

$$\frac{\psi_L^{(+)}(r)}{kr} = j_L(kr) + \int_0^\infty dr' r'^2 G_L^{(+)}(r,r',k^2)V(r')j_L(kr') + \cdots \tag{1.124}$$

If n iterative terms are included in Eq. (1.124), the approximation is termed the nth Born approximation. A similar Born–Neumann series solution is possible for all scattering equations. For weak potentials, $V \to 0$, and at high energies, $G_L^{(+)} \to 0$, this procedure leads to a convergent iterative scheme.

The partial-wave form of Eq. (1.97) is given by

$$\begin{aligned} t_L(p,p',k^2) =& V_L(p,p') + \sum_B \frac{f_{BL}(p)^* f_{BL}(p')}{k^2 - E_B} \\ &+ \frac{2}{\pi}\int_0^\infty q^2 dq \frac{t_L(p,q,q^2)t_L(p',q,q^2)^*}{k^2 - q^2 + i0}, \end{aligned} \tag{1.125}$$

where the summation now includes the bound states in the Lth partial wave. The residue at the bound-state pole is related to the bound-state wave function $\psi_{BL}(q)$ via

$$f_{BL}(p) = \frac{2}{\pi}\int_0^\infty q^2\,dq\,V_L(p,q)\psi_{BL}(q), \tag{1.126}$$

where $\psi_{BL}(q)$ satisfies the homogeneous version (1.73) of the Lippmann–Schwinger equation (1.63) or (1.66). Explicitly, the bound-state problem is given by

$$\psi_{BL}(p) = -\frac{2}{\pi(\alpha^2+p^2)}\int_0^\infty q^2\,dq\,V_L(p,q)\psi_{BL}(q), \tag{1.127}$$

where α^2 is the binding energy. The consideration above shows that the partial-wave t-matrix elements have the same unitarity cut and pole structure of the full t matrix (1.97).

If we take $p = p' = k$ in Eq. (1.125), the following on-shell partial-wave unitarity condition is obtained:

$$\Im[t_L(k,k,k^2)] = \frac{t_L(k,k,k^2) - t_L(k,k,k^2)^*}{2i} = -k|t_L(k,k,k^2)|^2, \tag{1.128}$$

which is often written as

$$\Im[t_L(k)] = -k|t_L(k)|^2 = -\frac{k}{4\pi}\sigma_L, \tag{1.129}$$

where the partial-wave cross section σ_L is defined by Eq. (1.41). Equation (1.128) is the on-shell partial-wave form of the unitarity relation (1.103). Equation (1.129) is the partial-wave projection of Eq. (1.42) and is the partial-wave optical theorem. The exact partial-wave t-matrix elements as well as any parametrization (1.47) with real phase shifts satisfy unitarity relation (1.128). If this condition is violated, the phase shifts are to be complex in general. An imaginary part of the phase shift would imply probability nonconservation, for example, by Eq. (1.36), which means that the number of particles is not conserved in the scattering process. This should be considered a serious fault of the theory, and we should try to preserve unitarity condition (1.128) in finding approximations to the t matrix.

From Eq. (1.125) we find that in the complex energy plane, the element $t_L(p,p',k^2)$ has the bound-state poles of Lth partial wave. From Eq. (1.128) we see that the partial-wave on-shell t matrix $t_L(k)$ has a square-root branch cut in the complex energy plane along the entire real energy axis. Hence, in this plane the on-shell t matrix is defined on two sheets, the first of which is called the *physical sheet.* Although for calculating physical observables—cross sections, phase shifts, and bound states—one needs the t matrix on the first sheet of the complex energy plane, the t matrix can be continued analytically to the second sheet and can have poles on the second sheet. The poles on the real energy axis on the second sheet at $E = -E_v$ correspond to virtual or quasi-bound states. With the increase in potential strength, they move along the real negative energy axis, come to the origin, then move to the first sheet of the complex energy plane and continue moving along the real negative energy axis [1]. The poles in the complex energy plane on the second sheet correspond to resonances and appear for complex-conjugate energies $E = a \pm ib$. Although such poles are not of direct consequence in the study of scattering and bound states, they may have a strong effect on the physical t matrix in the first sheet when $E_v \to 0$ or $b \to 0$. In the first case, the zero-energy t matrix tends to infinity, an effect known as *threshold enhancement*. In the second case, the t matrix at energy $E = a$ tends to infinity, which is known as a resonance at energy a and of width b. There is no unitarity cut in the complex $k = \sqrt{E}$ plane. In the transformation from the complex energy to the complex momentum plane, the entire first (second) sheet of the complex energy plane gets mapped onto the upper (lower)-half complex momentum plane. The resonances appear at pairs of k values $k = \pm k_\alpha - ik_\beta$ in the lower-half complex-k plane, and virtual (bound) states appear on the negative imaginary axis $k = -ik_v$ ($k = +ik_b$) of the same plane. The t matrix develops poles at these positions. In Section 2.8 we discuss numerical methods for determining the virtual states and resonances.

1.5 *K*-MATRIX DESCRIPTION

The t-matrix description of Section 1.4 involves a delicate treatment of a limiting $\pm i0$ prescription, which leads to complex algebra and could be inconvenient for numerical purposes. Also, in approximate treatments for the t matrix, such as the Born-Neumann series based on Eq. (1.123), optical theorem (1.42) and unitarity relation (1.128) are not expected to be satisfied.

An equivalent description of scattering could be made in terms of the K matrix. The K matrix is real and hence is numerically more convenient to deal with than the t matrix. Also, any real approximation to it preserves unitarity automatically or leads to real phase shifts. It is because of these two advantages that the K-matrix description is frequently used in realistic scattering problems.

Previously, we have included an infinitesimal imaginary part in the denominator of Eq. (1.54) so as to generate a meaningful Green's function. To introduce the K-matrix description, we consider the standing-wave free Green's function by taking the principal-value integral in Eq. (1.54), so that

$$G_0^{\mathcal{P}}(\mathbf{r}, \mathbf{r}', k^2) = \mathcal{P} \int \frac{d^3 q}{(2\pi)^3} \frac{\exp[i\mathbf{q} \cdot (\mathbf{r}' - \mathbf{r})]}{k^2 - q^2} \tag{1.130}$$

$$= -\frac{1}{4\pi} \frac{\cos(k|\mathbf{r}' - \mathbf{r}|)}{|\mathbf{r}' - \mathbf{r}|}. \tag{1.131}$$

The standing-wave Green's function $G_0^{\mathcal{P}}(\mathbf{r}, \mathbf{r}', k^2)$ is the real part of the Green's functions $G_0^{(\pm)}(\mathbf{r}, \mathbf{r}', k^2)$ of Eq. (1.58) because of Eq. (1.104). The standing-wave free Green's function is real and does not possess a unitarity branch cut along the real energy axis.

The outgoing-wave free Green's function (1.58) leads directly to the physical scattering boundary condition (1.23), as we have seen. The t-matrix elements derived with this Green's function are closely related to the physical scattering amplitude via Eq. (1.85). Green's function (1.131) also satisfies differential equation (1.60), but now corresponding to a standing-wave boundary condition at positive energies. In this case, the boundary condition, such as given by Eq. (1.61) for the outgoing-wave Green's function, is more complicated and is not directly related to the physical boundary condition (1.23) of the outgoing-wave wave function. Hence, we do not derive this asymptotic limit of the principal-value Green's function. The principal-value Green's function (1.131) does not lead directly to the scattering boundary condition (1.23), nor are the K-matrix elements derived with the principal-value Green's function related directly to the physical scattering amplitude. However, the K-matrix elements so derived can be used to calculate the t-matrix elements and/or scattering observables.

For numerical treatment one again employs a partial-wave projection, as in the outgoing-wave case. Using the partial-wave expansion for the Green's function in configuration space given by Eqs. (1.75) and (1.76), one obtains the following partial-wave standing-wave Green's function:

$$G_L^{\mathcal{P}}(r, r', k^2) = kj_L(kr_<)n_L(kr_>) \tag{1.132}$$

for $k^2 > 0$. Here $<$ ($>$) denotes the smaller (larger) of the variables r and r', and j_L, and n_L are the usual spherical Bessel and Neumann functions. As expected, the standing-wave Green's function (1.132) is the real part of the outgoing-wave Green's function (1.77). As $r \to \infty$ the standing-wave Green's function (1.132) has the following asymptotic behavior:

$$\lim_{r\to\infty} G_L^{\mathcal{P}}(r, r', k^2) \to -kj_L(kr')\frac{\cos(kr - L\pi/2)}{r}. \tag{1.133}$$

In close analogy with the outgoing-wave case, one can now define the standing-wave wave function and its partial-wave projection. The standing-wave scattering wave function again has the partial-wave projection (1.33) and satisfies the partial-wave Schrödinger differential equation (1.34). The complete wave function and its partial-wave projection satisfy the following Lippmann–Schwinger equations:

$$\psi_{\mathbf{k}}^{\mathcal{P}}(\mathbf{r}) = \phi_{\mathbf{k}}(\mathbf{r}) + \int d^3r'\, G_0^{\mathcal{P}}(\mathbf{r}, \mathbf{r}', k^2)V(\mathbf{r}')\psi_{\mathbf{k}}^{\mathcal{P}}(\mathbf{r}'), \tag{1.134}$$

$$\frac{\psi_L^{\mathcal{P}}(r)}{kr} = j_L(kr) + \int_0^\infty dr' r'^2 \frac{G_L^{\mathcal{P}}(r, r', k^2)V(r')\psi_L^{\mathcal{P}}(r')}{kr'}, \tag{1.135}$$

respectively.

As the standing-wave wave function $\psi_{\mathbf{k}}^{\mathcal{P}}(\mathbf{r})$ is not easily related to the physical wave function $\psi_{\mathbf{k}}^{(+)}(\mathbf{r})$, it is not convenient to study its asymptotic behavior. The asymptotic behavior of the partial-wave wave function $\psi_L^{\mathcal{P}}(r)/kr$ is directly related to the scattering phase shifts and we study it in the following. This wave function also satisfies condition (1.35), satisfied by other wave functions. Using the asymptotic behavior (1.133) of the standing-wave Green's function $G_L^{\mathcal{P}}(r, r', k^2)$ in Eq. (1.135), we obtain the following boundary condition for the wave function:

$$\lim_{r\to\infty} \psi_L^{\mathcal{P}}(r) \to \sin\left(kr - \frac{L\pi}{2}\right) + \tan\delta_L \cos\left(kr - \frac{L\pi}{2}\right) \tag{1.136}$$

$$\to \frac{\sin(kr - L\pi/2 + \delta_L)}{\cos\delta_L}, \tag{1.137}$$

provided that

$$\frac{1}{k}\int_0^\infty r\,dr\,j_L(kr)V(r)\psi_L^{\mathcal{P}}(r) = -\frac{\tan\delta_L}{k}. \tag{1.138}$$

The boundary conditions (1.36) and (1.37) are very similar and the phase shifts can be calculated from the asymptotic forms of both the outgoing- and standing-wave wave functions.

Partial-wave integral equation (1.135) has a compact kernel. However, for numerical computation, it is often convenient to use the K-matrix description of scattering. Equation (1.138) should be compared with Eq. (1.82) in the case of t matrix. Equation (1.82) defines the on-shell t-matrix element. We shall see in the following that Eq. (1.138) defines the on-shell K-matrix elements introduced below.

Using the standing-wave Green's function, one can define a real reactance or K matrix, in the same way as the t matrix, by

$$\langle\phi_{\mathbf{k}'}|K(k^2)|\phi_{\mathbf{k}}\rangle \equiv \langle\phi_{\mathbf{k}'}|V|\psi_{\mathbf{k}}^{\mathcal{P}}\rangle = \langle\psi_{\mathbf{k}'}^{\mathcal{P}}|V|\phi_{\mathbf{k}}\rangle. \tag{1.139}$$

Using the following formal integral equation satisfied by the wave function,

$$|\psi_{\mathbf{k}}^{\mathcal{P}}\rangle = |\phi_{\mathbf{k}}\rangle + G_0^{\mathcal{P}}(k^2)V|\psi_{\mathbf{k}}^{\mathcal{P}}\rangle, \tag{1.140}$$

the following K-matrix Lippmann–Schwinger equations are obtained:

$$K(k^2) = V + VG_0^{\mathcal{P}}(k^2)K(k^2), \tag{1.141}$$

$$K(k^2) = V + K(k^2)G_0^{\mathcal{P}}(k^2)V. \tag{1.142}$$

The derivation of Eqs. (1.141) and (1.142) follows steps similar to those of t-matrix equations (1.93) and (1.94), where the outgoing-wave free Green's function has to be substituted by the principal-value free Green's function.

The standing-wave Green's function is the real part of the outgoing-wave Green's function. But the K matrix is not the real part of the t matrix and the following inequality should be noted:

$$K(k^2) \neq \Re[t(k^2)] = V + VG^{\mathcal{P}}(k^2)V \neq K(k^2), \tag{1.143}$$

where $G^{\mathcal{P}}(k^2)$ is the full Green's function with the principal-value prescription. However, the K-matrix Lippmann–Schwinger equation can be solved numerically to yield the scattering observables.

In explicit notation, the momentum-space K-matrix Lippmann–Schwinger equation is given by

$$\langle \phi_{\mathbf{p}} | K(k^2) | \phi_{\mathbf{p}'} \rangle = \langle \phi_{\mathbf{p}} | V | \phi_{\mathbf{p}'} \rangle + \mathcal{P} \int d^3q \frac{\langle \phi_{\mathbf{p}} | V | \phi_{\mathbf{q}} \rangle \langle \phi_{\mathbf{q}} | K(k^2) | \phi_{\mathbf{p}'} \rangle}{k^2 - q^2}. \tag{1.144}$$

Again, Eq. (1.144) is essentially the t-matrix equation (1.106) with the principal-value free Green's function. A partial-wave projection of the K matrix is defined in the same way as in the case of the t matrix through Eqs. (1.116), (1.118), and (1.123). The partial-wave form of the K-matrix Lippmann–Schwinger equation is

$$K_L(p, p', k^2) = V_L(p, p') + \frac{2}{\pi} \mathcal{P} \int_0^\infty q^2\, dq \frac{V_L(p, q) K_L(q, p', k^2)}{k^2 - q^2}. \tag{1.145}$$

In contrast to the t-matrix elements, both the total and partial-wave K-matrix elements of (1.144) and (1.145) are real for real positive energies above the scattering threshold at $k^2 = 0$ and do not possess the unitarity cut. For real Hermitian potentials the K-matrix elements are real and Hermitian:

$$K(k^2) = K^\dagger(k^2), \tag{1.146}$$

$$\Im[K_L(k, k, k^2)] = 0. \tag{1.147}$$

These equations should be contrasted with the unitarity conditions (1.105) and (1.128) for the t matrix.

Next we would like to relate the partial-wave K-matrix elements with the partial-wave t-matrix elements. The partial-wave t-matrix Lippmann–Schwinger equation (1.121) can be rewritten as

$$t_L(p, p', k^2) = V_L(p, p') + \frac{2}{\pi} \mathcal{P} \int_0^\infty q^2 dq \frac{V_L(p, q) t_L(q, p', k^2)}{k^2 - q^2} - ik V_L(p, k) t_L(k, p', k^2), \tag{1.148}$$

where we have broken up the outgoing-wave Green's function into the following real principal-value and imaginary δ function parts according to Eq. (1.104). Using Eqs. (1.145) and (1.148), the following relation between the partial-wave K- and t-matrix elements are obtained:

$$t_L(p,p',k^2) = K_L(p,p',k^2) - ikK_L(p,k,k^2)t_L(k,p',k^2). \qquad (1.149)$$

For calculating the scattering observables, only the on-shell t-matrix elements are required. The on-shell partial-wave t-matrix elements $t_L(k)$ are related to the on-shell partial-wave K-matrix elements $K_L(k^2) \equiv K_L(k,k;k^2)$ by

$$t_L(k) = \frac{K_L(k^2)}{1 + ikK_L(k^2)}. \qquad (1.150)$$

In terms of the wave function, the on-shell partial-wave K-matrix elements are given by

$$K_L(k^2) = \int_0^\infty r^2 dr \frac{j_L(kr)V(r)\Psi^{\mathcal{P}}(r)}{kr}. \qquad (1.151)$$

Equation (1.151) is the partial-wave projection of Eq. (1.139).

Using parametrization (1.47) for the t matrix in Eq. (1.150), one obtains the following parametrization for the K matrix:

$$K_L(k^2) = -\frac{\tan \delta_L}{k}. \qquad (1.152)$$

This equation relates the partial-wave on-shell K-matrix elements to phase shifts, which are in turn related to the scattering observables such as cross section. Now we realize that requirement (1.138) is equivalent to Eqs. (1.151) and (1.152), and the quantity in Eq. (1.138) is related directly to the partial-wave on-shell K matrix. The advantage of using the K matrix for making approximate calculation is that any real approximation to it always leads to real phase shifts via Eq. (1.152) and to S and t matrices obeying conditions of unitarity (1.46) and (1.129). The implementation of unitarity on an approximation to the t matrix is less trivial.

The on-shell K matrix of (1.152) is a real analytic function of k^2 in the neighborhood of the scattering threshold, and for $L = 0$ one can make the following alternative Taylor series expansions [11]:

$$k \cot \delta_0 = -\frac{1}{a} + \frac{1}{2} r_1 k^2 + O(k^4), \tag{1.153}$$

$$k^{-1} \tan \delta_0 = -a + \frac{1}{6} r_2^3 k^2 + O(k^4). \tag{1.154}$$

In these expansions the quantity $a = -\lim_{k^2 \to 0} k^{-1} \tan \delta_0$, called the *scattering length*, plays a fundamental role in low-energy scattering. At low energies the scattering occurs predominantly in the S wave, and the scattering length gives a measure of scattering. The expansion (1.153) is commonly used in the analysis of scattering. This is valid for large $|a|$. For $|a| \to 0$ this expansion breaks down, as the first term on the right-hand side of Eq. (1.153) tends to diverge. Then one should consider expansion (1.154), which breaks down when $|a| \to \infty$. Depending on whether one is near the $|a| \to 0$ or $|a| \to \infty$ limit, one should use expansion (1.153), or (1.154), respectively. Expansion (1.153) [(1.154)] is exact for S-wave scattering, with a square-well potential with range $R = r_1$ $[r_2]$ in the limit $|a| \to 0$ $[\infty]$. This is why the parameters r_1 and r_2 are called effective ranges under the two conditions.

The S-matrix elements defined by (1.43) are the on-shell partial-wave projections of the S operator defined by (1.108). The relation of these S-matrix elements to the t- and K-matrix elements are given by

$$S_L(k) = 1 - 2ikt_L(k) \tag{1.155}$$

$$= \frac{1 - ikK_L(k^2)}{1 + ikK_L(k^2)}. \tag{1.156}$$

1.6 MULTICHANNEL SCATTERING

Scattering between two composite objects, which we call A and B, generally produce a multiplicity of final states involving many particles. In such cases, a mere description of the asymptotic boundary conditions, especially in the presence of multicluster final states, is a nontrivial task. The problem comes under control somewhat, when multicluster final states are altogether neglected. At low energies the multicluster final states are not energetically open, and this procedure seems to be reasonable. Even after neglecting the multicluster final states, an exact solution of most of these problems is out of the question and one has to rely on approximation techniques for solution. If only the two-cluster final states, corresponding to the possibilities

$$A + B \to A + B, \tag{1.157}$$

$$\to A^* + B^*, \tag{1.158}$$

$$\to C + D, \tag{1.159}$$

of elastic (including spin-flip transitions), inelastic excitation, and rearrangement collisions, respectively, are considered, one can derive multiparticle Lippmann–Schwinger equations in close analogy with single-channel potential scattering. However, these equations do not have Fredholm kernels and cannot be used for numerical purposed [5].

Approximate multichannel Fredholm scattering models can be derived from these multiparticle Lippmann–Schwinger equations. There are many ways of deriving such approximate equations. Essentially, by projecting the multiparticle Lippmann–Schwinger equations to two-cluster channel states, one can derive coupled reaction channel equations of nuclear physics or close-coupling (CC) equations of atomic physics [7,8]. For identical fermions this approach yields the resonating group method [9]. These multichannel models are usually applied for numerical studies.

1.6.1 Inadequacy of Multiparticle Lippmann–Schwinger Equations

The inadequacy of the multiparticle Lippmann-Schwinger equation is already explicit in the three-particle problem, which we discuss below. Suppose that the two-cluster scattering is initiated by projectile 1 and target 23, which is the bound state of particles 2 and 3. The full Hamiltonian H in this case, can be written as

$$H = H_0 + \sum_{i=1}^{3} V_i \tag{1.160}$$

$$= H_1 + V^1, \tag{1.161}$$

where H_0 is the total kinetic energy of the system and V_i is the potential between particles j and k, so that $i \neq j \neq k \neq i$, with $i, j, k = 1, 2, 3$. Here $H_1 = H_0 + V_1$ is the internal Hamiltonian for the initial channel state of particle 1 and bound state 23, and $V^1 = V_2 + V_3$ is the potential between projectile 1 and target 23. The partition (1.161) of the Hamiltonian is very similar to that in potential scattering. Here H_1 is some kind of generalized kinetic energy responsible for a correct description of the target wave function and the free relative motion between the target and the projectile. The potential V^1 is solely responsible for scattering.

Now one can develop the three-particle Lippmann–Schwinger equation in close analogy with potential scattering. The incident scattering state satisfies

$$H_1|\Phi_{\mathbf{k}_1}\rangle = E|\Phi_{\mathbf{k}_1}\rangle, \tag{1.162}$$

where $|\Phi_{\mathbf{k}_1}\rangle = |\phi_{B1}\phi_{\mathbf{k}_1}\rangle$. Here $|\phi_{B1}\rangle$ represents the target bound state 23 and $|\phi_{\mathbf{k}_1}\rangle$ is the usual plane wave of relative motion between the target and the projectile. As in potential scattering, the Lippmann–Schwinger equation now becomes

$$|\Psi^{(+)}_{\mathbf{k}_1}\rangle = |\Phi_{\mathbf{k}_1}\rangle + G_1^{(+)}(E)V^1|\Psi^{(+)}_{\mathbf{k}_1}\rangle, \tag{1.163}$$

where $G_1^{(+)}(E) \equiv (E - H_1 + i0)^{-1}$ is the outgoing-wave free Green's function.

The Lippmann–Schwinger equation (1.163) is formally very similar to the equation for potential scattering (1.63). But there are conceptual differences between the two. Equation (1.163) is in three-particle space, and the intermediate states in this equation are also in that space. Disconnected pieces in the following momentum-space matrix element of potential V_2 in three-particle space in the form of a δ function of infinite norm are responsible for the divergence of the Hilbert–Schmidt norm of the kernel of Eq. (1.163):

$$\langle \mathbf{p}_2, \mathbf{q}_2|V_2|\mathbf{p}_2', \mathbf{q}_2'\rangle = \langle \mathbf{q}_2|V_2|\mathbf{q}_2'\rangle\delta(\mathbf{p}_2 - \mathbf{p}_2'). \tag{1.164}$$

The potential V_2 of Eq. (1.164) is just a part of the potential V^1 of Eq. (1.163). In Eq. (1.164), $\mathbf{q}_2$, and $\mathbf{q}_2'$ are the relative momenta between particles 1 and 3, and $\mathbf{p}_2$ and $\mathbf{p}_2'$ are the relative momenta between particle 2 and the center of mass of particles 1 and 3. The δ function appears because the potential V_2 between particles 1 and 3 cannot change the relative momentum between particle 2 and the center of mass of particles 1 and 3.

The Hilbert–Schmidt norm of the kernel in three-particle space is infinite even if the two-particle potential V_2 between particles 1 and 3 satisfies Eq. (1.72) in two-particle space. For example, in three-particle space the Hilbert–Schmidt norm of V_2 will contain, among other things,

$$|V_2|^2_{HS} \sim \int d^3p_2\, d^3p'_2\, \delta^2(\mathbf{p}_2 - \mathbf{p}'_2) \to \infty, \tag{1.165}$$

because one gets a δ function from one potential term of the norm. Obviously, the norm in Eq. (1.165) is undefined. The inclusion of the Green's function $G_1^{(+)}$ of Eq. (1.163) in the norm of Eq. (1.165) does not save the situation. Hence, Eq. (1.163) has a kernel of infinite norm and is ill defined, so the kernel of the Lippmann–Schwinger equation (1.163) is not compact or Fredholm. Consequently, this equation does not possess a unique scattering solution. Related to this are the existence of homogeneous equations of scattering, which we describe below in the multiparticle case. Any numerical solution algorithm based on Eq. (1.163) must fail above the lowest scattering threshold [1].

The situation can only worsen in the real multiparticle problem. Nevertheless, we derive and discuss the multiparticle Lippmann–Schwinger equation below in order to exhibit the deficiencies more clearly and to show how such an equation can lead to a well-defined and meaningful solution if certain approximations are made to the multiparticle dynamics. Many of the multiparticle models actually in use are based on definite approximations to Lippmann–Schwinger equations.

Suppose that the scattering in a multiparticle system is initiated by objects A and B in partition a and in two-cluster channel α. It is natural to consider the partition (1.6) $H = H_\alpha + V^\alpha$ of the Hamiltonian. Here, V^α is the sum of pair potentials between the constituents of A and B and H_α is the sum of the kinetic energy and intracluster interactions of A and B. Consequently, H_α is responsible for the binding of A and B and for their free relative motion, and V_α is responsible for scattering. It is technically correct to label the Hamiltonian by partition index as in Eq. (1.6), as there can be many channels in a single partition. However, we assume only one channel in each partition in the following and use Greek letters α, β, ... to label both the channel and the partition to which the channel belongs.

In this case, the channel Green function and the incident state are defined, respectively, by

$$G_\alpha^{(+)}(E) \equiv (E - H_\alpha + i0)^{-1}, \tag{1.166}$$

$$(E - H_\alpha)|\Phi_{\mathbf{k}_\alpha}\rangle = 0, \tag{1.167}$$

where E, H_α, V^α, ... are expressed in units of $\hbar^2/2m_\alpha$, with m_α the reduced mass in channel α. Here $|\Phi_{\mathbf{k}_\alpha}\rangle$ represents the initial plane wave composed of the bound states, $|\phi_{B\alpha}\rangle$, of the clusters in channel α and a plane wave of relative motion between them:

$$|\Phi_{\mathbf{k}_\alpha}\rangle = |\phi_{B\alpha}\rangle|\phi_{\mathbf{k}_\alpha}\rangle. \tag{1.168}$$

If both initial clusters have internal structure, the function $|\phi_{B\alpha}\rangle$ is the product of the bound states of these two clusters. The wave function $|\Psi_{\mathbf{k}_\alpha}^{(+)}\rangle$ for scattering initiated in channel α at energy E satisfies the following Lippmann–Schwinger equations [4,5]:

$$|\Psi_{\mathbf{k}_\alpha}^{(+)}\rangle = |\Phi_{\mathbf{k}_\alpha}\rangle + G_\alpha^{(+)}(E)V^\alpha|\Psi_{\mathbf{k}_\alpha}^{(+)}\rangle \tag{1.169}$$

$$= G_\beta^{(+)}(E)V^\beta|\Psi_{\mathbf{k}_\alpha}^{(+)}\rangle, \tag{1.170}$$

where β is a two-cluster channel in partition b. The appearance of the homogeneous Lippmann–Schwinger scattering equation (1.170) in the multiparticle case has no analogy in potential scattering. By multiplying Eqs. (1.169) and (1.170) by $E - H_\alpha$ and $E - H_\beta$, respectively, equivalence of these scattering equations and the Schrödinger equation can be established.

The scattering wave function $|\Psi_{\mathbf{k}_\alpha}^{(+)}\rangle$ should possess asymptotically the incoming wave $|\Phi_{\mathbf{k}_\alpha}\rangle$ in channel α and spherically outgoing wave in all open channels. It is this asymptotic boundary condition that determines that Eq. (1.170) for $|\Psi_{\mathbf{k}_\alpha}^{(+)}\rangle$ should not have any incoming wave. [The equivalence of Eq. (1.170) with the Schrödinger equation requires that the only possible incoming wave in this equation be $|\Phi_{\mathbf{k}_\beta}\rangle$. But this incoming wave is not physically acceptable.] A partial configuration representation of Eqs. (1.169) and (1.170) can be written as

$$\Psi_{\mathbf{k}_\alpha}^{(+)}(\mathbf{r}_\alpha) = \Phi_{\mathbf{k}_\alpha}(\mathbf{r}_\alpha) + \int d^3r'_\alpha \, d^3r''_\alpha \, G_\alpha^{(+)}(E; \mathbf{r}_\alpha, \mathbf{r}'_\alpha) \times V^\alpha(\mathbf{r}'_\alpha, \mathbf{r}''_\alpha)\Psi_{\mathbf{k}_\alpha}^{(+)}(\mathbf{r}''_\alpha), \tag{1.171}$$

$$\Psi_{\mathbf{k}_\alpha}^{(+)}(\mathbf{r}_\beta) = \int d^3r'_\beta \, d^3r''_\beta \, G_\beta^{(+)}(E; \mathbf{r}_\beta, \mathbf{r}'_\beta)V^\beta(\mathbf{r}'_\beta, \mathbf{r}''_\beta)\Psi_{\mathbf{k}_\alpha}^{(+)}(\mathbf{r}''_\beta), \tag{1.172}$$

where $\mathbf{r}_\beta$ ($\mathbf{r}_\alpha$) is the asymptotic separation between the fragments in channel β (α).

At infinity, one can generate asymptotic configuration for different channels by taking the appropriate channel separation to tend to infinity. This leads to the following asymptotic behavior in the presence of two-cluster open channels β:

$$\Psi_{\mathbf{k}_\alpha}^{(+)}(\mathbf{r}) \to \sum_\beta \frac{\phi_{B\beta}}{(2\pi)^{3/2}} \left[\delta_{\alpha\beta} \exp(i\mathbf{k}_\alpha \cdot \mathbf{r}_\alpha) + f_{\beta\alpha}(\hat{\mathbf{k}}_\beta, \hat{\mathbf{k}}_\alpha) \frac{\exp(ik_\beta r_\beta)}{r_\beta}\right]. \tag{1.173}$$

The quantity $f_{\beta\alpha}(\hat{\mathbf{k}}_\beta, \hat{\mathbf{k}}_\alpha)$ is the amplitude for transition from channel α to β:

$$f_{\beta\alpha}(\hat{\mathbf{k}}_\beta, \hat{\mathbf{k}}_\alpha) = -2\pi^2 \langle \phi_{\mathbf{k}_\beta} | V^\beta | \Psi_{\mathbf{k}_\alpha}^{(+)} \rangle, \tag{1.174}$$

and $\hat{\mathbf{k}}_\alpha, \ldots$ are unit vectors in the direction of $\mathbf{k}_\alpha$. Equations (1.173) and (1.174) are generalizations of single-channel equations (1.23) and (1.74) to the multichannel case. Only in two-cluster channel α is there an incoming plane wave in addition to an outgoing spherical wave. In all other open two-cluster channels $\beta \neq \alpha$, one has only spherical outgoing waves. As energy increases, more and more two-cluster channels are open, and one has more outgoing waves in Eq. (1.173). In writing the boundary condition (1.173), we have assumed the absence of multicluster breakup channels, where more than two fragments may appear in the final state. In the presence of such channels the boundary condition (1.173) becomes more complex and one should have appropriate outgoing waves in each of the open breakup channels.

Equation (1.170) suggests the following equation with channel indices α and β changed:

$$|\Psi_{\mathbf{k}_\beta}^{(+)}\rangle = G_\alpha^{(+)}(E)V^\alpha|\Psi_{\mathbf{k}_\beta}^{(+)}\rangle, \qquad \beta \neq \alpha, \tag{1.175}$$

where $|\Psi_{\mathbf{k}_\beta}^{(+)}\rangle$ is the wave function for scattering initiated in channel β. The scattering equations are homogeneous when the channel labels of the wave function and the Green's function are different.

From Eqs. (1.169) and (1.175) we find that the homogeneous version of Eq. (1.169) is satisfied by a scattering wave function in other channels at any energy E above the scattering threshold [5]. As the general solution of an inhomogeneous equation may include an arbitrary admixture of nontrivial solution of its homogeneous version, the solution of Eq. (1.169) need not be unique. This is in contrast to the two-particle case, where the only nonuniqueness is associated with the appearance of a two-particle bound state, which is usually not realized in the scattering region. Hence, unlike in potential scattering, Eq. (1.169) alone does not constrain the physics sufficiently so as to incorporate the correct boundary condition (1.173) in the solution and does not have a unique solution.

The implication of this result, although not always emphasized, is of concern. It implies that any attempt to calculate the exact scattering state from a direct solution of Eq. (1.169) alone should fail, because this equation does not constrain the solution correctly. There are two alternatives to pursue. (1) One could completely abandon the multiparticle Lippmann–Schwinger equation and formulate compact scattering equations with unique solutions. These are the multiparticle connected-kernel scattering equations of the Faddeev–Yakubovskii type [6]. (2) The other possibility is to approximate the dynamics to such a degree in solving Eq. (1.169) that the resultant equation yields a unique prediction. The simplest example of this is the first Born approximation, $|\Psi_{\mathbf{k}_\alpha}^{(+)}\rangle \simeq |\Phi_{\mathbf{k}_\alpha}\rangle$, or the second Born approximation,

$$|\Psi_{\mathbf{k}_\alpha}^{(+)}\rangle \simeq |\Phi_{\mathbf{k}_\alpha}\rangle + G_\alpha^{(+)}(E)V^\alpha|\Phi_{\mathbf{k}_\alpha}\rangle, \tag{1.176}$$

to Eq. (1.169). Other approaches involve approximations that, in effect, reduce Eq. (1.169) to an integral equation or a coupled system of integral equations with the degrees of freedom and uniqueness properties of the two-body case. Many widely used calculational techniques fall in this category. We describe below a commonly used approximation scheme, known as the close-coupling (CC) scheme, based on the multiparticle Lippmann–Schwinger equations (1.169) and (1.170).

1.6.2 Close-Coupling Equations

In many problems of interest there could be an infinite number of two-cluster channels. This is true in electron hydrogen atom scattering, where each excited state of the hydrogen atom forms a two-cluster channel. The CC equations with Fredholm kernel couple a finite number of preselected two-cluster channels, and the remaining channels are neglected.

Let us consider the asymptotic boundary condition for the wave function $|\Psi^{(+)}_{\mathbf{k}_\alpha}\rangle$. As the relative separation r_α in channel α tends to infinity, $|\Psi^{(+)}_{\mathbf{k}_\alpha}\rangle$ tends to an incident plane wave and an outgoing spherical wave. However, as the relative separation in another two-fragment channel β tends to infinity, $|\Psi^{(+)}_{\mathbf{k}_\alpha}\rangle$ tends to an outgoing spherical wave in that channel with no incident plane wave. It is convenient to break up wave function $|\Psi^{(+)}_{\mathbf{k}_\alpha}\rangle$ into components so as to distribute these asymptotic behaviors.

Suppose that we select N channels denoted by $\gamma = \alpha, \beta, \sigma, \ldots$ in the model. The Schrödinger equation for scattering initiated in channel α is approximated as

$$(E - H)|\Psi^{(+)}_{\mathbf{k}_\alpha}\rangle \approx (E - H)\sum_{\gamma=1}^{N}|\Psi^{(+)}_{\mathbf{k}_\alpha\gamma}\rangle = 0, \tag{1.177}$$

where the full wave function is broken into components $|\Psi^{(+)}_{\mathbf{k}_\alpha\gamma}\rangle$. The complete asymptotic behavior of the full wave function is distributed over these components, each component carrying a specific behavior.

The component $|\Psi^{(+)}_{k_\alpha\alpha}\rangle$ tends to an incident plane wave and an outgoing spherical wave as the relative separation in channel α tends to infinity; it does not carry the outgoing spherical waves, as the relative separations in other channels tend to infinity. The component $|\Psi^{(+)}_{\mathbf{k}_\alpha\beta}\rangle$ tends to an outgoing spherical wave with no incident plane wave, as the relative separation in channel $\beta\,(\neq \alpha)$ tends to infinity.

Equation (1.177) can be rewritten as the following set of N coupled equations

$$(E - H_\gamma)|\Psi^{(+)}_{\mathbf{k}_\alpha\gamma}\rangle = \sum_{\beta=1}^{N} U_{\gamma\beta}|\Psi^{(+)}_{\mathbf{k}_\alpha\beta}\rangle, \tag{1.178}$$

with

$$U_{\gamma\beta} = [V^\gamma - (E - H_\gamma)\bar{\delta}_{\gamma\beta}], \tag{1.179}$$

where $\bar{\delta}_{ij} \equiv 1 - \delta_{ij}$ is the anti-Kronecker δ function. The passage from Eq. (1.177) to Eq. (1.178) is quite arbitrary. There are other ways of defining wave-function components and introducing coupling between them. The coupling scheme of Eq. (1.178) is called the *Fock coupling scheme*. By operating on Eq. (1.178) with $\gamma = \alpha$ ($\gamma = \sigma \neq \alpha$), by Green's functions $G^{(+)}_\alpha$ ($G^{(+)}_\sigma$), respectively, we obtain the following set of equations:

$$|\Psi^{(+)}_{\mathbf{k}_\alpha\alpha}\rangle = |\Phi_{\mathbf{k}_\alpha}\rangle + G^{(+)}_\alpha(E)\sum_{\beta=1}^{N} U_{\alpha\beta}|\Psi^{(+)}_{\mathbf{k}_\alpha\beta}\rangle, \tag{1.180}$$

$$|\Psi^{(+)}_{\mathbf{k}_\alpha\sigma}\rangle = G^{(+)}_\sigma(E)\sum_{\beta=1}^{N} U_{\sigma\beta}|\Psi^{(+)}_{\mathbf{k}_\alpha\beta}\rangle, \quad \sigma \neq \alpha. \tag{1.181}$$

The incident plane wave has been included explicitly in the equation for $|\Psi^{(+)}_{\mathbf{k}_\alpha\alpha}\rangle$, as this component should contain the incident plane wave and the outgoing spherical wave in channel α. The component $|\Psi^{(+)}_{\mathbf{k}_\alpha\sigma}\rangle$ ($\sigma \neq \alpha$) asymptotically has the outgoing spherical wave in channel σ and no incident plane wave. Equations (1.180) and (1.181) still have momentum δ functions in the kernel and hence are not of Fredholm type.

Fredholm integral equations can be written by considering bound-state approximation for Green's function(s) and wave function(s). In this approximation, the Green's function is taken as

$$G^{(+)}_\sigma(E) \approx \int d^3q_\sigma \frac{|\phi_{B\sigma}\rangle\langle\phi_{B\sigma}|}{E + e_\sigma - q_\sigma^2 + i0}. \tag{1.182}$$

Here e_σ, the total binding energy of the clusters, and the energy variables in the denominator of Eq. (1.182) are expressed in units of $\hbar^2/2m_\sigma$ with m_σ the reduced mass in channel σ. The on-shell value of the channel wave vector is

$$k_\sigma \equiv (E + e_\sigma)^{1/2}, \tag{1.183}$$

whether or not the channel is actually open. The last possibility is realized only if

$$E + e_\sigma \geq 0 \qquad \text{(open channel)}. \tag{1.184}$$

Then approximation (1.182) to the Green's function can be rewritten as

$$G_\sigma^{(+)}(E) \approx |\phi_{B\sigma}\rangle g_\sigma^{(+)}(k_\sigma^2)\langle\phi_{B\sigma}|, \tag{1.185}$$

where

$$g_\sigma^{(+)}(k_\sigma^2) = (k_\sigma^2 - h_\sigma + i0)^{-1}, \tag{1.186}$$

where h_σ is the Hamiltonian for relative motion in channel σ in units of $\hbar^2/2m_\sigma$. The bound-state approximation for the wave function implies taking

$$|\Psi_{\mathbf{k}_\sigma}^{(+)}\rangle \approx \sum_{\beta=1}^{N} |\Psi_{\mathbf{k}_\sigma\beta}^{(+)}\rangle \approx \sum_{\beta=1}^{N} |\phi_{B\beta}\rangle|\psi_{\mathbf{k}_\sigma\beta}^{(+)}\rangle, \tag{1.187}$$

where

$$|\psi_{\mathbf{k}_\sigma\beta}^{(+)}\rangle = \langle\phi_{B\beta}|\Psi_{\mathbf{k}_\sigma\beta}^{(+)}\rangle. \tag{1.188}$$

If these approximations for the Green's function and the wave function are introduced in Eqs. (1.180) and (1.181), the following equations are obtained:

$$|\psi_{\mathbf{k}_\alpha\alpha}^{(+)}\rangle = |\phi_{\mathbf{k}_\alpha}\rangle + g_\alpha^{(+)} \sum_{\beta=1}^{N} \langle\phi_{B\alpha}|U_{\alpha\beta}|\phi_{B\beta}\rangle|\psi_{\mathbf{k}_\alpha\beta}^{(+)}\rangle, \tag{1.189}$$

$$|\psi_{\mathbf{k}_\alpha\sigma}^{(+)}\rangle = g_\sigma^{(+)} \sum_{\beta=1}^{N} \langle\phi_{B\sigma}|U_{\sigma\beta}|\phi_{B\beta}\rangle|\psi_{\mathbf{k}_\alpha\beta}^{(+)}\rangle, \quad \sigma \neq \alpha. \tag{1.190}$$

After introducing approximations (1.185) and (1.187) to non-Fredholm Equations (1.180) and (1.181), well-defined Fredholm equations (1.189) and (1.190) have been obtained. There are no infinities in the norm of the kernels of the set of equations (1.189) and (1.190). Because of the projection onto the cluster bound states, the Green's functions $g_\sigma^{(+)}$ are of the two-cluster or two-particle type and the effective potentials $\langle\phi_{B\sigma}|U_{\sigma\beta}|\phi_{B\beta}\rangle$ appearing in Eqs. (1.189) and (1.190) are compact. Approximations (1.185) and (1.187) have, in effect, destroyed the multiparticle nature of the dynamics and have essentially led to a coupled set of two-particle equations (1.189) and (1.190). These equations are known as the *CC equations for wave-function components.*

For momentum-space treatment it is convenient to derive a coupled set of CC integral equations with Fredholm kernel for the t-matrix elements for transition from channel α to channel β, defined by

$$\langle\Phi_{\mathbf{k}_\beta}|T_{\beta\alpha}|\Phi_{\mathbf{k}_\alpha}\rangle = \langle\Phi_{\mathbf{k}_\beta}|V^\beta|\Psi_{\mathbf{k}_\alpha}^{(+)}\rangle \tag{1.191}$$

$$= \langle\Psi_{\mathbf{k}_\beta}^{(-)}|V^\alpha|\Phi_{\mathbf{k}_\alpha}\rangle. \tag{1.192}$$

Equations (1.191) and (1.192) are generalizations of potential scattering t matrix (1.84). Because of the asymmetry between initial and final channels α and β, we have two possibilities, (1.191) and (1.192). The first (second) is called the *post* (*prior*) *t matrix*, as it involves the interaction potential in the final (initial) or post (prior) channel. If exact wave functions are used in Eqs. (1.191) and (1.192), both yield the same t matrix. However, if approximate wave functions are used, each leads to a distinct t matrix.

If we operate on Eq. (1.189) with $\langle\phi_{B\rho}|U_{\rho\alpha}|\phi_{B\alpha}\rangle$ and on Eq. (1.190), for all σ, with $\langle\phi_{B\rho}|U_{\rho\sigma}|\phi_{B\sigma}\rangle$ and add all N equations, we get

$$\sum_{\sigma=1}^{N}\langle\phi_{B\rho}|U_{\rho\sigma}|\phi_{B\sigma}\rangle|\psi_{\mathbf{k}_\alpha\sigma}^{(+)}\rangle = \langle\phi_{B\rho}|U_{\rho\alpha}|\phi_{B\alpha}\rangle|\phi_{\mathbf{k}_\alpha}\rangle$$
$$+\sum_{\sigma=1}^{N}\langle\phi_{B\rho}|U_{\rho\sigma}|\phi_{B\sigma}\rangle \tag{1.193}$$
$$\times\, g_\sigma^{(+)}\sum_{\beta=1}^{N}\langle\phi_{B\sigma}|U_{\sigma\beta}|\phi_{B\beta}\rangle|\phi_{\mathbf{k}_\alpha\beta}^{(+)}\rangle.$$

Let us introduce the two-cluster effective potential $V_{\rho\alpha}$ and the t matrix $\mathcal{T}_{\rho\alpha}$ connecting the channels ρ and α, respectively, by

$$\langle \phi_{B\rho} | U_{\rho\alpha} | \phi_{B\alpha} \rangle = V_{\rho\alpha}, \tag{1.194}$$

$$\sum_{\sigma=1}^{N} \langle \phi_{B\rho} | U_{\rho\sigma} | \phi_{B\sigma} \rangle | \psi^{(+)}_{\mathbf{k}_\alpha \sigma} \rangle = \mathcal{T}_{\rho\alpha} | \phi_{\mathbf{k}_\alpha} \rangle. \tag{1.195}$$

Consistent with present approximation (1.187), the on-shell elements of the t matrix of Eq. (1.195) are identical with the on-shell elements of the t matrix defined in (1.191) and (1.192):

$$\langle \Phi_{\mathbf{k}_\beta} | T_{\beta\alpha} | \Phi_{\mathbf{k}_\alpha} \rangle = \langle \Phi_{\mathbf{k}_\beta} | \mathcal{T}_{\beta\alpha} | \Phi_{\mathbf{k}_\alpha} \rangle. \tag{1.196}$$

The on-shell momenta for different channels are defined by Eq. (1.183). These two t matrices of Eq. (1.196) are identical on shell, as the effective potential appearing on the left-hand side of Eq. (1.195) has the energy-dependent term $(E - H_\rho)\bar{\delta}_{\rho\sigma}$, which vanishes on shell. In terms of the effective potential and t matrices of Eqs. (1.194) and (1.195), Eq. (1.193) can be rewritten as

$$\mathcal{T}_{\rho\alpha} = V_{\rho\alpha} + \sum_{\sigma=1}^{N} V_{\rho\sigma} g_\sigma^{(+)} \mathcal{T}_{\sigma\alpha} \tag{1.197}$$

Equation (1.197) is the multichannel two-cluster Lippmann–Schwinger equation with compact kernel and unique solution. This equation is commonly known as the *CC equation for the t matrix*.

The effective CC potential (1.194) is called the *post form* of the CC potential, as the two-particle potentials in V^ρ of $U_{\rho\alpha}$ refer to the final or the post channel ρ. The CC potential is not symmetric in channel indices ρ and α. The potential

$$\hat{V}_{\rho\alpha} = \langle \phi_{B\rho} | [V^\alpha - (E - H_\alpha)\bar{\delta}_{\alpha\rho}] | \phi_{B\alpha} \rangle, \tag{1.198}$$

is known as the *prior form* of the CC potential, as the two-particle potentials in it refer to the initial or prior channel. Starting from the time-reversed multiparticle Lippmann–Schwinger equation, one can derive Eq. (1.197), but with the post CC potential replaced by the prior CC potential. Both the prior and post CC potentials carry an energy-dependent off-shell part. If exact cluster bound states are used, the on-shell matrix element of both prior and post potentials are the same and energy independent:

$$\langle \phi_{\mathbf{k}_\rho} | V_{\rho\alpha} | \phi_{\mathbf{k}_\alpha} \rangle = \langle \phi_{\mathbf{k}_\rho} | \hat{V}_{\rho\alpha} | \phi_{\mathbf{k}_\alpha} \rangle \tag{1.199}$$

$$= \langle \phi_{\mathbf{k}_\rho} | \langle \phi_{B\rho} | V^\rho | \phi_{B\alpha} \rangle | \phi_{\mathbf{k}_\alpha} \rangle \tag{1.200}$$

$$= \langle \phi_{\mathbf{k}_\rho} | \langle \phi_{B\rho} | V^\alpha | \phi_{B\alpha} \rangle | \phi_{\mathbf{k}_\alpha} \rangle. \tag{1.201}$$

In practice, in most cases the function $|\phi_{B\alpha}\rangle$ is taken to be approximate bound-state wave function in channel α. As a small finite number of bound-state functions $|\phi_{B\alpha}\rangle$ with the complete neglect of scattering state functions is far from forming a complete set, actual convergence of the CC calculational scheme is slow [8,12].

1.6.3 Partial-Wave Multichannel *K* and *S* Matrices

As in potential scattering, for obtaining a numerical solution of the CC scattering equation (1.197), the technique of partial-wave projection is useful. After partial-wave expansion of the CC equation (1.197), one constructs equations with conserved total angular momentum J and parity P. If the scattering wave function is expanded in terms of cluster s-wave bound states, the number of coupled equations equals the number of expansion functions for all partial waves J. Also, in all cases, the number of coupled equations for total angular momentum $J = 0$ is equal to the number of expansion functions employed. However, if cluster states with higher partial waves are used in the expansion, for $J \neq 0$ the number of equations and number of expansion functions are different in general, governed by the angular momenta sum rule. For each p-wave (d-wave) expansion function, the number of coupled equations for $J > 0$ is greater than the number of expansion functions by one (two), and so on.

If partial-wave projections, in conserved total angular momentum J, of the t matrix and the potential operators are taken exactly as in the case of potential scattering, namely, by Eqs. (1.116) and (1.118), the partial-wave CC equations have the following form:

$$t^J_{\beta\alpha}(p_\beta, k_\alpha, E) = V^J_{\beta\alpha}(p_\beta, k_\alpha) + \frac{2}{\pi} \sum_{\sigma=1}^{N} \int_0^\infty V^J_{\beta\sigma}(p_\beta, q_\sigma) \frac{q_\sigma^2 \, dq_\sigma}{k_\sigma^2 - q_\sigma^2 + i0} t^J_{\sigma\alpha}(q_\sigma, k_\alpha, E). \tag{1.202}$$

The partial-wave projection is straightforward when in the expansion of the wave function only s-wave cluster states are used [8]. However, if excited non-s-wave cluster states are used, the partial-wave projection is more involved [8]. In both cases a partial-wave CC equation of type (1.202) results. Often, it is convenient to work in terms of a dimensionless t matrix $\tilde{t}$ defined by

$$\tilde{t}^J_{\beta\alpha}(p_\beta, q_\alpha, E) = \sqrt{p_\beta q_\alpha}\, t^J_{\beta\alpha}(p_\beta, q_\alpha, E).$$

This t matrix satisfies

$$\tilde{t}^J_{\beta\alpha}(p_\beta, k_\alpha, E) = \tilde{V}^J_{\beta\alpha}(p_\beta, k_\alpha) + \frac{2}{\pi}\sum_{\sigma=1}^{N}\int_0^\infty \tilde{V}^J_{\beta\sigma}(p_\beta, q_\sigma)\frac{q_\sigma\, dq_\sigma}{k_\sigma^2 - q_\sigma^2 + i0}\tilde{t}^J_{\sigma\alpha}(q_\sigma, k_\alpha, E), \tag{1.203}$$

where $\tilde{V}$ is defined in a similar fashion.

As in potential scattering, the desired asymptotic limit for the wave function is obtained with the t matrices of Eqs. (1.192) and (1.197), provided that

$$f_{\beta\alpha}(\hat{\mathbf{k}}_\beta, \hat{\mathbf{k}}_\alpha) = -2\pi^2\langle\phi_{\mathbf{k}_\beta}|\mathcal{T}_{\beta\alpha}|\phi_{\mathbf{k}_\alpha}\rangle, \tag{1.204}$$

where the scattering amplitude $f_{\beta\alpha}$ is defined by Eq. (1.174).

For s-wave cluster states the full t matrix is given by

$$\langle\phi_{\mathbf{k}_\beta}|\mathcal{T}_{\beta\alpha}|\phi_{\mathbf{k}_\alpha}\rangle = \frac{1}{2\pi^2}\sum_J (2J+1)\frac{1}{\sqrt{k_\alpha k_\beta}}\tilde{t}^J_{\beta\alpha}(k_\beta, k_\alpha, E)P_J(\cos\theta). \tag{1.205}$$

In the presence of cluster states in non-s waves, one has the slightly more involved relation (1.217), given below.

The t matrix of Eq. (1.202) possesses scattering unitarity cuts for all open channels included in the CC model. A real K matrix involving principal-value integrals can be introduced as in the case of potential scattering. A real Hermitian dimensionless K matrix is introduced by

$$K_{\beta\alpha}(p_\beta, k_\alpha, E) = \tilde{V}_{\beta\alpha}(p_\beta, k_\alpha) + \frac{2}{\pi}\sum_{\sigma=1}^{N}\mathcal{P}\int_0^\infty \tilde{V}_{\beta\sigma}(p_\beta, q_\sigma)\frac{q_\sigma\, dq_\sigma}{k_\sigma^2 - q_\sigma^2}K_{\sigma\alpha}(q_\sigma, k_\alpha, E), \tag{1.206}$$

where the integration over the outgoing-wave Green's functions are replaced by principal-value integrals. Now, exactly as in the case of potential scattering, the K matrix can be related to the t matrix. The generalization of Eq. (1.149) to the multichannel case is given by

$$\tilde{t}_{\beta\alpha}(k_\beta, k_\alpha, E) = K_{\beta\alpha}(k_\beta, k_\alpha, E) - i\sum_{\sigma=1}^{N} K_{\beta\sigma}(k_\beta, k_\sigma, E)\tilde{t}_{\sigma\alpha}(k_\sigma, k_\alpha, E), \tag{1.207}$$

where we have considered only the on-shell version. This relation can be rewritten as

$$\sum_{\sigma=1}^{N}[\delta_{\beta\sigma} - iK_{\beta\sigma}(k_\beta, k_\sigma, E)]\tilde{t}_{\sigma\alpha}(k_\sigma, k_\alpha, E) = K_{\beta\alpha}(k_\beta, k_\alpha, E)., \tag{1.208}$$

where $\delta_{\beta\delta}$ is the Kronecker delta function: $\delta_{\beta\delta} = 1(0)$ for $\beta = \delta(\neq \delta)$. Hence, one has to invert a complex matrix to find the multichannel t matrix from the multichannel K matrix. The complex cut structure of the t matrix appears from this matrix inversion. This procedure yields t-matrix elements obeying conditions of unitarity for any real approximation to the K-matrix elements and is useful for a numerical treatment. In operator form the relation between the multichannel t and K matrices is given by

$$\tilde{t} = \frac{K}{I + iK}, \tag{1.209}$$

and the S matrix is given by

$$S = I - 2ikt = I - 2i\tilde{t} = \frac{I - iK}{I + iK}. \tag{1.210}$$

Equation (1.208) is the explicit matrix element of Eq. (1.209) in channel space.

The multichannel S matrix is symmetric and unitary:

$$S_{\alpha\beta} = S_{\beta\alpha}, \qquad \sum_{\sigma} S^*_{\alpha\sigma} S_{\sigma\beta} = \delta_{\alpha\beta}, \tag{1.211}$$

or, in operator form,

$$S = S^T, \qquad SS^\dagger = S^\dagger S = I. \tag{1.212}$$

The multichannel S, t, and K matrices are related to each other by Eqs. (1.209) and (1.210) and determine the observables.

In the presence of many open channels, the S matrix in the open-channel space is not diagonal. Nor does the parametrization

$$S_{\alpha\alpha} = \exp(2i\delta_{\alpha\alpha}) \tag{1.213}$$

of the diagonal elements lead to real phase shifts $\delta_{\alpha\alpha}$. However, the S matrix in the open-channel space can be diagonalized by an unitarity transformation U to lead to a diagonal unitary matrix $S' = USU^\dagger$. Then each of the diagonal elements of S' must have the form

$$S'_{\alpha\alpha} = \exp(2i\delta'_{\alpha\alpha}), \tag{1.214}$$

where $\delta'_{\alpha\alpha}$ are real and are called the *eigenphases*. Although S' does not determine the physical observables directly, the eigenphases behave like phase shifts in potential scattering.

1.6.4 Observables

One can define the Born approximation in the CC multichannel case in close analogy with the single-channel potential scattering Born approximation given by Eq. (1.101). If we use the effective post potential of Eq. (1.194), the Born approximation to the scattering amplitude is given by

$$f^B_{\beta\alpha}(\hat{\mathbf{k}}_\beta, \hat{\mathbf{k}}_\alpha) = -\frac{1}{4\pi}\frac{2m_\beta}{\hbar^2}\int d^3r \exp(-ik_\beta \cdot r_\beta)\phi^*_{B\beta} U_{\beta\alpha}\phi_{B\alpha} \exp(-ik_\alpha \cdot r_\alpha). \tag{1.215}$$

Here d^3r represents integration over coordinates of all the constituents of the target and projectile, the two exponentials denote the free relative motion in the incident and final channels, and the variables of the partition bound states in the initial and final channels are not explicitly shown. The complex conjugate on the bound state is shown explicitly because for non-s-wave bound states the angular part is usually complex.

In the presence of nonzero spins, this transition amplitude is related to the asymptotic behavior of the reaction amplitude for transition $\alpha s m_s \to \beta t m_t$ by [8,13]

$$\Psi_{\text{reac}}(\beta t) \to \phi_{\beta t} \frac{\exp(ik_\beta r_\beta)}{r_\beta} \sum_{m_t} f_{\beta t m_t;\alpha s m_s}(\theta\phi)\chi(t, m_t). \tag{1.216}$$

The spin and angular momentum states of the initial (final) channel are taken to be s, m_s, l (t, m_t, l'), so that the conserved total angular momentum state is J, M in each channel, $\chi(t, m_t)$ is the spin function in final channel β, $\phi_{\beta t}$ is the final bound cluster state, and v_α and v_β are relative velocities in these channels. The detector selects particles traveling in directions θ and ϕ with spin t and m_t. The reaction amplitude is related to the scattering S matrix by

$$\begin{aligned} f_{\beta t m_t;\alpha s m_s}(\theta\phi) = \\ \frac{1}{(k_\alpha k_\beta)^{1/2}} \sum_{J=0}^{\infty} \sum_{M=-J}^{J} \sum_{l=|J-s|}^{J+s} \sum_{l'=|J-t|}^{J+t} \sum_{\mu=-l'}^{l'} \\ i^{l-l'+1}\pi^{1/2}(2l+1)^{1/2} Y_{l'\mu}(\theta\phi) \\ \times \langle ls0m_s|JM\rangle\langle l't\mu m_t|JM\rangle(\delta_{\beta\alpha}\delta_{st}\delta_{l'l} - S^J_{\beta tl';\alpha sl}). \end{aligned} \tag{1.217}$$

The vector addition coefficients vanish unless $M = m_s$ and $\mu = M - m_t = m_s - m_t$.

The differential cross section for transition from $\alpha s m_s \to \beta t m_t$ for a collision in which the final objects emerge within the solid angle $d\Omega$ in the direction $\theta\phi$ with respect to the incident beam in the CM system is given by

$$\frac{d\sigma_{\beta t m_t;\alpha s m_s}}{d\Omega_\beta} = \frac{k_\beta}{k_\alpha}\frac{m_\alpha}{m_\beta}|f_{\beta\alpha}(\mathbf{k}_\beta, \mathbf{k}_\alpha)|^2, \tag{1.218}$$

where m_α and m_β are reduced masses in channels α and β, respectively. The ratio of masses is relevant in rearrangement channels and drops out in elastic and inelastic channels. In the following we consider the usual case of inelastic scattering with $m_\alpha = m_\beta$. The cross section for the $\alpha s \to \beta t$ collision with an unpolarized beam is obtained by averaging over the incident spin directions m_s and summing over the final spin directions m_t:

$$d\sigma_{\beta t;\alpha s} = (2s+1)^{-1} \sum_{m_s, m_t} d\sigma_{\beta t m_t;\alpha s m_s}, \tag{1.219}$$

where $(2s+1)$ is the spin degeneracy.

Finally, the differential cross section for the $\alpha \to \beta$ collision without regard to the channel spins s and t is obtained by averaging over initial spin s and summing over final spin t,

$$d\sigma_{\beta;\alpha} = \sum_{s=|I-i|}^{I+i} \sum_{t=|I'-i'|}^{I'+i'} \frac{2s+1}{(2I+1)(2i+1)} d\sigma_{\beta t;\alpha s}, \tag{1.220}$$

where I (I') and i (i') are the spins of the fragments in the initial (final) channel α (β), and the fraction represents the statistical weight of the channel spin s.

The total cross section for transition from channel α to β is given by

$$\sigma_{\beta\alpha}(\mathbf{k}_\alpha) = \frac{k_\beta}{k_\alpha} \frac{m_\alpha}{m_\beta} \int d\Omega_\beta |f_{\beta\alpha}(\mathbf{k}_\beta, \mathbf{k}_\alpha)|^2. \tag{1.221}$$

For the time-reversed process, the total cross section is given by

$$\sigma_{\alpha\beta}(\mathbf{k}_\beta) = \frac{k_\alpha}{k_\beta} \frac{m_\beta}{m_\alpha} \int d\Omega_\alpha |f_{\alpha\beta}(\mathbf{k}_\alpha, \mathbf{k}_\beta)|^2, \tag{1.222}$$

so that, for $m_\alpha = m_\beta$, the cross sections for the direct and time-reversed processes are related by

$$k_\alpha^2 \sigma_{\beta\alpha}(\mathbf{k}_\alpha) = k_\beta^2 \sigma_{\alpha\beta}(\mathbf{k}_\beta). \tag{1.223}$$

Relation (1.223) is known as the *principle of detailed balance*. The cross sections averaged over all possible incident $(\mathbf{k}_\alpha)$ is given by

$$\sigma_{\beta\alpha} = (4\pi)^{-1} \int d\Omega_\alpha \sigma_{\beta\alpha}(\mathbf{k}_\alpha). \tag{1.224}$$

This averaged quantity is measured in laboratory when the target is a volume of atomic gas with random orientations.

1.7 NOTES AND REFERENCES

[1] See, for example, Newton (1982).
[2] For a review see, for example, Adhikari and Kowalski (1991).
[3] See, for example, Truhlar et al. (1990).
[4] Lippmann and Schwinger (1950), Lippmann (1956).
[5] Adhikari and Kowalski (1991), Glöckle (1970a,b, 1983), Epstein (1957), Foldy and Tobocman (1957), Gerjuoy (1958, 1971).
[6] Faddeev (1965), Yakubovskii (1967), Lovelace (1964a).
[7] Seaton (1953, 1973), Burke (1965, 1968), Burke and Seaton (1971).
[8] Geltman (1969), Bransden (1983), Burke (1977).
[9] Tang et al. (1978), Wildermuth and Tang (1977).
[10] See, for example, Newton (1982), Joachain (1975), and Goldberger and Watson (1964).
[11] Adhikari (1985).
[12] Callaway (1978a).
[13] Nesbet (1980), Blatt and Biedenharn (1952).

CHAPTER 2

NUMERICAL METHODS

2.1 INTRODUCTION

In Chapter 1 we presented a mathematical formulation of the quantum theory of scattering. This mathematical framework for the problem incorporates different virtues, such as asymptotic condition and unitarity, in terms of the S, t, and K matrices. Specifically, one has a formulation in terms of well-defined integral equations for both the scattering wave function and for the t- or K-matrix elements. Next comes the task of finding efficient means for obtaining numerical solutions. Most of the remaining parts of the book are devoted to this topic, with illustrations from nuclear, atomic, and molecular physics and physical chemistry.

The general approaches to finding a complete numerical solution of Fredholm scattering equations [1,2] are relatively straightforward for single-channel potential scattering or for simple two-particle scattering with a real Hermitian central short-range potential. We describe these approaches in some detail first. For these problems one must select a specific computational method and find the numerical solution. Virtually, all solution procedures convert the Fredholm integral equation to a well-defined set of linear algebraic equations with relatively small dimensions, which is solved numerically.

In realistic scattering problems, due to the presence of one or more of the following features, the situation could be far more complicated than the ideal scattering model noted above.

1. There is the limiting prescription $\lim_{\epsilon \to 0}$ and/or the evaluation of a principal-value integral in the momentum-space Green's function in order to incorporate the correct boundary condition. This calls for certain care.
2. Also, in atomic and molecular physics and physical chemistry, there could be a large number of open two-body scattering channels in addition to the open three- and multiparticle channels at low to medium energies. Even if the three- and multiparticle channels are neglected and the actual problem is modeled as a two-cluster multichannel process, the existence of many two-body channels may increase the dimensions of the problem beyond control. In addition, the effective interaction could be quite complicated and the coupling between different channels could be strong.
3. As we shall see in the following, some formulations, such as the Γ matrix, are not time-reversal symmetric.
4. In some cases, for example, in close-coupling (CC) approximation employing p-wave basis functions, the potentials are purely absorptive or imaginary.
5. In molecular scattering problems the solution depends on the orientation of the target molecules and a tedious averaging has to be performed over various orientations of the target. In the presence of many of these complications, the numerical task could be quite intricate and a brute-force computing may not lead to satisfactory convergent results. An intelligent method of solution or a variational principle for solution has proved to be an asset in these cases.

The time-independent scattering integral equations with some short-range well-behaved potentials could be solved numerically. The well-behaved short-range nature of the interaction allows the development of a convergent partial-wave projection scheme and subsequent truncation in angular momentum. This procedure always leads to a single (or a set of coupled) one-dimensional Fredholm integral equation(s) of the generic form

$$Y_{ij}^E(x,y) = B_{ij}^E(x,y) + \sum_l \int_0^\infty dz\, \mathcal{K}_{il}^E(x,z) Y_{lj}^E(z,y), \tag{2.1}$$

where x, y, and z are continuum momentum variables, E is the parametric energy, and i, j, and l are the discrete channel variables. The channel variables specify the different initial and final states, their angular momenta, and other quantum numbers. The variable x has an active role in the solution process [i.e., the functional dependence of the unknown Y on x is exploited in Eq. (2.1)]. Both the parametric energy E and the variable y play a passive role in the solution process, which means that fixed values of these parameters are preselected before solving the scattering problem. Nevertheless, the functional dependencies of the Born term B, the kernel $\mathcal{K}$, and the solution Y on energy E or momentum y could be singular. Usually, in scattering problems, $Y_{ij}^{E}(x, y)$ also satisfies a set of integral equations that have the following conjugate form:

$$Y_{ij}^{E}(x, y) = \tilde{B}_{ij}^{E}(x, y) + \sum_{l} \int_{0}^{\infty} dz\, Y_{il}^{E}(x, z) \tilde{\mathcal{K}}_{lj}^{E}(z, y). \tag{2.2}$$

In Eq. (2.2), y now assumes an active role.

In the case of single-channel potential scattering, after suppressing the passive variables, one has the following standard form of scattering integral equation:

$$Y(x) = B(x) + \int_{0}^{\infty} dz\, \mathcal{K}(x, z)\, Y(z), \tag{2.3}$$

which is recognized as a standard textbook Fredholm integral equation [1,2].

Equations of types (2.1) and (2.3) appear in many multiparticle collision problems besides simple potential scattering [3]. Many of the models are approximate and involve the reduction of the multiparticle problems to a set of coupled effective two-body problems. A frequently used model is based on the CC equations of Section 1.6.2. These equations may have highly complicated Born terms and kernels, in addition to having a large number of coupled channels, which invariably lead to numerical complications. However, these difficulties are much less than those encountered in the proper treatment of multiparticle dynamics. Below the multiparticle thresholds, multiparticle dynamics plays a passive role in a collision process, and two-cluster models yield reliable results. Above the multiparticle thresholds the multiparticle dynamics plays an active role in the scattering process, and their neglect may lead to serious theoretical defects in the solution, such as flux nonconservation or loss of unitarity due to open multiparticle channels. However, in the absence of tractable multichannel models, effective two-body models have been used successfully at higher energies.

The formal Fredholm solution to Eq. (2.3) ceases to exist when its homogeneous version, with $B = 0$, permits a solution. This means that the operator $(I - \mathcal{K})$ does not have an inverse. The eigenfunction–eigenvalue problem

$$Y(x) = \int dz\, \mathcal{K}(x, z) Y(z) \tag{2.4}$$

corresponds to normalizable localized states in potential scattering. Below the lowest scattering threshold, these states are the bound states. They do not exist for potential scattering with local Hermitian potential. For potential scattering with nonlocal potentials, they may appear as continuum bound states at certain discrete energies [4]. The homogeneous version of disconnected-kernel multiparticle scattering equations allows scattering solutions at all energies for rearrangement channels [3]. Then Eq. (2.3) ceases to be of the Fredholm type and cannot be solved numerically.

Next, it remains to establish that the scattering integral equations we derived in Chapter 1 are really of the Fredholm type, so that they may yield unique solutions. The kernel $\mathcal{K}$ of these equations has to be $\mathcal{L}^2$, or compact, or Hilbert–Schmidt for these equations to be of the Fredholm type. In the scattering region in the t-matrix Lippmann–Schwinger equation there exists a limiting $\epsilon \to 0$ procedure to be performed in the parametric energy in the kernel over the scattering cut so that the kernel $\mathcal{K}$ is complex in the scattering region. Also, if ϵ is set equal to zero in the parametric energy, the norm of the kernel does not exist. Hence, it is not obvious that the kernel is $\mathcal{L}^2$ or compact in the scattering region. However, Weinberg has demonstrated that a sufficient condition for compactness for such scattering equations is to have $\text{Tr}(\mathcal{K}\mathcal{K}^\dagger) < \infty$ [1,2]. This condition is satisfied by short-range pair potentials with only two-cluster channels. This is the case for most of the scattering problems that we consider. Usually, more analytical work is needed to establish compactness of the scattering integral equations in the presence of multicluster channels [3]. In the two-particle attractive Coulomb problem, an infinite number of bound states are present, which is a sufficient condition for making the kernel of the scattering equation noncompact.

One of the various equivalent definitions of compactness is that such operators can be approximated uniformly by operators of finite ranks. These finite-rank approximations of the kernel (and the Born term) of the scattering integral equation will be of great help numerically. They will be shown to lead to different numerical methods for the solution of the scattering integral equation, such as the degenerate kernel scheme, separable expansion scheme, and many types of variational principles [3].

In Section 2.2 a general solution strategy for solving Lippmann–Schwinger scattering integral equations for the t or K matrix is given. In Section 2.3 we specialize to the case of the K matrix and related principal-value treatment. In Section 2.4 we describe a solution procedure in terms of an auxiliary nonsingular real integral equation that does not involve any principal-value treatment. The principal-value treatment appears at a later stage in the form of a single integral over solution of the nonsingular equation. This approach has certain advantages over the conventional t- or K-matrix approaches. In Section 2.5 we describe approximate iterative solution methods for the scattering integral equations. In Section 2.6 we describe the solution procedure for scattering with a complex potential. In Section 2.7 we describe a multi-channel generalization of the nonsingular equations presented in Section 2.4. Finally, in Section 2.8 we describe a method for calculating bound states and resonances which uses momentum-space integral equations similar to those used for scattering.

2.2 SOLUTION STRATEGY

It should be noted that there is a preferred momentum k in the scattering integral equation, the *on-shell momentum*, which is related to the parametric energy E by $E = k^2$. The physical or on-shell amplitude is related to observables, such as the phase shift or cross section, and corresponds to taking both the initial and final momenta equal to the on-shell momentum. However, the scattering integral equation couples the on-shell amplitude with infinitely many off-shell amplitudes defined for all continuous momenta between 0 and ∞. This coupling makes solution of the scattering equation more complicated.

Any numerical procedure to solve the continuous Fredholm integral equation (2.3) starts by approximating it by a set of discrete linear algebraic equations by means of a quadrature rule. The original Fredholm equation is termed *continuous*, as the momentum variables are continuous functions. On the other hand, the set of linear algebraic equations are defined on some discrete momentum mesh. The solution is then effected on this mesh. If needed, the solution for other values of momentum can be found by some interpolation technique or by using the integral equation once again. This last observation is relevant, as the on-shell momentum is not allowed to be one of the discrete momenta, and the on-shell amplitude is not one of the unknowns of the set of algebraic equations. Additional use of the integral equation or some interpolation of the solution is needed to get the on-shell amplitude.

If the domain of integration is subdivided into N mesh points $\{x_i, i = 1, ..., N\}$, then Eq. (2.3) reduces to the following set:

$$Y_i = B_i + \sum_{j=1}^{N} \mathcal{K}_{ij} \Delta_j Y_j, \tag{2.5}$$

where $Y_i = Y(x_i)$, $B_i = B(x_i)$, $\mathcal{K}_{ij} = \mathcal{K}(x_i, x_j)$, and $\Delta_j = dx_j$. If we have an efficient quadrature rule for discretization, the solution of Eq. (2.5) approaches $Y(x)$ uniformly as $N \to \infty$. An account of different quadrature rules, such as Gauss, Laguerre, Hermite, and so on, are given in Abramowitz and Stegan [5]. Evidently, a finite value of N is used in calculations. The magnitude of N depends on the complexity of the scattering equation, the quadrature rule employed, and the desired degree of precision. Equation (2.5) can be rewritten in the form of the following algebraic set of equations:

$$\sum_{j=1}^{N} M_{ij} Y_j = B_i, \tag{2.6}$$

with

$$M_{ij} = \delta_{ij} - \mathcal{K}_{ij} \Delta_j, \tag{2.7}$$

where δ_{ij} is the Kronecker delta function: $\delta_{ij} = 1\ (0)$ for $i = j\ (\neq j)$. The discrete solution Y_i of Eq. (2.6) can be obtained by inverting the matrix M. Hence, the discretization procedure leads to the solution at the discretization points. Once these solutions are obtained, a convenient interpolation, using Eq. (2.5) again, yields the unknown function Y at any point not a discretization point. In such an iterative interpolation one uses Eq. (2.5) with x_i as the point where one needs the unknown function.

As most of the scattering problems have the usual Lippmann–Schwinger form, it is worthwhile to consider the single-channel form (1.121) of this equation. It is assumed that both the potential and the t matrix are time-reversal symmetric [e.g., $V(p,p') = V(p',p)$], in addition, the potential is assumed to be real. However, the condition of real potential can be removed. Also, the kernel of Eq. (1.121) is assumed compact, and this equation is Fredholm.

We recall that $i0$ in the denominator of Eq. (1.121) implies a limiting procedure $\epsilon \to 0$. As the kernel of this equation is singular at $q = k$ once ϵ is set equal to zero, special care is needed in discretization. This singularity can be handled by decomposing $(k^2 - q^2 + i0)^{-1}$ into its real principal-value and imaginary δ-function parts according to Eq. (1.104). Because of the complex nature of the outgoing-wave Green's function, even for a real potential, the kernel of Eq. (1.121) is complex in the scattering region.

A straightforward but not the most efficient or especially intelligent way to find the solution of Eq. (1.121) starts by discretizing it directly. This procedure generates a set of complex linear algebraic equations which is then solved by matrix inversion or otherwise. One procedure to avoid complex algebra is to use Eq. (1.104) and break up Eq. (1.121) into its real and imaginary parts as follows [6]:

$$\Re[t(p,p',k^2)] = V(p,p') + \frac{2}{\pi}\mathcal{P}\int_0^\infty \frac{q^2\,dq}{k^2 - q^2} V(p,q)\Re[t(q,p',k^2)] \quad (2.8)$$

$$+ \sqrt{k}V(p,k)\Im[t(k,p',k^2)],$$

$$\Im[t(p,p',k^2)] = \frac{2}{\pi}\mathcal{P}\int_0^\infty \frac{q^2 dq}{k^2 - q^2} V(p,q)\Im[t(q,p',k^2)]$$

$$- \sqrt{k}V(p,k)\Re[t(k,p',k^2)], \quad (2.9)$$

where the potential has been assumed to be real and Hermitian. In Eqs. (2.8) and (2.9) and in the following we have suppressed the angular momentum label. In this way one has real algebra, but the dimension is increased by a factor of 2. Numerical treatment of real matrices is often convenient and leads to more precise results than those obtained with complex algebra.

Another procedure of avoiding complex algebra without increasing the dimension of the equation is to deal with the K matrix, which satisfies Eq. (1.145), where the integral over the outgoing-wave Green's function $(k^2 - q^2 + i0)^{-1}$ is replaced by its principal-value part. Then the phase shifts are obtained directly from the on-shell K-matrix elements; or one can also construct the t-matrix elements by using Eq. (1.149), with a knowledge of the K-matrix elements. These solution procedures, however, has to deal with principal-value integrals, and this requires special care which we describe in the next section.

The singularity problem can be avoided by the contour deformation technique [3]. Then instead of integrating along the positive real momentum axis in Eq. (1.121), one integrates along a deformed contour C in the lower-half complex momentum plane. One possible contour is a slightly rotated straight line beginning at the origin and going to ∞ in the lower-half complex momentum plane. The result of integration remains unchanged provided that no singularities of $V(p,q)t(q,p',k^2)$ are crossed in rotating the line $(0,\infty)$ to C. This method involves complex algebra and is not particularly advantageous for simple potential scattering. However, it is very suitable for multiparticle collisions with moving singularities in the complex potential that depend on both p and p' [3].

A better and simpler way of handling the fixed-point singularity in Eq. (1.121) is to express its solution $t(p,p',k^2)$ in terms of the solution of an auxiliary real nonsingular integral equation of the same dimension as the original one, but without having to deal with the principal-value prescription. In the end one has to perform a principal-value integral over the solution of the nonsingular equation while calculating the observables. We describe this approach in Section 2.4. The application of this technique to collision integral equations has proved to be extremely useful. However, there are an infinite number of ways of performing this nonsingular reduction, with each way leading to a distinct nonsingular equation to solve [3]. Other criteria must be used to prefer a particular reduction scheme.

Once a method for handling a fixed-point singularity has been chosen, there remains the problem of solving the Fredholm integral equation. One can write the formal solution of such an equation as an infinite series involving the traces of different powers of the kernel [1,2]. But such a formal solution is hardly of any use from a numerical point of view. The simplest practical procedure is to solve the set of (real or complex) linear algebraic equations (2.5) exactly without further approximation.

An entirely different strategy is to seek controllable approximations while solving Eq. (2.5). This will be of great help in realistic problems when there are many coupled channels. One such method is to consider the iterative Born–Neumann series (Section 2.5) of Eq. (2.1) written schematically as

$$Y = B + \mathcal{K}B + \mathcal{K}^2 B + \mathcal{K}^3 B + \cdots \tag{2.10}$$

The quantity B is called the *first Born term* or simply the *Born term*, $B + \mathcal{K}B$ is the *second Born term*, and so on. Such a series converges when the kernel is sufficiently weak. This happens when the energy is reasonably high or the interaction potential is weak enough. Then one can use only a small number of terms in Eq. (2.10) for calculating the t matrix. In many problems of atomic and molecular physics a small number of terms of the Born–Neumann series have been used with great success at medium energies for the past 40 or so years. Even the first Born approximation may lead to reasonable result. If the Born–Neumann series diverges, the information contained in the first few terms of the Born–Neumann series can be summed effectively by means of the Padé or other techniques to produce the solution [7].

Variational approaches have proved to be very useful for obtaining a successively convergent approximation scheme for realistic scattering problems in nuclear, atomic, and molecular physics. They are applicable at all energies for both weak and strong interactions and generate a variety of approximation schemes for a general Fredholm integral equation, such as the method of separable expansion, the degenerate kernel scheme, the method of moments, and so on. Of these the separable expansion scheme is specially useful for representing the two- and three-particle amplitudes that appear in the kernel of few-particle scattering integral equations. This procedure simplifies certain types of few-particle scattering equations by reducing their dimensionality and has frequently been used in nuclear few-particle problems [3]. Variational approaches have also been used with great success in problems of atomic and molecular physics over the last three decades. The problems studied include the scattering of atoms and molecules by photons, electrons, positrons, and diatoms. In most of these problems the use of variational principles has been fundamental in dealing with the mathematical complexity. We devote a significant portion of this book to describing these variational methods.

In practice, a solution strategy is selected depending on the complexity of the problem, available computing power, and ultimate use of the t-matrix elements. In a realistic problem of molecular dynamics, the number of coupled channels could be several hundreds, and with the computing power available, an approximate variational calculation is welcome. The effort needed to improve on the variational solution by attempting direct solution of the set of algebraic equations may be prohibitive and not worthwhile. On the other hand, if one is interested in calculating nucleon–nucleon scattering observables, a direct precise solution of the scattering integral equation is called for. However, if one is interested in using the two- and three-nucleon t-matrix elements in a four-nucleon scattering calculation, it might be advantageous to use the approximate variational separable expansion scheme, because of the ultimate simplification of the four-nucleon scattering equations [8].

The range of momentum integration in Eq. (1.121) covers the entire phase space (e.g., 0 to ∞), so one encounters an infinite integral. The Fredholm nature of the problem guarantees that the kernel decays rapidly to zero as the momentum variables tend to infinity. It is often useful from a practical point of view to transform the infinite integral to a finite integral. As the usual Gauss quadrature points often used to discretize the integrals are given between 0 and 1 or between -1 and $+1$, it is convenient to transform the infinite integral to a finite integral between 0 and 1 or between -1 and $+1$. In principle, any transformation can be used. Two trivial transformations mapping $0 < q < \infty$ into $0 < x < 1$ are

$$q = c\left(\ln\frac{1+x}{1-x}\right), \tag{2.11}$$

$$q = c\frac{x}{1-x}, \tag{2.12}$$

where the images of points $x = 0, 1$ are $q = 0, \infty$. The transformation

$$q = c\frac{1+x}{1-x} \tag{2.13}$$

maps $0 < q < \infty$ into $-1 < x < 1$, where the images of points $x = -1, 0, 1$ are $q = 0, c, \infty$. The differential dq in Eq. (1.121) is obtained directly from the transformation and is given by

$$dq = \frac{2c}{1 - x^2} \tag{2.14}$$

$$= \frac{c}{(1 - x)^2} \tag{2.15}$$

$$= \frac{2c}{(1 - x)^2}, \tag{2.16}$$

in the case of transformations (2.11), (2.12), and (2.13), respectively. One can have an infinite class of such transformations, each distributing the integration points in a different fashion. It is not a priori clear which distribution is most convenient for one particular integral. Actually, that is decided by trial. These mappings imply a maximum value of q in momentum space which is efficiently controlled by the parameter c. This corresponds to a cutoff in the infinite integral. Otherwise, the choice of parameter c is entirely arbitrary. Generally, the form of the transformation function and its parameters, such as c, are dictated by special features of the kernel and should be chosen to get the most accurate numerical result with a given number of mesh points. Once the numerical value of c is decided upon, the number of quadrature points should be increased to achieve the desired precision.

2.3 *K* MATRIX AND PRINCIPAL-VALUE TREATMENT

The K-matrix approach has often been used in solving scattering problems. For real Hermitian potentials this solution procedure then deals with real algebra and hence implies less memory and processing time and a more stable result in computation. This approach has been used with great success in scattering problems of nuclear, atomic, and molecular physics. However, this approach involves an integral equation with a principal-value prescription in the kernel. This requires special care, which we discuss in this section.

The principal-value treatment is also required in a t-matrix description of scattering. The implied $\epsilon \to 0$ limit in the Green's function and the kernel of the t-matrix integral equation (1.121) is numerically troublesome and needs special care. If one puts the imaginary part in this outgoing-wave Green's function equal to zero, the integral diverges due to a pole singularity at $q = k$. However, the integration appearing in Eq. (1.121) is perfectly well defined in the limit $\epsilon \to 0$, consistent with the outgoing-wave nature of the Green's function.

One possible numerical way of discretization is to use the following decomposition of an integral $\mathcal{I}$ over the outgoing-wave Green's function into its real principal-value and imaginary δ-function parts:

$$\mathcal{I} \equiv \frac{2}{\pi}\int_0^\infty \frac{q^2 dq f(q)}{k^2 - q^2 + i0} = \frac{2}{\pi}\mathcal{P}\int_0^\infty \frac{q^2 dq f(q)}{k^2 - q^2} - ikf(k), \tag{2.17}$$

where $f(q)$ is a smooth integrable function of q. Equation (2.17) follows with the use of Eq. (1.104). There are many ways of dealing with the principal-value part. In a numerical treatment, if the integration points around the singularity at $q = k$ are symmetrically chosen on both sides of the pole of the integrand, one obtains the principal-value part. Sloan [9] pointed out that if one employs mapping (2.13) with $c = k$ in order to transform the infinite q integral in (2.17) to a finite integral in x between -1 to 1 and chooses the integration quadrature points symmetrically about $x = 0$, the result immediately yields the principal-value integral. The usual Gauss quadrature points between -1 and 1 are symmetrical about 0 and can be used directly for this purpose.

Then the discretized version of Eq. (1.121) is written as

$$t_{ij} = V_{ij} + \frac{2}{\pi}\sum_{n=1}^{N} \frac{V_{in} p_n^2 \Delta_n}{p_0^2 - p_n^2} t_{nj} - ip_0 V_{i0} t_{0j} \tag{2.18}$$

where, for example, $t_{ij} = t(p_i, p_j)$, $V_{ij} = V(p_i, p_j)$, $\Delta_n = dp_n$, $p_0 = k$, and $p_i, i = 1, 2, ..., N$ are the discretization mesh points. Then the momentum points between 0 and ∞ are transformed to x points between -1 and $+1$ using mapping (2.13) with $c = k$. Once the integration mesh points exclude $x = 0$ and are chosen symmetrically about $x = 0$, as in mapping (2.13), consistency with principal-value prescription is achieved. The resultant complex matrix equation (2.18) is then solved numerically by matrix inversion or otherwise. However, there are numerical problems with very large complex matrices, and this procedure may not yield particularly stable and precise results.

It might be worthwhile to use a set of real linear equations. This can be achieved by working with the K-matrix integral equation (1.145) in place of the t-matrix equation (1.121). The discretized version of this equation is given by

$$K_{ij} = V_{ij} + \frac{2}{\pi} \sum_{n=1}^{N} \frac{V_{in} p_n^2 \Delta_n}{p_0^2 - p_n^2} K_{nj}, \tag{2.19}$$

where, for example, $K_{ij} = K(p_i, p_j)$. The integration points are to be chosen as in Eq. (2.18), in accordance with the principal-value prescription of Sloan. This real set of equations is then solved, for example, by matrix inversion. The K-matrix elements are then used to calculate the physical observables, such as the cross section or the phase shift.

The Sloan method requires the choice $c = k$ in Eq. (2.13) for an evaluation of the principal-value integrals. However, c was introduced as a free parameter of the mapping to be fixed by the range of the momentum-space kernel. A small (large) c is to be preferred when in the momentum space the kernel is short-ranged (long-ranged). Such a c will make the discretization of the integral more efficient with fewer integration points. Once c is taken to be k, flexibility in the choice of c is gone and the mapping (2.13) may turn out to be inefficient for discretization. For example, at low (high) energies $k \to 0$ ($k \to \infty$) and half of the integration points accumulate near $k = 0$ ($k \to \infty$). In both cases this results in an uneconomical distribution of mesh points, and the accuracy of the Sloan mapping could be low. This could be remedied by increasing the number of mesh points or by employing a more complex distribution of mesh points. The real auxiliary nonsingular equations of the next section for solving the scattering problem seems to be very suitable for avoiding the principal-value prescription. There c could be chosen independent of k and this will give an extra flexibility in distributing the mesh points.

One can also remove the principal-value prescription in an integral by use of the identity

$$\mathcal{P} \int_0^\infty \frac{dq}{k^2 - q^2} = 0. \tag{2.20}$$

Using this identity the principal-value integral of (2.17) can be rewritten as

$$\frac{2}{\pi} \mathcal{P} \int_0^\infty dq \frac{q^2 f(q)}{k^2 - q^2} = \frac{2}{\pi} \int_0^\infty dq \frac{q^2 f(q) - k^2 f(k)}{k^2 - q^2}. \tag{2.21}$$

As the integral on the right-hand side of this equation is nonsingular, the principal-value label $\mathcal{P}$ has been dropped from the integral.

By using identity (2.20), one can rewrite the K-matrix equation (1.145) in the following form [10,11]:

$$K(p,p',k^2) = V(p,p') + \frac{2}{\pi}\int_0^\infty \frac{dq[q^2V(p,q)K(q,p',k^2) - k^2V(p,k)K(k,p',k^2)]}{k^2 - q^2}. \quad (2.22)$$

The integral in this equation is nonsingular and there is no need for the principal-value prescription. The discretization of the integrals in Eqs. (2.21) and (2.22) can be done without paying any attention to the principal-value prescription provided that the point $q = k$ is not chosen accidentally as a mesh point. The discretized version of Eq. (2.22) is given by the following set of equations:

$$K_{ij} = V_{ij} + \frac{2}{\pi}\sum_{n=1}^{N} \Delta_n \frac{p_n^2 V_{in}K_{nj} - p_0^2 V_{i0}K_{0j}}{p_0^2 - p_n^2}, \quad (2.23)$$

$$K_{0j} = V_{0j} + \frac{2}{\pi}\sum_{n=1}^{N} \Delta_n \frac{p_n^2 V_{0n}K_{nj} - p_0^2 V_{00}K_{0j}}{p_0^2 - p_n^2}, \quad (2.24)$$

where the index 0 denotes on-shell quantities: $p_0 = k$, $K_{0j} = K(k,p_j)$, and so on. In Eq. (2.22), to remove the principal-value integral we have introduced an extra half-on-shell K matrix element $K(k,p',k^2)$. We have included an extra equation for the half-on-shell K-matrix element (2.24) to have a complete set of equations. Although the sum in Eqs. (2.23) and (2.24) runs from 1 to N, there are $N+1$ unknowns including the on-shell quantities. This closed set of $N+1$ real equations can be solved numerically. Nevertheless, the manifest asymmetry of Eq. (2.22) with respect to the on-shell point k may limit the numerical accuracy of the solution [11].

2.4 Γ MATRIX AND AUXILIARY NONSINGULAR EQUATION

Many of the above-mentioned treatments for the fixed-point singularity involve a principal-value prescription in the integral equation. As we have seen in Section 2.3, the principal-value treatment may require special care, which may limit the numerical accuracy under many situations. A more efficient way of dealing with the fixed-point singularity is to express the t or K matrix in terms of the solution of an auxiliary nonsingular Fredholm integral equation which does not involve the $i0$ prescription or principal-value integral. The principal-value prescription reappears in some algebraic relation in the form of an integral over known functions that relates the t or K matrices with the solution of the real auxiliary nonsingular equation. In such an integral the principal-value prescription can be handled reliably, for example, via Eq. (2.20) [12].

The Fredholm reduction of the singular equation is performed by introducing a real subtraction function $\gamma(\bar{k}, q)$, such that $\gamma(\bar{k}, \bar{k}) = 1$. Then the Lippmann–Schwinger equation (1.121) can be rewritten as

$$t(p, p', k^2) = V(p, p') + \frac{2}{\pi} V(p, \bar{k}) \int_0^\infty \frac{q^2 dq \gamma(\bar{k}, q)}{k^2 - q^2 + i0} t(q, p', k^2)$$

$$+ \frac{2}{\pi} \int_0^\infty q^2 dq A(p, q, k^2) t(q, p', k^2) \tag{2.25}$$

where

$$A(p, q, k^2) = \mathcal{V}(p, q) \frac{1}{k^2 - q^2} \tag{2.26}$$

with

$$\mathcal{V}(p, q) \equiv [V(p, q) - V(p, \bar{k}) \gamma(\bar{k}, q)]. \tag{2.27}$$

The subtraction function γ is supposed to be a smooth function of the momentum variables. The $\bar{k}$ dependencies of operators A and $\mathcal{V}$ are not explicitly shown. The kernel A is nonsingular at positive energies provided that $\bar{k} = k$. Then the fixed-point singularity at $q = k$ is canceled by the zero of the potential $\mathcal{V}(p, q)$. It will be assumed that at positive energies, $\bar{k} = k$. All the equations of this section are also valid at complex and negative energies, where there is no fixed-point singularity for real momenta and there is no reason to stick to $\bar{k} = k$. We shall see that negative-energy equations of are special interest for calculating binding energies and bound-state wave functions. This is why we keep the provision $\bar{k} \neq k$ in the present equations. At negative energies both $\bar{k} = (|E|)^{1/2}$ and $= c$, where c is an energy-independent constant, can be used. The $i0$ limit in the Green function of Eq. (2.26) is irrelevant and has been dropped.

The t matrix can be expressed in terms of the solution, called the Γ *matrix*, of the following real auxiliary nonsingular Fredholm integral equation:

$$\Gamma(p, p', k^2) = V(p, p') + \frac{2}{\pi}\int_0^\infty q^2\, dq\, A(p, q, k^2)\Gamma(q, p', k^2). \quad (2.28)$$

The Γ matrix is not time-reversal symmetric: $\Gamma(p, p', k^2) \neq \Gamma(p', p, k^2)$. However, this is not a problem in numerical calculation. The Γ matrix can easily be related to scattering and bound-state solutions.

The formal manipulation needed to relate the t matrix with the Γ matrix becomes very transparent in operator form. In operator form Eqs. (2.25) and (2.28) are written as

$$t = V + \bar{V}\mathcal{G}_0 t + At, \quad (2.29)$$

$$\Gamma = V + A\Gamma, \quad (2.30)$$

with $\bar{V}(p, q) = V(p, \bar{k})$, and $\mathcal{G}_0(p, q) = \delta(q - p)(k^2 - q^2 + i0)^{-1} \gamma(\bar{k}, q)$. Eliminating the potential V between Eqs. (2.29) and (2.30), the t matrix can be expressed in terms of the Γ matrix by

$$t = \Gamma + \bar{\Gamma}\mathcal{G}_0 t, \quad (2.31)$$

with $\bar{\Gamma}(p, q, k^2) = \Gamma(p, \bar{k}, k^2)$. The kernel $\bar{\Gamma}\mathcal{G}_0$ of Eq. (2.31) is separable: $\langle p|\bar{\Gamma}\mathcal{G}_0|q\rangle = \Gamma(p, k, k^2)(k^2 - q^2 + i0)^{-1}\gamma(k, q)$, which means that functional dependence on the momenta p and q are separated. An integral equation such as (2.31) with a separable kernel has an analytic solution (Section 3.1.2). The momentum-space matrix element of Eq. (2.31) is given by

$$t(p, p', k^2) = \Gamma(p, p', k^2) + \Gamma(p, \bar{k}, k^2)\mathcal{I}(\bar{k}, p', k^2), \quad (2.32)$$

with

$$\mathcal{I}(\bar{k}, p', k^2) \equiv (2/\pi)\int_0^\infty q^2\, dq (k^2 - q^2 + i0)^{-1}\gamma(\bar{k}, q)t(q, p', k^2) \quad (2.33)$$

$$= \frac{(2/\pi)\int_0^\infty q^2\, dq(k^2 - q^2 + i0)^{-1}\gamma(\bar{k}, q)\Gamma(q, p', k^2)}{1 - (2/\pi)\int_0^\infty q^2\, dq(k^2 - q^2 + i0)^{-1}\gamma(\bar{k}, q)\Gamma(q, \bar{k}, k^2)}. \quad (2.34)$$

In deriving Eq. (2.34) we have multiplied Eq. (2.32) from the left by $\mathcal{G}_0$, integrated over the appropriate momentum variables, and solved the resultant equation for $\mathcal{I}(\bar{k}, p', k^2)$. The preceding equations of this section are also valid for the K matrix provided that we replace the limiting $i0$ prescription by the principal-value prescription in appropriate integrals. The limiting $i0$ prescription of the t-matrix equation or the principal-value prescription of the K-matrix equation does not appear in the real Γ-matrix equation. However, it appears in Eq. (2.34) in integrals over known functions, where it can be handled by Eqs. (1.104) and/or (2.21).

If the system has a bound state, the t matrix develops a pole at the corresponding energy. This bound-state pole of the t matrix is given by the vanishing of the denominator in Eq. (2.34) at a negative bound-state energy $k^2 = -\alpha^2$, so that

$$1 = -\frac{2}{\pi}\int_0^\infty q^2 dq \frac{\gamma(\bar{k}, q)\Gamma(q, \bar{k}, k^2)}{\alpha^2 + q^2}. \tag{2.35}$$

This gives the condition of the bound state. We shall see in Section 2.8 that this condition could be used for producing a practical method for determining the bound-state energy(ies) and wave function(s). The approach could be modified for the determination of virtual states and resonances.

Although the formulation is valid for a general $\bar{k}$, in the following we assume explicitly that $\bar{k} = k$; this removes the fixed-point singularity of the scattering equation and makes the Γ matrix real at positive energies. From Eq. (2.32) one has the following two half-shell t-matrix elements:

$$t(p, k, k^2) = \frac{\Gamma(p, k, k^2)}{\Gamma(k, k, k^2)} t(k), \tag{2.36}$$

with

$$\begin{aligned} t(k) &\equiv t(k, k, k^2) \\ &= \frac{\Gamma(k, k, k^2)}{1 - (2/\pi)\int_0^\infty q^2\, dq (k^2 - q^2 + i0)^{-1} \gamma(k, q)\Gamma(q, k, k^2)}, \end{aligned} \tag{2.37}$$

$$t(k, p', k^2) = \Gamma(k, p', k^2) + \Gamma(k, k, k^2)\mathcal{I}(k, p'). \tag{2.38}$$

For calculating the scattering phase shifts, one can use Eq. (2.37) and parametrization (1.47). One can alternatively use the following equivalent equation for the on-shell K matrix:

$$K(k^2) \equiv K(k,k,k^2)$$
$$= \frac{\Gamma(k,k,k^2)}{1 - (2/\pi)\mathcal{P}\int_0^\infty q^2\,dq(k^2 - q^2)^{-1}\gamma(k,q)\Gamma(q,k,k^2)} \tag{2.39}$$

and its phase-shift parametrization (1.152).

If we eliminate $\mathcal{I}(k,p')$ between Eqs. (2.32) and (2.38) and use Eq. (2.36) and the fact that the t matrix is time-reversal symmetric, $t(p,k,k^2) = t(k,p,k^2)$, the following symmetric form for the t matrix is obtained:

$$t(p,p',k^2) = [f(p,k)t(k,k,k^2)f(p',k)] + \{R(p,p',k^2)\}, \tag{2.40}$$

with

$$R(p,p',k^2) = \left\{\Gamma(p,p',k^2) - \frac{\Gamma(p,k,k^2)\Gamma(k,p',k^2)}{\Gamma(k,k,k^2)}\right\}, \tag{2.41}$$

$$f(p,k) = \frac{\Gamma(p,k,k^2)}{\Gamma(k,k,k^2)} = \frac{t(p,k,k^2)}{t(k,k,k^2)} = \frac{K(p,k,k^2)}{K(k,k,k^2)}. \tag{2.42}$$

The universal function $f(p,k)$ is called the *half-shell function* and $R(p,p',k^2)$ the *residual function.* Although the Γ matrix is not symmetric, in Eq. (2.40) the two groups of terms in brackets and braces are each symmetric [13].

The dynamical Γ-matrix formulation should be considered entirely equivalent to a t- or K-matrix description of scattering at both positive and negative energies. The real nonsingular Γ-matrix equation (2.28) is first solved numerically. The t matrix is next calculated via Eqs. (2.40) to (2.42). If one of the momentum variables is on-shell, the quantity in the brackets of Eq. (2.40) is exact because of Eq. (2.36). The fully off-shell residual term R is zero if one of the momentum variables is on-shell: $R(k,k,k^2) = R(p,k,k^2) = R(k,p',k^2) = 0$.

The potential term $\mathcal{V}(p,q)$ that appears in the kernel A of the Γ-matrix formulation is zero for right-sided half-on-shell values of momentum: $\mathcal{V}(p,k) = 0$. This potential involves a difference between two potentials and is weaker than the original interaction potential. Hence, the Γ-matrix formulation not only leads to a real nonsingular Fredholm description of scattering but is also a good starting point for perturbative or iterative treatment of scattering. We shall see later that the Γ-matrix formulation also provides an alternative approach for constructing variational principles for scattering.

The function γ is yet undetermined. There are a couple of interesting choices for this function which deserve some comment. The choice

$$\gamma(k,q) = \frac{q^L}{k^L} \tag{2.43}$$

for the Lth partial wave leads to a convergent Neumann series solution for the Γ matrix for any local Hermitian potential independent of its strength [14]. This choice also leads to convergent Neumann series solution for many nonlocal potentials.

The choice [15]

$$\gamma(k,q) = \frac{V(k,q)}{V(k,k)} \tag{2.44}$$

has some interesting consequences. It leads to the following two real nonsingular Fredholm equations for the half-shell and the residual functions f and R:

$$f(p,k) = \gamma(k,p) + \frac{2}{\pi}\int_0^\infty q^2 dq A(p,q,k^2) f(q,k), \tag{2.45}$$

$$R(p,p',k^2) = \mathcal{V}(p,p') + \frac{2}{\pi}\int_0^\infty q^2\, dq\, \mathcal{V}(p,q)(k^2-q^2)^{-1} \times R(q,p',k^2), \tag{2.46}$$

$$= \mathcal{V}(p,p') + \frac{2}{\pi}\int_0^\infty q^2\, dq\, A(p,q,k^2) R(q,p',k^2) \tag{2.47}$$

where now

$$\mathcal{V}(p,p') = V(p,p') - \frac{V(p,k)V(k,p')}{V(k,k)} \tag{2.48}$$

Equations (2.28), (2.45), and (2.47) have the same kernel. In this case

$A(k, q, k^2) = \mathcal{V}(k, q)/(k^2 - q^2) = 0$ and consequently, $\Gamma(k, p', k^2) = V(k, p')$. The residual term R now satisfies the Lippmann–Schwinger equation (2.47) with the subtracted potential (2.48). Because of the subtracted potential, no $i0$ limit or principal-value prescription is required in the Green's function. As this equation has the form of a Lippmann–Schwinger equation, successive subtractions can be introduced in this equation as in the Γ-matrix formulation in order to lead to an efficient iterative calculational scheme [3,13].

The integral equations (2.45) and (2.47) can be solved exactly for a square-well potential in terms of spherical Bessel functions [16]. These solutions are useful for studying analytical behavior. The technique used in the Γ-matrix formulation is a particular version of a general class of subtraction schemes [3].

2.5 SOLUTION BY ITERATION

There are several alternatives to solving either the singular or nonsingular integral equation of scattering by matrix inversion. Apart from the use of different variational methods for the solution of these equations, which we describe in detail in later chapters, high-precision numerical results can be obtained via solution by iteration with comparatively little numerical effort.

The iterative solution of Lippmann–Schwinger equations (1.93) and (1.94) yields the Born–Neumann series

$$t = V + VG_0^{(+)}V + (VG_0^{(+)})^2V + (VG_0^{(+)})^3V + \cdots \tag{2.49}$$

Equation (2.49) is a special case of the general iterative Neumann series solution (2.10) of a Fredholm equation. The first term of this series is the Born or first Born term, and the sum of first n terms is the nth Born term. Calculation of the t matrix using Eq. (2.49) requires only successive matrix vector multiplications. The advantage of the iterative method lies in the small accumulation of numerical error, called *round-off error*. Such errors are usually large in other methods, which limits the final precision. If the iterative series (2.49) converges, precision of the final result can be increased by increasing the number of iterations and the number of mesh points in discretizing the scattering integral equation. For a given scattering problem, the iterative method can thus be used to yield numerical results superior in precision to that obtained in other methods by several orders of magnitude.

A solution scheme based on the iterative Born–Neumann series has been used successfully in conjunction with close-coupling equation (1.197) in problems of atomic and molecular physics [17]. The first term of Eq. (1.197) or the Born term in this approach is given by Eq. (1.215). The problems studied by this approach include those dealing with electron–atom and electron–molecule scattering. At medium energies, the Born approximation (1.215) often yields reasonable result. At very low energies, however, scattering depends on the details of the dynamics, and the lowest-order terms of the iterative Born–Neumann series do not lead to satisfactory results. Calculation of higher-order Born terms becomes increasingly complicated and is not as useful in atomic and molecular scattering problems because of coupling to other channels.

Despite the above-mentioned advantages, the iterative Born–Neumann series is not frequently used in the numerical solution of scattering problems in nuclear and particle physics. There the potential is sufficiently strong that the scattering equation converges only at very high energies. At such energies the validity of the physical model breaks down, due to the significant role played by many neglected dynamical effects, such as meson and quark degrees of freedom, and relativistic effects.

Actually, for many scattering problems, the iterative series diverges at low energies. It converges only for very weak potentials or at very high energies. However, divergence of the iterative Born–Neumann series (2.49) does not mean that the Lippmann–Schwinger equation does not have a solution. It only reflects the failure of the Born–Neumann iterative series to find this solution. This is similar to the fact that the sum of a divergent geometric progression cannot be found by naively summing the terms of the series.

Following Weinberg [1], we present next a description of the criteria of convergence of the iterative Born–Neumann series. The condition of divergence of this series is related to the existence of an eigenvalue of the kernel of the Lippmann–Schwinger equation greater than 1 in magnitude. Let us consider the following eigenfunction–eigenvalue problem for the kernel $\mathcal{K}(E)$ of the Lippmann–Schwinger equation:

$$\mathcal{K}(E)|\chi\rangle \equiv VG_0^{(+)}(E)|\chi\rangle = \eta(E)|\chi\rangle, \tag{2.50}$$

where $\eta(E)$ is the complex eigenvalue at a specific energy E above the scattering threshold. At negative energies, both the kernel and the eigenvalues are real. The Born–Neumann series converges if all the eigenvalues are less than unity in magnitude. Weinberg made a careful study of these eigenvalues and we present an account of his study. For an attractive potential at large negative energies, all the eigenvalues are smaller than unity and each becomes larger monotonically as the energy approaches zero from a large negative value. As the energy becomes equal to the ground-state energy, $E = -B_0$, the largest eigenvalue becomes unity. At the energy of the nth excited bound state, $E = -B_n$, the nth eigenvalue becomes unity, and $(n-1)$ eigenvalues corresponding to the other bound states are larger than unity. The existence of n bound states of a system guarantees n eigenvalues of the kernel greater than 1 at the lowest scattering threshold, and consequently, the Born–Neumann series diverges [1,2]. This series also diverges for strongly repulsive potentials. In these cases the divergence of the Born–Neumann series is related to the existence of an eigenvalue(s) smaller than -1 at the lowest scattering threshold. Such an eigenvalue(s) does (do) not correspond to a bound state(s) of the original potential V but to the sign-changed potential $-V$. The latter possibility is realized for realistic nucleon–nucleon potentials and for interatomic potentials due to a strongly repulsive core at short distances. These potentials may also have added convergence difficulty, due to the actual bound states of these systems.

At the lowest scattering threshold, the Green's function is real and so is η. Typically, if the largest (in magnitude) eigenvalue η of the kernel of the Lippmann–Schwinger equation at the lowest scattering threshold satisfies [1,18]

$$|\eta| > 1, \tag{2.51}$$

the Born–Neumann series diverges at this energy. In the presence of bound states with a partly attractive and partly repulsive potential, several eigenvalues of the kernel may satisfy (2.51) at the lowest scattering threshold. At small positive energies the eigenvalues become complex. However, Weinberg [1] showed that Eq. (2.51) is still satisfied and the Born–Neumann series diverges. As energy is increased past a sufficiently large energy $E = E_c$, all the eigenvalues satisfy

$$|\eta| < 1, \tag{2.52}$$

and the Born–Neumann series converges.

One way to circumvent these difficulties with the iterative solution is to separate the divergent feature in a auxiliary problem that can be solved analytically. Then the iteration technique can be applied to the remainder of the problem. In this approach we decompose the kernel $\mathcal{K}$ of the Lippmann–Schwinger equation (1.93) into a separable part $\mathcal{K}_S$ and a nonseparable remainder $\mathcal{K}_R$:

$$\mathcal{K}(E) = \mathcal{K}_S(E) + \mathcal{K}_R(E). \tag{2.53}$$

Then the Lippmann–Schwinger equation (1.93) can be rewritten as

$$t = \Gamma_R V + \Gamma_R \mathcal{K}_S(E) t, \tag{2.54}$$

where

$$\Gamma_R = 1 + \mathcal{K}_R(E)\Gamma_R. \tag{2.55}$$

Since $\mathcal{K}_S$ is separable, if Γ_R is known, Eq. (2.54) can be solved analytically. Any integral equation with a separable kernel is analytically solvable, as has been demonstrated in Section 3.1.2. The objective is to choose $\mathcal{K}_S$ as such a good approximation to $\mathcal{K}$ that $\mathcal{K}_R$ is small and the iterative series for Γ_R converges, even when the iterative series for the original Lippmann–Schwinger equation diverges. For a compact $\mathcal{K}$, it is always possible, in principle, to find a $\mathcal{K}_S$ such that the eigenvalues of $\mathcal{K}_R$ are all less than 1. In practice, however, the criteria for such a choice are not always obvious [3].

One method for determining an appropriate $\mathcal{K}_S$ is provided by the Γ-matrix formulation of Section 2.4. This formulation essentially corresponds to the choice

$$\mathcal{K}_S(p, p', k) = \frac{2}{\pi} \frac{V(p, k)\gamma(k, p')p'^2}{k^2 - p'^2 + i0}, \tag{2.56}$$

so that

$$\mathcal{K}_R(p, p', k) = A(p, p', k^2), \tag{2.57}$$

where $\mathcal{K}_R$ corresponds to a difference between two terms $(\mathcal{K} - \mathcal{K}_S)$, with $A(p, p', k^2)$ defined by (2.26). With these choices for $\mathcal{K}_R$ and $\mathcal{K}_S$, Eqs. (2.54) and (2.55) become very similar to the Γ-matrix equations (2.29) and (2.30). Hence, as in the Γ-matrix method, the residual kernel $\mathcal{K}_R$ is weaker than the original one and Eq. (2.55) should have better convergence properties than the original Lippmann–Schwinger equation (1.93). One cannot assure in general that the iteration solution corresponding to the $\mathcal{K}_R$ of Eq. (2.57) really converges. The criteria of convergence depend on the potential, the energy under consideration, and choice of the function γ. The answer is always affirmative for a local potential with the choice (2.43), independent of the strength of the potential and energy under consideration. In the case of nonlocal potentials the choice (2.43) also leads to good convergence under many situations. However, for a given local or nonlocal potential, it is possible to find other choices for γ that lead to faster convergence than choice (2.43) [3].

The subtraction procedure using the Γ-matrix formulation relates the solution of the original Lippmann–Schwinger t- or K-matrix equation to that of the nonsingular Γ-matrix equation. This amounts to a "weakening" of the Lippmann–Schwinger kernel and ensures that the iteration series for the Γ matrix either converges or at best diverges slower than the original equation (1.93) or (1.94). If additional subtractions are made to weaken kernel(s) of successive auxiliary equations, the corresponding rates of convergence progressively increase. This last approach has been used in the solution of realistic scattering problems in nuclear physics [19,20].

Even when the Born–Neumann series diverges, the successive terms of this series contain information about exact solution of the problem. Similarly, the first few terms of a divergent geometric series can be used to sum this series. However, a general procedure for extracting this solution is not known. There are advantages in finding the solution of a scattering problem with a divergent Born–Neumann series (2.49) for the Lippmann–Schwinger equation via the convergent iterative series for the physically motivated Γ-matrix equation. This procedure utilizes the first few terms of the original iterative series for finding the solution. The technique of Padé approximants is also a mathematical procedure for approximating the sum of a divergent series utilizing the first few terms of this series [4]. This technique is specially suited for scattering problems and uses the first few terms of the Born–Neumann series of a scattering equation to construct its solution numerically irrespective of whether the series is slowly convergent or even divergent. The technique of Padé approximants can be applied either to the integral equation for the t matrix or the Γ matrix. This method has been used exhaustively for this purpose in a variety of scattering problems, and we refer the interested readers to appropriate places [7].

2.6 SCATTERING WITH COMPLEX POTENTIALS

In most of the discussions presented so far, we assumed that the potential is real and Hermitian. However, in an actual scattering problem, one often encounters a complex potential. Such potentials account for absorption in scattering. Transition of the incident flux from the elastic to other channels, not included in the scattering model, makes the effective potential for scattering complex. The elastic scattering phase shifts obtained by solving a dynamical set of equations in the presence of many channels, for example, close-coupling (CC) equations (1.202), are real (complex) below (above) the lowest inelastic threshold with real effective potentials in this equation. Hence, if one studies the elastic scattering of the same system as the foregoing multichannel CC model via a model potential, this potential has to be complex above the lowest inelastic threshold in order to generate the complex phase shifts of the CC model and incorporate correctly the loss of flux to inelastic channels. Also, if higher angular-momentum cluster states are included in the CC model, the coupling potentials in the CC model could be complex.

In the presence of a complex potential, the partial-wave wave function $\psi_L^{(+)}$ of Eq. (1.34) becomes intrinsically complex with a complex phase shift δ_L. Then, because of asymptotic property (1.36), the probability amplitude $|\psi_L^{(+)}|^2$ acquires an exponentially decaying part at infinity, due to absorption of flux. For this to happen, the phase shift δ_L should have a positive imaginary part. A negative imaginary part will correspond to a exponentially growing solution at infinity.

The partial-wave Lippmann–Schwinger equation for the t matrix (1.121) and its phase-shift parametrization (1.47) continue the same in the presence of the complex potential and complex phase shift. Once the complex phase shifts are obtained, the scattering amplitude and cross sections can be calculated with the help of Eqs. (1.49), (1.51) and so on. In the following we present an account of the solution of Eq. (1.121) in the presence of complex potentials.

The complex t-matrix equation can now be discretized as in Eq. (2.18), which can be solved directly . With a real potential the first two terms on the right-hand side of Eq. (2.18) are real. For a complex potential, both these terms are complex and the Lippmann–Schwinger equation is intrinsically complex. However, in Eqs. (1.121) and (2.18) we need the principal-value prescription in the integration over the outgoing-wave Green's function. The principal-value integral should be evaluated as in the case of real potentials. As the potential is intrinsically complex, it is not possible to write real auxiliary scattering equations such as the K- or Γ-matrix equation. However, one can formally write the K- and the Γ-matrix equations for complex potentials as in the case of real potentials. There is no advantage in using the K-matrix approach in this case, and the K matrix is complex in this case. However, there could be some advantage in using the Γ-matrix equation for complex potentials. Although complex algebra cannot be avoided in the Γ-matrix approach, there is no principal-value prescription in this approach. Also, as by construction the Γ-matrix equation has a weaker kernel, it may lead to a convergent Born–Neumann series solution.

One way to avoid complex algebra in the presence of complex potentials is to break up the Lippmann–Schwinger equation into its real and imaginary parts, as in Eqs. (2.8) and (2.9). For a complex potential, the real and imaginary parts of the t matrix satisfy [6]

$$\begin{aligned}\Re[t(p,p',k^2)] = &\\ &\Re[V(p,p')] + \sqrt{k}\{\Re[V(p,k)]\Im[t(k,p',k^2)] + \Im[V(p,k)]\Re[t(k,p',k^2)]\}\\ &+ \frac{2\mathcal{P}}{\pi}\int_0^\infty \frac{q^2dq\{\Re[V(p,q)]\Re[t(q,p',k^2)] - \Im[V(p,q)]\Im[t(q,p',k^2)]\}}{k^2-q^2},\end{aligned} \tag{2.58}$$

$$\begin{aligned}\Im[t(p,p',k^2)] = &\\ &\Im[V(p,p')] - \sqrt{k}\{\Re[V(p,k)]\Re[t(k,p',k^2)] - \Im[V(p,k)]\Im[t(k,p',k^2)]\}\\ &+ \frac{2\mathcal{P}}{\pi}\int_0^\infty \frac{q^2dq\{\Re[V(p,q)]\Im[t(q,p',k^2)] + \Im[V(p,k)]\Re[t(k,p',k^2)]\}}{k^2-q^2}.\end{aligned} \tag{2.59}$$

By solving the set of real equations (2.58) and (2.59), one can find the complex t matrix. This set of real equations for the t matrix in the presence of complex potential is no more difficult to solve than the corresponding equations (2.8) and (2.9) for real potentials. The principal-value integrals in Eqs. (2.58) and (2.59) are to be dealt with exactly in the same fashion as those in the case of real potentials: for example, by the Sloan prescription [9].

In the next two chapters we develop variational methods for the solution of scattering problems. These methods use an approximate trial wave function or form factor for calculating an improved variational estimate of these quantities. The phase shifts are calculated from this improved result. The variational methods of Chapter 3 need to incorporate the asymptotic property (1.36) of the scattering wave function in the trial wave function. As the asymptotic properties of the wave function gets completely altered in the presence of complex potentials, the methods of Chapter 3 are not appropriate for complex potentials. Most of the variational methods presented in Chapter 4 involve the free outgoing-wave Green's function and does not need to incorporate the asymptotic property (1.36) of the scattering wave function or the form factors in the trial quantities. The potentials and the Green's function of these variational methods construct the correct asymptotic properties of the wave function with a $\mathcal{L}^2$ basis set. These variational methods of Chapter 4 can be applied to scattering problems with complex potential.

2.7 MULTICHANNEL Γ-MATRIX FORMULATION

We have seen that the Γ-matrix formulation of Section 2.4 is an advantage in the case of potential scattering. Realistic scattering problems are formulated in terms of multichannel scattering equations: for example, the CC equations (1.202). These multichannel scattering equations have the usual Lippmann–Schwinger form, but now the various variables have the channel indices over and above the momentum labels.

The Γ-matrix formulation can be generalized for multichannel scattering. One possible multichannel generalization of the partial-wave Lippmann–Schwinger equation (1.121) is the coupled set of CC equations (1.202) [19]. We shall use the following off-shell continuation of this set of equations for developing the multichannel Γ-matrix formulation

$$t_{\beta\alpha}(p_\beta, p'_\alpha, E) = V_{\beta\alpha}(p_\beta, p'_\alpha) + \frac{2}{\pi}\sum_{\sigma=1}^{N}\int_0^\infty q_\sigma^2 dq_\sigma V_{\beta\sigma}(p_\beta, q_\sigma)$$

$$\times \frac{1}{k_\sigma^2 - q_\sigma^2 + i0} t_{\sigma\alpha}(q_\sigma, p'_\alpha, E). \tag{2.60}$$

Here α, β, σ and so on, denote N distinct channels, k_σ is the on-shell momentum for channel σ defined by (1.183), and the σ-channel

Green's function $(k_\sigma^2 - q_\sigma^2 + 10)^{-1}$ now contains the fixed-point singularity of the σ channel.

Equation (2.60) involves N different multichannel Green's functions via coupling of the dynamics to other channels. Each of these Green's functions contributes to a different branch cut. It would be of advantage if we could eliminate these singularities corresponding to different channels and introduce a nonsingular equation of scattering in the multichannel case, and in the following we do the same.

In close analogy with the single-channel Γ-matrix equation (2.28), the multichannel Γ-matrix elements now satisfy the following auxiliary set of coupled equations:

$$\Gamma_{\beta\alpha}(p_\beta, p'_\alpha, E) = V_{\beta\alpha}(p_\beta, p'_\alpha) + \frac{2}{\pi}\sum_{\sigma=1}^{N}\int_0^\infty q_\sigma^2\, dq_\sigma\, A_{\beta\sigma}(p_\beta, q_\sigma, E)t_{\sigma\alpha}(q_\sigma, p'_\alpha, E), \tag{2.61}$$

with the nonsingular kernel $A_{\beta\sigma}(p_\beta, q_\sigma, E)$ defined by

$$A_{\beta\sigma}(p_\beta, q_\sigma, E) = [V_{\beta\sigma}(p_\beta, q_\sigma) - V_{\beta\sigma}(p_\beta, k_\sigma)\gamma_\sigma(k_\sigma, q_\sigma)](k_\sigma^2 - q_\sigma^2)^{-1}, \tag{2.62}$$

where the diagonal function $\gamma_\sigma(k_\sigma, q_\sigma)$ satisfies

$$\gamma_\sigma(k_\sigma, k_\sigma) = 1. \tag{2.63}$$

At the on-shell points $q_\sigma = k_\sigma$, the multichannel kernel $A_{\beta\sigma}(p_\beta, q_\sigma, E)$ of Eq. (2.62) is nonsingular, and hence the Γ matrix of Eq. (2.61) is real and nonsingular. It remains to relate the solution of the nonsingular Γ-matrix equation (2.61) with that of the t-matrix equation (2.60).

Using the nonsingular kernel $A_{\beta\sigma}(p_\beta, q_\sigma, E)$, Eq. (2.60) can be rewritten as

$$t_{\beta\alpha}(p_\beta, p'_\alpha, E) = V_{\beta\alpha}(p_\beta, p'_\alpha) + \frac{2}{\pi}\sum_{\sigma=1}^{N}\int_0^\infty q_\sigma^2\, dq_\sigma\, A_{\beta\sigma}(p_\beta, q_\sigma, E)t_{\sigma\alpha}(q_\sigma, p'_\alpha, E) + \sum_{\sigma=1}^{N} V_{\beta\sigma}(p_\beta, k_\sigma)\frac{2}{\pi}\int_0^\infty \frac{q_\sigma^2\, dq_\sigma\, \gamma_\sigma(k_\sigma, q_\sigma)}{k_\sigma^2 - q_\sigma^2 + i0} t_{\sigma\alpha}(q_\sigma, p'_\alpha, E). \tag{2.64}$$

As in the single-channel case, the t- and Γ-matrix equations can be written in operator forms (2.29) and (2.30), respectively. The relation between the t- and the Γ-matrix equations is given again by Eq. (2.31). In operator forms (2.29), (2.30), and (2.31), both the momentum labels and channel indices have been dropped. When the proper momentum labels and channel indices are introduced in Eq. (2.31), the t-matrix elements are expressed in terms of the auxiliary Γ-matrix elements via

$$t_{\beta\alpha}(p_\beta, p'_\alpha, E) = \Gamma_{\beta\alpha}(p_\beta, p'_\alpha, E) + \sum_{\sigma=1}^{N} \Gamma_{\beta\sigma}(p_\beta, k_\sigma, E) I_{\sigma\alpha}(k_\sigma, p'_\alpha, E), \tag{2.65}$$

with

$$I_{\sigma\alpha}(k_\sigma, p'_\alpha, E) = \frac{2}{\pi}\int_0^\infty q_\sigma^2\, dq_\sigma (k_\sigma^2 - q_\sigma^2 + i0)^{-1} \gamma_\sigma(k_\sigma, q_\sigma) t_{\sigma\alpha}(q_\sigma, p'_\alpha, E). \tag{2.66}$$

From Eqs. (2.65) and (2.66), it can be seen that $I_{\beta\alpha}$ are solutions of

$$I_{\beta\alpha}(k_\beta, p'_\alpha, E) = d_{\beta\alpha}(k_\beta, p'_\alpha, E) + \sum_{\sigma=1}^{N} d_{\beta\sigma}(k_\beta, k_\sigma, E) I_{\sigma\alpha}(k_\sigma, p'_\alpha, E), \tag{2.67}$$

with

$$d_{\sigma\alpha}(k_\sigma, p'_\alpha, E) = \frac{2}{\pi}\int_0^\infty q_\sigma^2\, dq_\sigma (k_\sigma^2 - q_\sigma^2 + i0)^{-1} \gamma_\sigma(k_\sigma, q_\sigma) \Gamma_{\sigma\alpha}(q_\sigma, p'_\alpha, E). \tag{2.68}$$

Using the solution $\Gamma_{\sigma\alpha}(q_\sigma, p'_\alpha, E)$ of the Γ-matrix equation (2.61), the matrices $d_{\beta\alpha}(k_\beta, p'_\alpha, E)$ and $I_{\beta\alpha}(k_\beta, p'_\alpha, E)$ are first calculated, which when substituted in Eq. (2.65) yields the t matrix.

As in the single-channel case, the final result given by Eq. (2.65) can again be written in a symmetric form. The two half-on-shell and on-shell versions of this equation are written as

$$t_{\beta\alpha}(p_\beta, k_\alpha, E) = \sum_{\sigma=1}^{N} \Gamma_{\beta\sigma}(p_\beta, k_\sigma, E)[\delta_{\sigma\alpha} + I_{\sigma\alpha}(k_\sigma, k_\alpha, E)], \tag{2.69}$$

$$t_{\beta\alpha}(k_\beta, p'_\alpha, E) = \Gamma_{\beta\alpha}(k_\beta, p'_\alpha, E) + \sum_{\sigma=1}^{N} \Gamma_{\beta\sigma}(k_\beta, k_\sigma, E) I_{\sigma\alpha}(k_\sigma, p'_\alpha, E), \tag{2.70}$$

$$t_{\beta\alpha}(k_\beta, k_\alpha, E) = \sum_{\sigma=1}^{N} \Gamma_{\beta\sigma}(k_\beta, k_\sigma, E)[\delta_{\sigma\alpha} + I_{\sigma\alpha}(k_\sigma, k_\alpha, E)]. \quad (2.71)$$

Eliminating $I_{\sigma\alpha}$ between Eqs. (2.69) and (2.71), one gets

$$t_{\beta\alpha}(p_\beta, k_\alpha, E) = \sum_{\sigma,\rho=1}^{N} \Gamma_{\beta\sigma}(p_\beta, k_\sigma, E)\Theta_{\sigma\rho}(k_\sigma, k_\rho, E)t_{\rho\alpha}(k_\rho, k_\alpha, E), \quad (2.72)$$

where Θ is a matrix whose momentum variables take only the on-shell values and is defined by

$$\sum_{\sigma=1}^{N} \Gamma_{\beta\sigma}(k_\beta, k_\sigma, E)\Theta_{\sigma\rho}(k_\sigma, k_\rho, E) = \sum_{\sigma=1}^{N} \Theta_{\beta\sigma}(k_\beta, k_\sigma, E)\Gamma_{\sigma\rho}(k_\sigma, k_\rho, E) = \delta_{\beta\rho}. \quad (2.73)$$

Equations (2.70) and (2.73) yield the following formal solution for $I_{\sigma\alpha}$:

$$I_{\sigma\alpha}(k_\sigma, p'_\alpha, E) = \sum_{\rho=1}^{N} \Theta_{\sigma\rho}(k_\sigma, k_\rho, E)[t_{\rho\alpha}(k_\rho, p'_\alpha, E) - \Gamma_{\rho\alpha}(k_\rho, p'_\alpha, E)]. \quad (2.74)$$

Now recalling that $t_{\rho\alpha}(k_\rho, p'_\alpha, E) = t_{\alpha\rho}(p'_\alpha, k_\rho, E)$ and using Eqs. (2.69) and (2.74), the following symmetric form for the t-matrix elements is obtained:

$$\begin{aligned} t_{\beta\alpha}(p_\beta, p'_\alpha, E) = \\ &\left[\sum_{\sigma,\rho,\mu,\nu=1}^{N} \Gamma_{\beta\sigma}(p_\beta, k_\sigma, E)\Theta_{\sigma\rho}(k_\sigma, k_\rho, E)\Gamma_{\alpha\nu}(p'_\alpha, k_\nu, E)\Theta_{\nu\mu}(k_\nu, k_\mu, E)\right. \\ &\left. \times\, t_{\mu\rho}(k_\mu, k_\rho, , E)\right] + \Bigg\{\Gamma_{\beta\alpha}(p_\beta, p'_\alpha, E) \\ &- \sum_{\sigma,\rho=1}^{N} \Gamma_{\beta\sigma}(p_\beta, k_\sigma, E)\Theta_{\sigma\rho}(k_\sigma, k_\rho, E)\Gamma_{\rho\alpha}(k_\rho, p'_\alpha, E)\Bigg\}. \end{aligned} \quad (2.75)$$

Equation (2.75) is the desired multichannel generalization of the single-channel equation (2.40). The quantity in brackets in Eq.

(2.75) is exact half-on-shell. The last term in the braces is the fully off-shell real residual term and is zero half-on-shell.

2.8 SCATTERING APPROACH TO BOUND STATE AND RESONANCE

There are numerous methods in configuration and momentum spaces for solving the bound-state problem. The usual configuration space methods are suitable for local potentials and become cumbersome for nonlocal potentials. Yet most of the realistic physical problems involve nonlocal potentials. Both in scattering and bound-state problems, most of the momentum-space numerical techniques do not require much modification to treat nonlocal potentials. Instead of describing them all, we consider some that are closely related to the momentum-space methods of this chapter for solving scattering integral equations. One of them is very useful for virtual states and resonances. Essentially, the same numerical program with minor modification can be used for the calculation of scattering, bound state, virtual state, and resonance.

After a partial-wave projection, the bound-state wave function $\psi_{BL}(p)$ satisfies the homogeneous integral equation (1.127), where $B = \alpha^2$ is the binding energy. This can be considered as the $E = -\alpha^2$ case of the general homogeneous integral equation

$$\psi(p, E) = \frac{2}{\pi(E - p^2)} \int dq\, q^2 V(p, q)\psi(q, E), \tag{2.76}$$

which has a nontrivial solution only at the bound-state energies. We have suppressed the angular momentum label in Eq. (2.76), which is the partial-wave momentum-space form of Eq. (1.173). The bound-state energies are presumably negative and there is no singularity in the Green's function $(E - p^2)^{-1}$ of Eq. (2.76) at negative energies.

Let us consider a practical method [21] for solving Eq. (2.76) for the ground state of a purely attractive potential. Schematically, Eq. (2.76) can be written as

$$\psi(E) = \mathcal{K}(E)\psi(E), \tag{2.77}$$

where $\mathcal{K}$ is the kernel of integral equation (2.76). Let us consider the eigenfunction-eigenvalue problem (2.50) at negative energies. For a purely attractive potential and at negative energies, all the eigenvalues

$\eta(E)$ of the kernel of Eq. (2.50) are real and positive. When the largest of these positive eigenvalues equals unity, Eq. (2.50) becomes identical with Eq. (2.77) and one has the ground state. If the kernel $\mathcal{K}$ has only one eigenvalue η equal to or slightly greater than unity at specific negative energy E, a consideration of the iterative Neumann series

$$\psi = \sum_n \psi_n = \sum_n \mathcal{K}^n(E)\phi_0, \tag{2.78}$$

where ϕ_0 is any function, shows that [21]

$$\lim_{n\to\infty} \frac{\psi_{n+1}}{\psi_n} = \eta. \tag{2.79}$$

Equation (2.79) follows straightforwardly in the diagonal representation of the matrix $\mathcal{K}$ and is valid in general. At the bound-state energy, $\eta = 1$. Hence, one should vary the energy and see at what energy the condition $\eta = 1$ is satisfied to a desired accuracy. Then this energy corresponds to the bound-state energy, and $\lim_{n\to\infty} \psi_n$ is the bound-state wave function.

The method above can be extended to calculate the ground state of a partly attractive and partly repulsive potential, where the kernel has negative (repulsive) eigenvalues larger than 1 in magnitude. These repulsive eigenvalues correspond to bound states of the sign-changed potential $-V$. Then the ground state corresponds to unit eigenvalue, but at this energy there are other states with eigenvalues greater than unity in magnitude. Hence, direct use of the iterative scheme above will pick up the largest (in modulus) eigenvalue η of the kernel $\mathcal{K}$, not the unit eigenvalue corresponding to the ground state. Eigenvalue(s) greater than 1 in magnitude could be eliminated in some situations by subtracting the largest negative eigenvalue(s) from the kernel [18] before iteration. For example, if the kernel has a single repulsive eigenvalue $-\eta_1(\eta_1 > 1)$, then in the kernel $\mathcal{K} + I\eta_1$ the largest (in modulus) eigenvalue will correspond to the ground state. So the iteration of this modified kernel will select the ground state via Eq. (2.79). However, one then has to apply the method before and after each such subtraction, which makes the method less attractive.

An alternative approach, free of the above-mentioned difficulties, valid for ground and excited states and also for partly attractive and repulsive potentials, is next described using the subtraction procedure used in the Γ-matrix approach [22]. The potential is decomposed into a separable part and a weak nonseparable part V^R:

$$V(p,q) = f_1(p)f_2(q) + V^R(p,q). \tag{2.80}$$

It is assumed that the separable term $f_1(p)\,f_2(q)$ is a good approximation of the potential, so that the residual term $V^R(p,q) \equiv V(p,q) - f_1(p)\,f_2(q)$ is weak. Next the wave-function normalization is chosen to be

$$\frac{2}{\pi}\int dq\, q^2 f_2(q)\psi(q,E) = 1. \tag{2.81}$$

Substituting Eq. (2.80) into Eq. (2.76) and using the normalization condition (2.81), we can transform the homogeneous bound-state equation (2.76) into the following inhomogeneous equation:

$$\psi(p,E) = \frac{f_1(p)}{E-p^2} + \frac{2}{\pi(E-p^2)}\int dq\, q^2 V^R(p,q)\psi(q,E). \tag{2.82}$$

If the potential $V^R(p,q)$ does not support a bound state and is weak enough, the integral equation (2.82) should permit iterative solution.

The homogeneous Schrödinger equation (2.76) has a solution for certain negative energies corresponding to the finite number of bound states of the system. However, the inhomogeneous equation (2.82) has a unique solution for all energies except those for which its homogeneous form admits a nontrivial solution. This corresponds to a bound state of potential V^R. It is assumed that the bound states of V^R, if they exist, are not in the same energy region as the bound states of the full potential V. By construction, any solution of Eq. (2.82) that satisfies the normalization condition (2.81) is also a solution to the Schrödinger equation (2.76). The strategy is to guess an energy, solve Eq. (2.82), and see if normalization condition (2.81) is satisfied to desired precision. If the answer is affirmative, the trial energy is the bound-state energy and the solution of Eq. (2.82) is the bound-state wave function. If the answer is negative, the process is repeated for a new trial energy. One of the advantages of this method is that it is applicable to both ground and excited states.

Before performing a numerical calculation one has to make a choice for the separable term $f_1(p)f_2(q)$ and the residual potential V^R in Eq. (2.80). From a discussion of Section 2.4 related to the Γ matrix we see that a good choice is

$$V^R(p,q) = V(p,q) - V(p,\bar{k})\gamma(\bar{k},q), \tag{2.83}$$

where $\bar{k}$ is some arbitrary positive momentum. Then Eqs. (2.81) and (2.82) become

$$\frac{2}{\pi}\int dq\, q^2\gamma(\bar{k},q)\psi(q,E) = 1, \tag{2.84}$$

$$\psi(p,E) = \frac{V(p,\bar{k})}{E-p^2} + \frac{2}{\pi(E-p^2)}\int dq\, q^2 V^R(p,q)\psi(q,E), \tag{2.85}$$

respectively. Comparing the bound-state conditions (2.35) and (2.84), we have the following relation between the bound-state wave function and the negative-energy Γ matrix:

$$\psi(p,B) = -\frac{1}{B+p^2}\Gamma(p,\bar{k},k^2). \tag{2.86}$$

The Γ matrix satisfies an inhomogeneous integral equation and contains full information about the solution of the Schrödinger equation at all energies and can be used for a determination of bound-state energies and wave functions. Thus for bound states, one avoids the solution of usual homogeneous momentum space integral equations corresponding to the diagonalization of a matrix and solving an eigenfunction-eigenvalue problem. Numerically, it is easier to solve an inhomogeneous integral equation than an eigenfunction-eigenvalue problem.

We recall from Section 1.4 that the t matrix has a square-root unitarity cut along the real positive energy axis, in addition to discrete bound-state poles at negative energies. In the complex energy plane, the t matrix is defined on two sheets. The physical scattering energies and bound-state poles lie on the first sheet of energy. For a local Hermitian potential, the resonance and virtual state appear as poles of the t matrix at complex and negative energies, respectively, on the second sheet of the complex energy plane. For finding these poles, one needs to continue the t matrix analytically through the unitarity cut of the complex energy plane onto the second sheet [2].

Essentially, the Γ (K) matrix can be used for finding the resonances and virtual states [23]. For this purpose, one needs to calculate the Γ

(K) matrix at complex and negative energies and then calculate the t matrix at these energies on the second sheet of the complex energy plane. The Γ (K) matrix is a real function of energy and does not have the unitarity cut. Hence, it has the same value on both sheets of the t matrix in the complex energy plane.

The analytic continuation can be performed via the Γ or K matrices. First, we describe this using the Γ-matrix approach of Section 2.4. To have an analytic Γ at all energies, the appropriate value of $\bar{k}$ is the on-shell wave number k, and we stick to this choice in the analytic continuation. This cut and the two-sheet structure of the t matrix appears via integral $\mathcal{I}(k)$ appearing in Eq. (2.37), having the following structure:

$$\mathcal{I}_I(k) = \frac{2}{\pi}\int_0^\infty q^2 dq \frac{F(q^2)}{k^2 - q^2 + i0}, \tag{2.87}$$

where the function

$$F(q^2) \equiv \gamma(k,q)\Gamma(q,k,k^2) \tag{2.88}$$

is assumed to be analytic and does not possess the unitarity cut. We have assumed that in addition to the Γ matrix, the γ function is also real. The suffix I denotes the first sheet of energy. In Eq. (2.87), the q integration path is along the positive momentum axis, and integral $\mathcal{I}_I(k)$ has a square-root branch point at zero energy in the complex energy plane and a branch cut along the positive real energy axis. Integral (2.87) can be rewritten as

$$\mathcal{I}_I(k) = \frac{2}{\pi}\int_0^\infty dq \frac{q^2F(q^2) - k^2F(k^2)}{k^2 - q^2} - ik_IF(k^2), \tag{2.89}$$

where the real principal-value and the imaginary parts of the integral have been separated by using Eq. (1.104). In Eq. (2.89), k_I is the k value on the first sheet of energy (e.g., with a positive imaginary part). The principal-value part of the integral in Eq. (2.89) has been rewritten using Eq. (2.21), and the principal-value prescription has been dispensed with. Thus we have separated the part analytic in k^2 from the part that generates the branch cut. The branch cut is generated by the ik term in Eq. (2.89). In expression (2.89), the explicit k dependence has been separated and its real and imaginary parts are separately analytic in k. If one wants to calculate the function $\mathcal{I}(k)$ on the second sheet of energy, one should substitute the appropriate k value in the ik term of this expression corresponding to the second sheet. The remaining integral term is to be calculated on the first sheet. The appropriate k value on the second sheet of energy should be of the form $k = \pm k_\alpha - ik_\beta$ with a negative imaginary part. The first (second) sheet of the complex energy plane is defined by k values with positive (negative) imaginary part. For a resonance, k_α is nonzero and k_β small, and for a virtual state, $k_\alpha = 0$ and k_β positive. The quantity k_α^2 is called the energy and k_β^2 the width of the resonance. The bound state appears on the first sheet for which $k_\alpha = 0$ and k_β negative. Suppose that we are looking for a resonance or virtual-state pole of the t matrix on the second energy sheet. On the second sheet of energy the integral $\mathcal{I}(k)$ of Eq. (2.89) becomes

$$\mathcal{I}_{II}(k) = \frac{2}{\pi}\int_0^\infty dq \frac{q^2F(q^2) - k^2F(k^2)}{k^2 - q^2} - ik_{II}F(k^2), \tag{2.90}$$

where we have to use an appropriate k value on the second sheet in place of k_{II}. If there is a bound or virtual state at energy $k^2 = -\alpha^2$, for the determination of these states one should use $k_I = i\alpha$ in Eq. (2.89) (bound state) and $k_{II} = -i\alpha$ in Eq. (2.90) (virtual state), respectively. For a resonance at $k_{II} = k_\alpha - ik_\beta$ one should use this k_{II} in Eq. (2.90). Using this $\mathcal{I}$ of Eq. (2.90), one should calculate the t matrix on the second sheet via Eq. (2.37). The zero of the denominator in Eq. (2.37) on the second sheet of energy corresponds to a resonance or virtual state. The last term of Eq. (2.89) or (2.90), in general, requires the knowledge of $\Gamma(q, k, k^2)$ for complex momentum k. Once $\Gamma(q, k, k^2)$ is known for real momentum variables, one can calculate it for complex momentum variables q by using the Γ-matrix equation (2.28) again and evaluating the right-hand side by quadrature.

Similar continuation of the t matrix to the second sheet of energy can be made via the use of the K matrix. On the first sheet of energy the partial-wave on-shell t matrix is related to the partial-wave on-shell K matrix via

$$t_I(k) = \frac{K(k^2)}{1 + ik_IK(k^2)}, \tag{2.91}$$

where on the right-hand side $K(k^2)$ is analytic in k^2, and only the ik

term has to be continued to the second sheet. On the second sheet

$$t_{II}(k) = \frac{K(k^2)}{1 + ik_{II}K(k^2)}. \tag{2.92}$$

Equations (2.91) and (2.92) can be used for determining bound and virtual states and resonances. For a bound [virtual] state at $k^2 = -\alpha^2$, $k_I = i\alpha$ $[k_{II} = -i\alpha]$ and the denominator of Eq. (2.91) [(2.92)] should vanish. A resonance corresponds to a vanishing of the denominator of (2.92) at a complex energy on the second sheet, so that $k_{II}(\equiv \pm k_\alpha - ik_\beta)$ has a negative imaginary part.

2.9 NOTES AND REFERENCES

[1] For a discussion of compactness of the kernel of the scattering integral equation, see, for example, Weinberg (1963, 1964) and Lovelace (1964a,b).

[2] See, for example, Newton (1982).

[3] Adhikari and Kowalski (1991).

[4] Bagchi et al. (1977), Bagchi and Mulligan (1979).

[5] Abramowitz and Stegun (1965).

[6] Ghosh (1992).

[7] See, for example, Baker and Gammel (1970) and Adhikari and Kowalski (1991).

[8] Fonseca (1986a,b).

[9] Sloan (1968). See Redish ans Stricker-Bauer (1987) for an application of this method to a realistic scattering problem.

[10] Haftel and Tabakin (1970).

[11] For a limitation of this approach, see Brown and Jackson (1976), Glöckle (1983), and Redish and Stricker-Bauer (1987).

[12] The first application of Fredholm reduction to a two-particle scattering integral equation appears in Kowalski and Feldman (1961, 1963a,b), and Kowalski (1965).

[13] Tomio and Adhikari (1980a,b, 1981, 1995), Blaszak and Fuda (1973), Whiting and Fuda (1976).

[14] This was proved in Coester (1971) and Blaszak and Fuda (1973).

[15] Kowalski (1965).

[16] Kowalski and Feldman (1961, 1963a,b).

[17] Bransden (1983), Burke (1977).

[18] Glöckle (1983).

[19] Adhikari (1979), Tomio and Adhikari (1981).

[20] Ishikawa (1987), Hořacek and Sasakawa (1983, 1984).

[21] Malfliet and Tjon (1968, 1970).

[22] This method is first proposed in Adhikari and Tomio (1981) and Adhikari and Kowalski (1991).

[23] Adhikari et al. (1982), Adhikari and Tomio (1982).

CHAPTER 3

VARIATIONAL PRINCIPLES FOR ON-SHELL AMPLITUDES

3.1 GENERAL VARIATIONAL PRINCIPLES

3.1.1 What Is a Variational Principle?

A variational principle is a certain algebraic stationary expression for the unknown solution(s) of a problem in terms of some unknown quantities. If exact solutions are used for these unknown quantities, the variational principle is an identity. Using approximations for these unknowns leads to an improved result. If these approximations have small errors, the variational principle leads to deviations that are second order in errors. A variational principle is extremely powerful for successive improvement in the solution of a problem when an approximate solution is known. A variational principle can be formed for the solution of an algebraic problem, for finding an inverse, for solving a quantum-mechanical eigenvalue problem, or for solving an integral equation. We illustrate it first for two algebraic problems [1].

Inverse of a Number. A variational principle for the calculation of the inverse of a number c can be written from the following identity, valid for arbitrary numbers $(c^{-1})_{t1}$ and $(c^{-1})_{t2}$:

$$c^{-1} = (c^{-1})_{t1} + (c^{-1})_{t2} - (c^{-1})_{t1}c(c^{-1})_{t2} + [c^{-1} - (c^{-1})_{t1}]c[c^{-1} - (c^{-1})_{t2}]. \tag{3.1}$$

If one takes the arbitrary numbers $(c^{-1})_{t1}$ and $(c^{-1})_{t2}$ to be approximations to c^{-1}, the last term in Eq. (3.1) is second order in the differences $c^{-1} - (c^{-1})_{t1}$ and $c^{-1} - (c^{-1})_{t2}$, so that one can write the following variational principle for the inverse of c, omitting this last term:

$$[c^{-1}] = (c^{-1})_{t1} + (c^{-1})_{t2} - (c^{-1})_{t1}c(c^{-1})_{t2}. \tag{3.2}$$

The square brackets in Eq. (3.2) and below indicate that the expression is stationary. If the exact value for c^{-1} is used in place of trial estimates $(c^{-1})_{t1}$ and $(c^{-1})_{t2}$, the three terms on the right-hand side of Eq. (3.2) are equal and this equation is an identity. Most of the variational principles we discuss in scattering theory are based on trilinear relations such as (3.2) with a similar stationary property. The stationary property of Eq. (3.2) could be verified by considering the independent estimates $(c^{-1})_{t1} = (c^{-1})_E + \Delta_1$ and $(c^{-1})_{t2} = (c^{-1})_E + \Delta_2$, where $(c^{-1})_E$ is the exact inverse and Δ_1 and Δ_2 are small independent deviations. With these trial estimates the inverse of Eq. (3.2) is given by

$$[c^{-1}] = (c^{-1})_E - (\Delta_1)c(\Delta_2). \tag{3.3}$$

The deviation in Eq. (3.3) is quadratic in Δ's. The linear terms in Δ have canceled out.

Equation (3.2) is the desired variational principle, which is based on identity (3.1). In the following we consider many variational principles. Usually, these variational expressions will be based on identities, such as Eq. (3.1). However, these identities will not be written down explicitly in most cases.

As an application of the variational principle (3.2), calculation of the inverse of 3 could be considered; the result is 0.3333333333.... The approximate guesses $(c^{-1})_{t1} = (c^{-1})_{t2} = 0.3$ in Eq. (3.2) lead to the improved result 0.33, which when introduced in the two guesses of this equation leads to an even more improved result, 0.3333. When the result 0.3333 is substituted for two approximate guesses, we obtain 0.33333333. This procedure can be continued indefinitely until the exact result or a result up to desired precision is obtained. A small error in the result is reduced quadratically after each iteration.

Solution of a Linear Algebraic Equation. The solution x of a linear algebraic equation

$$Ax = B \tag{3.4}$$

can be expressed in the form of the following variational principle:

$$[x] = x_t + (A^{-1})_t B - (A^{-1})_t A x_t, \tag{3.5}$$

where x_t and $(A^{-1})_t$ are trial estimates. The example with $A = 3$ and $B = 2$ leads to the exact result $x = 0.666666...$ and may be used to illustrate the advantage of using the variational principle. When used in the variational principle (3.5), the guesses $x_t = 0.6$ and $(A^{-1})_t = 0.3$ lead to the improved result $x = 0.66$. An improved estimate for A^{-1} could be found from $A^{-1} = x/B$ or by the use of variational principle (3.2). In either case, use of these improved results leads to a more precise estimate for x.

3.1.2 Variational Principles for Fredholm Integral Equations

The variational principles have proved to be extremely useful for solving a variety of mathematical problems in physics, including the scattering integral equations. Different types of variational principles have been considered for quantum scattering problems. Formulation of these variational principles is best understood in the context of the general Fredholm equation (2.3), and here some useful variational principles are considered. This equation has the following form:

$$F(x) = B(x) + \mathcal{K}F(x), \tag{3.6}$$

where $F(x)$ is the unknown, $B(x)$ is the known Born term, and the kernel $\mathcal{K}$ satisfies

$$\mathcal{K}F(x) = \int_0^\infty dy\, \mathcal{K}(x, y)F(y). \tag{3.7}$$

The formal solution of Eq. (3.6) is given by

$$F(x) \equiv JB(x) = \int_0^\infty dy J(x, y)B(y), \tag{3.8}$$

with

$$J = (I - \mathcal{K})^{-1}. \tag{3.9}$$

The straightforward numerical attempt to solve Eq. (3.6) proceeds via construction of the operator J of Eq. (3.9). If there is a certain error in this construction, the same error will propagate to the solution (3.8). The expression (3.8) is not variational.

A variational expression for the solution can be developed from the following operator identity:

$$F = F_t + J_t B - J_t(I - \mathcal{K})F_t + (J - J_t)(I - \mathcal{K})(F - F_t), \qquad (3.10)$$

where J_t and F_t are independent trial estimates of J and F. Neglecting the last term in this identity, which is second order in the differences, a simple variational expression for the solution can be written as

$$[F(x)] = F(x) + JB(x) - J(I - \mathcal{K})F(x), \qquad (3.11)$$

where the subscript t on the trial functions has been dropped. Expression (3.11) is stationary with respect to small variations of J and $F(x)$ around their exact values. With the use of Eqs. (3.8) and (3.9), the variational property of Eq. (3.11) can be verified. With the trial quantities

$$F_t(x) = F_E(x) + \Delta F(x), \qquad (3.12)$$

$$J_t = J_E + \Delta J, \qquad (3.13)$$

the expression (3.11) becomes

$$\begin{aligned}[F(x)] &= F_E(x) + \Delta F(x) + J_E B(x) + \Delta J B(x) \\ &\quad - (J_E + \Delta J)(I - \mathcal{K})[F_E(x) + \Delta F(x)] \\ &= F_E(x) - \Delta J(I - \mathcal{K})\,\Delta F(x),\end{aligned} \qquad (3.14)$$

where in deriving Eq. (3.14), Eqs. (3.8) and (3.9) have been used. The subscript E denotes *exact* and t denotes *trial*. In Eq. (3.14) the linear terms in errors Δ have canceled out and only the quadratic term survives.

The variational property above can be exploited to develop a successive improvement scheme for numerical calculation in the following fashion. For this purpose the trial functions are chosen as

$$F_t(x) = \sum_{i=1}^{n} a_i u_i(x), \qquad (3.15)$$

$$J_t(x, y) = \sum_{j=1}^{n} b_j(x) v_j(y), \qquad (3.16)$$

where $u_i(x)$ and $v_j(y)$ are two sets of arbitrarily chosen functions. With these trial functions the expression (3.11) becomes

$$[F(x)] = \sum_{i=1}^{n} a_i u_i(x) + \sum_{j=1}^{n} b_j(x)(v_j|B) - \sum_{j=1}^{n} b_j(x) \sum_{i=1}^{n} a_i[(v_j|u_i) - (v_j|\mathcal{K}u_i)], \tag{3.17}$$

where

$$(A|B) = \int_0^\infty dx\, A(x)B(x). \tag{3.18}$$

Now by demanding that expression (3.17) be stationary with respect to small variations of a_i and $b_j(x)$ around their exact values, the following solutions for the coefficients are obtained:

$$a_i = \sum_{j=1}^{n} D_{ij}^{(n)}(v_j|B), \tag{3.19}$$

$$b_j(x) = \sum_{i=1}^{n} u_i(x) D_{ij}^{(n)}, \tag{3.20}$$

with

$$(D^{-1})_{ji} = (v_j|u_i) - (v_j|\mathcal{K}u_i). \tag{3.21}$$

It is to be noted that after inversion of Eq. (3.21), the D matrix is n-dependent, shown explicitly by the label n in Eqs. (3.19) and (3.20). These results for a_i and $b_j(x)$ when substituted into Eq. (3.17) leads to the following variational estimate:

$$[F_n(x)] = \sum_{i,j=1}^{n} u_i(x) D_{ij}^{(n)}(v_j|B). \tag{3.22}$$

This result can also be obtained nonvariationally just by substituting Eq. (3.19) into Eq. (3.15). Equation (3.22) is closely related to the Schwinger variational principle of scattering, as we shall see later [2]. This is also the well-known result of the method of moments [3].

The method of moments of rank n usually postulates a solution to Eq. (3.6) of the form (3.15) and finds a_i by demanding that this solution is exact in the space spanned by the functions $v_j(x)$, or that

$$(v_j|F_t) = (v_j|B) + (v_j|\mathcal{K}F_t) \tag{3.23}$$

be satisfied. This requirement leads to solution (3.19) for the coefficient a_i. When this solution is substituted in Eq. (3.15), the variational result (3.22) is obtained. If the trial functions $u_i(x)$ are chosen to be orthonormal in a suitable space of $\mathcal{L}^2$ functions, the variational nature of the result guarantees successive improvement as the number of functions n is increased.

Equation (3.22) can also be derived if one considers the following operator form of Eq. (3.6):

$$F = B + KF. \tag{3.24}$$

Equation (3.6) should be understood as the x-space projection of Eq. (3.24), so that $F(x) \equiv (x|F)$, and so on. The formal solution of this equation could be written as

$$F = I \times \frac{1}{I - K} \times B, \tag{3.25}$$

where I is the identity operator. If we introduce the presumably complete set of states $\sum_{i=1}^{\infty} |u_i)(u_i|$ and $\sum_{j=1}^{\infty} \quad |v_j)(v_j|$ in places marked by the two $\times$'s, respectively, the variational result (3.22) is obtained provided that only a finite number of functions n are included in the expansion.

Variational principle (3.11) can also be written in an equivalent fractional form:

$$[F(x)] = \frac{F(x)JB(x)}{J(I - \mathcal{K})F(x)}. \tag{3.26}$$

The variational nature of Eq. (3.26) can be verified using the trial functions (3.12) and (3.13).

In fact, an infinite number of such variational expressions can be written down for the solution $F(x)$ of Eq. (3.6). Another variational expression can be developed from the following operator identity:

$$F = B + Q_t B + KF_t - Q_t(I - \mathcal{K})F_t + (Q - Q_t)(I - \mathcal{K})(F - F_t) \tag{3.27}$$

where the subscript t denotes trial estimates and $Q = \mathcal{K}(I - \mathcal{K})^{-1}$. Neglecting the last term in this expression, which is second order in differences, the following variational principle is obtained:

$$[F(x)] = B(x) + QB(x) + \mathcal{K}F(x) - Q(I - \mathcal{K})F(x). \tag{3.28}$$

Explicitly,

$$Q(x, y) = \int_0^\infty dz\, \mathcal{K}(x, z)J(z, y). \tag{3.29}$$

In Eq. (3.28) the quantities Q and $F(x)$ are considered unknowns. Equation (3.28) is a stationary expression for the difference $(F - B)$, and should yield good results when this difference is small compared to the full solution F.

The stationary property of Eq. (3.28) can be verified by considering the trial functions $Q_t = Q_E + \Delta Q$ and $F_t(x) = F_E(x) + \Delta F(x)$. Then Eq. (3.28) becomes

$$[F(x)] = B(x) + Q_E B(x) + \Delta Q B(x) + \mathcal{K}F_E(x) + \mathcal{K}\,\Delta F(x)$$
$$- (Q_E + \Delta Q)(I - \mathcal{K})[F_E(x) + \Delta F(x)] \tag{3.30}$$
$$= F_E(x) - (\Delta Q)(I - \mathcal{K})\Delta F(x), \tag{3.31}$$

which is quadratic in error, consistent with the stationary property.

Again the stationary property can be turned to good advantage by taking the trial function $F_t(x)$ of Eq. (3.15) and

$$Q_t(x, y) = \sum_{j=1}^{n} b_j(x)v_j(y). \tag{3.32}$$

Substituting these trial functions in Eq. (3.28) and requiring that the resultant expression be stationary with respect to small variations of a_i and $b_j(x)$ around their exact values, the following expressions for the coefficients result:

$$a_i = \sum_{j=1}^{n} D_{ij}^{(n)}(v_j|B), \tag{3.33}$$

$$b_j(x) = \sum_{i=1}^{n} \mathcal{K}u_i(x)D_{ij}^{(n)}. \tag{3.34}$$

When these expressions for a_i and $b_j(x)$ are substituted into Eqs. (3.32) and (3.15) and the results substituted into Eq. (3.28), the following variational estimate is obtained:

$$F_n(x) = B(x) + \sum_{i,j=1}^{n} \mathcal{K}u_i(x)D_{ij}^{(n)}(v_j|B). \tag{3.35}$$

Equation (3.35) is closely related to Kohn [4] and Newton [5] variational principles of scattering, as we shall see later. Also, when substituted into Eq. (3.15), Eq. (3.33) leads to the following nonvariational estimate:

$$F_n(x) = \sum_{i,j=1}^{n} u_i(x)D_{ij}^{(n)}(v_j|B). \tag{3.36}$$

The present nonvariational expression (3.36) is now recognized to be the variational expression (3.22) and hence may also be termed variational.

Expression (3.35) is the well-known result of the degenerate kernel method [3]. This method proceeds through construction of a degenerate kernel $\mathcal{K}_n$ of rank n using functions $u_i(x), i = 1, 2, ..., n$, and finding an approximate solution $F_n(x)$ satisfying

$$F_n(x) = B(x) + \mathcal{K}_n F_n(x). \tag{3.37}$$

For a specific choice of $\mathcal{K}_n$ a specific solution is obtained. For example, let us choose

$$\mathcal{K}_n = \sum_{i,j=1}^{n} \mathcal{K}u_i(x)C_{ij}^{(n)}v_j(y), \tag{3.38}$$

where $(C^{-1})_{ji} = (v_j|u_i)$. With this choice, Eq. (3.37) leads to the solution (3.35). For a different choice of $\mathcal{K}_n$, one can have a different result.

Variational result (3.35) can also be derived if we consider the following formal solution of Eq. (3.24):

$$F = B + \mathcal{K} \times \frac{1}{I - \mathcal{K}} \times B \tag{3.39}$$

and introduce the presumably complete set of states $\sum_{i=1}^{n} |u_i)(u_i|$ and $\sum_{j=1}^{n} |v_j)\ (v_j|$ in place of the two $\times$'s, respectively. This particular derivation of the solution of the Fredholm equations works in all situations. However, from this derivation the variational property of the result is usually not realized. The variational property of the solutions in the case of the usual derivation of the method of moments and the degenerate kernel method [3] is also never explicit.

Variational principle (3.28) can also be written in the equivalent fractional form:

$$[F(x)] = B(x) + \frac{QB(x)\mathcal{K}F(x)}{Q(I-\mathcal{K})F(x)}. \tag{3.40}$$

The variational nature of Eq. (3.40) can be verified directly.

Finally, it is instructive to write the following variational principle:

$$[F(x)] = B(x) + \mathcal{K}B(x) + RB(x) + \mathcal{K}^2F(x) - R(I-\mathcal{K})F(x), \tag{3.41}$$

where $R = \mathcal{K}^2(I-\mathcal{K})^{-1}$. In Eq. (3.41) R and $F(x)$ are unknown and should be approximated in calculations.

Employing the trial functions (3.15) and

$$R_t(x,y) = \sum_{j=1}^{n} b_j(x)v_j(y) \tag{3.42}$$

in Eq. (3.41) and demanding that the resultant expression be stationary with respect to small variations of a_i and $b_j(x)$, these constants can be found as before, and the following variational estimate is obtained:

$$[F_n(x)] = B(x) + \mathcal{K}B(x) + \sum_{i,j=1}^{n} \mathcal{K}^2u_i(x)D_{ij}^{(n)}(v_j|B), \tag{3.43}$$

where D is defined by Eq. (3.21). This result is closely related to the similar consideration by Takatsuka and McKoy in relation to the Lippmann–Schwinger equation [6].

3.2 KOHN VARIATIONAL PRINCIPLES

The *Rayleigh–Ritz variational method* for bound states is based on a variational functional in terms of the unknown wave function and the energy. The trial wave function in this variational functional is expanded in terms of a set of $\mathcal{L}^2$ functions, and the variational property then reduces the original eigenfunction–eigenvalue problem in the full Hilbert space to one in the limited space spanned by the basis functions. This procedure yields both the binding energy and the eigenfunction via solution of an algebraic set of homogeneous equations of small dimension.

The observables for potential scattering by a central potential are determined by the phase shifts. These parameters could be determined by solving either a differential or an integral equation. For multichannel problems involving noncentral potentials or composite particles, the exact solution is much more difficult, as one encounters a set of coupled differential or integrodifferential equations, which can also be cast in the form of a set of coupled integral equations. Algebraic variational methods of the type discussed in Section 3.1 provides great advantage for solving these scattering problems not only for potential scattering but also for multichannel problems.

The Rayleigh–Ritz variational functional can be extended to scattering processes. One can write a stationary expression for this functional at positive energies. By expanding the trial scattering wave function in a basis set with appropriate boundary conditions and by making use of the stationary property, one reduces the original scattering problem to an inhomogeneous algebraic set of equations of small dimension. The solution of this set of equations yields the scattering observables. This particular implementation of the variational principle is known as the *variational basis-set method.* For collision, the energy is known, and the trial wave function should have a non-$\mathcal{L}^2$ oscillating part at infinity as in Eq. (1.36). Once some of the basis functions are taken consistent with this boundary condition and others as $\mathcal{L}^2$ functions, the algebraic basis-set methods yield the scattering observables.

The first of these methods for the single-channel problem, due to Hulthén in 1944 [7], involved the solution of a quadratic equation. This method can lead to unphysical (complex) phase shifts. In 1948, independently, Hulthén and Kohn suggested a method that does not have the drawback of the original *Hulthén method* [4,8]. The second Hulthén method is also known as the *Rubinow* or *inverse Kohn method.* These methods could be generalized to multichannel problems. After some preliminary applications, these methods were studied critically by Schwartz in 1961 [9]. He found that the phase shift of the Kohn variational method often exhibits spurious resonancelike behavior where the phase shift rapidly increases (resonance) or decreases (antiresonance) by π as the energy is varied. The exact solution did not have these resonances. Later it has been found that most of the algebraic variational methods exhibit such spurious behavior. These behaviors are referred to below as anomalous behaviors or anomalies. Similar behavior can also appear in the Rubinow method. This unpleasant feature of the Kohn and Rubinow methods induced people to find other alternatives.

In 1967 Harris suggested another algebraic method, which increased interest in these variational approaches [10]. In a critical study of these methods, Nesbet in 1968 showed the origin of the anomalies in these methods, and several anomaly-free methods were suggested [11,12]. He also showed the relation among various methods and generalized them to multichannel problems. The works of Nesbet increased interest in the algebraic variational methods, and since then they have been applied to many physical problems of interest. Several new algebraic variational methods having certain advantages were suggested later.

In the remaining part of this section we present a general formalism for deriving Kohn, Rubinow and Hulthén variational principles which is used later for deriving different basis-set methods. In Section 3.3 we derive different algebraic basis-set methods based on these variational principles to be used in actual numerical calculations. We also present the variational R-matrix method, which has often been used for solving scattering problems. In Section 3.4 we describe the closely related Harris methods for scattering calculations. These methods are not based on a variational principle but are aimed to reduce the error in Kohn methods. In Section 3.5 we discuss the appearance of spurious singularities or anomalies in these calculations. Some ways of avoiding these anomalies are suggested. A generalization of the Kohn, Rubinow, and Hulthén basis-set methods as well as the variational R-matrix method to multichannel scattering is given in Section 3.6. Finally, in Sections 3.7 and 3.8, several anomaly-free methods for multichannel scattering are discussed.

3.2.1 Formulation

For the discrete (bound) states of the Schrödinger equation, the Rayleigh–Ritz variational principle makes use of the stationary property of the following functional:

$$[I_L] \equiv \langle \psi_L^t | (H - E) | \psi_L^t \rangle = \int_0^\infty \psi_L^t(r)(H - E)\psi_L^t(r)\, dr, \qquad (3.44)$$

where $\psi_L^t(r)$ is an approximation to the exact bound-state solution $\psi_L(r)$ of the following Schrödinger equation in the Lth partial wave:

$$(H - E)|\psi_L\rangle \equiv \left[-\frac{d^2}{dr^2} + V_L(r) + \frac{L(L+1)}{r^2} - k^2 \right] \psi_L(r) = 0. \quad (3.45)$$

The sub- or superscript t on functions in Eqs. (3.44) and (3.45) and in the remainder of the book denotes a trial estimate. The variational functional (3.44) can also be used for scattering. In the following we study the properties of this functional for both scattering and bound states. If the exact wave functions $\psi_L(r)$ are used in the functional (3.44), the result is identically zero by use of the Schrödinger equation (3.45).

Let us first study the condition of the stationary property of the functional (3.44) for the following trial function $\psi_L^t(r)$:

$$\psi_L^t(r) \to \psi_L(r) + \Delta\psi_L(r), \quad (3.46)$$

where the trial function is assumed to satisfy boundary condition (1.36) of scattering or (1.39) for bound states. Here we consider the scattering solution ψ_L with standing-wave boundary condition though the superfix $\mathcal{P}$ on the wave function is suppressed. In both scattering and bound-state problems, the first-order variation of the functional I_L of (3.44) is given by

$$\Delta I_L = \int_0^\infty \Delta\psi_L(r)(H - E)\psi_L(r)\, dr + \int_0^\infty \psi_L(r)(H - E)\Delta\psi_L(r)\, dr \quad (3.47)$$

$$= 2\int_0^\infty \Delta\psi_L(r)(H - E)\psi_L(r)\, dr + \lim_{r\to\infty} \left[\Delta\psi_L(r)\frac{d\psi_L(r)}{dr} - \psi_L(r)\frac{d\Delta\psi_L(r)}{dr} \right]. \quad (3.48)$$

Unlike in the space spanned by $\mathcal{L}^2$ functions, in the space of unnormalizable scattering functions, the differential kinetic energy operator in the Hamiltonian is not Hermitian. Hence, care must be taken if the differential operator in the kinetic energy of the Hamiltonian operates to the left. An integration by parts over the second derivative of the kinetic energy term (3.47) has been performed to obtain Eq. (3.48) from Eq. (3.47). In the space of $\mathcal{L}^2$ functions, the Hamiltonian operator is Hermitian and the last term on the right-hand side of Eq. (3.48) is zero for any bound-state $\mathcal{L}^2$ trial function satisfying condition (1.35) or (1.38). The condition (1.39) is not necessary for the vanishing of this term, which contributes only for scattering functions.

First, let us consider the bound-state problem. In the bound-state problem, if both trial $\psi_L^t(r)$ and $\Delta\psi_L(r)$ satisfy boundary condition (1.35), the first-order variation (proportional to Δ) of the functional I_L is zero:

$$\Delta I_L = 0. \tag{3.49}$$

Only the second-order variation (proportional to Δ^2) survives. The variational property (3.49) follows from (3.48) by using the Hermitian nature of the Hamiltonian in the space of $\mathcal{L}^2$ functions.

In application of the Rayleigh–Ritz variational principle for bound states, a variational trial function is constructed in the form

$$\psi_L^t(r) = \sum_{i=1}^{n} b_i^{(\nu)} \chi_i(r), \tag{3.50}$$

where ν labels the eigenvalue. The basis functions $\chi_i(r)$ are $\mathcal{L}^2$ and satisfy Eq. (1.35). The use of Eq. (3.50) in Eqs. (3.44) and (3.49) leads to the following eigenfunction–eigenvalue problem

$$\sum_{j=1}^{n} \mathcal{M}_{ij} b_j^{(\nu)} = 0, \qquad i = 1, 2, ..., n, \tag{3.51}$$

where

$$\mathcal{M}_{ij} \equiv \langle \chi_i | (H - E) | \chi_j \rangle = \int_0^\infty \chi_i(r)(H - E_\nu)\chi_j(r)\, dr. \tag{3.52}$$

The solution, $\chi_i(r)$, of this eigenfunction–eigenvalue problem in the space of $\mathcal{L}^2$ functions yields both the binding energy and the wave function.

The variational functional (3.44) can also be used for the scattering problem. Calculational schemes based on this functional were initiated with the studies of Hulthén and Kohn about 50 years ago and proved to be very useful for the calculation of scattering phase shifts in potential scattering, although the idea behind these studies can be extended to multichannel scattering. The exact scattering wave function satisfies Eq. (1.35) and the asymptotic boundary condition (1.136). The use of different approximate trial wave functions in Eqs. (3.44) and (3.48) leads to different variational principles. Several possibilities are considered in the following.

Kohn considered trial function (3.46), satisfying boundary conditions (1.35) [or (1.38)] and (1.136), where δ_L is the approximate phase shift. The trial function (3.46), consequently, satisfies

$$\lim_{r\to\infty} \Delta\psi_L(r) \to \Delta(\tan\delta_L)\cos\left(kr - \frac{L\pi}{2}\right). \tag{3.53}$$

In this case, the following first-order variation of the functional (3.44) is obtained from (3.48):

$$\Delta I_L = \lim_{r\to\infty}\left[\Delta\psi_L(r)\frac{d\psi_L(r)}{dr} - \psi_L(r)\frac{d\,\Delta\psi_L(r)}{dr}\right]. \tag{3.54}$$

If we use asymptotic properties (1.136) and (3.53), this limit can be evaluated, so that Eq. (3.54) becomes

$$\Delta I_L = k\,\Delta(\tan\delta_L). \tag{3.55}$$

For the exact wave function, with the exact phase shift, $I_L = 0$. Hence, Eq. (3.55) leads to the Kohn variational principle

$$[\tan\delta_L]_K = \tan\delta_L - \frac{I_L(\tan\delta_L)}{k}. \tag{3.56}$$

The quantity ΔI_L of Eq. (3.55) is only the first-order change in I_L. The total change in I_L is the sum of the first-order change given by (3.55) and the second-order change and is written as

$$I_{Lt} - I_L = k\,\Delta(\tan\delta_L) + \int_0^\infty \Delta\psi_L(r)(H-E)\,\Delta\psi_L(r)\,dr. \tag{3.57}$$

The exact functional I_L of Eq. (3.57) is zero. Hence, Eq. (3.57) leads to the following identity:

$$[\tan \delta_L]_K = \tan \delta_L - \frac{I_L(\tan \delta_L)}{k} + k^{-1} \int_0^\infty \Delta\psi_L(r)(H - E)\Delta\psi_L(r)\, dr. \tag{3.58}$$

Equation (3.58) is the Kato [13] identity, from which the Kohn variational principle can be derived. For example, if we ignore the last term in this identity, we obtain the Kohn variational principle (3.56).

If we use trial wave functions with different asymptotic properties, alternative variational principles can be derived. For example, let us consider trial function (3.46), satisfying the following asymptotic properties:

$$\psi_L^t(r) \to \cos\left(kr - \frac{L\pi}{2}\right) + \cot \delta_L \sin\left(kr - \frac{L\pi}{2}\right), \tag{3.59}$$

$$\Delta\psi_L(r) \to \Delta(\cot \delta_L) \sin\left(kr - \frac{L\pi}{2}\right), \tag{3.60}$$

where δ_L is the approximate phase shift. Condition (3.59) is consistent with (1.136). The first-order variation of the functional (3.44) is given by Eq. (3.54). If we use asymptotic properties (3.59) and (3.60), the limit in Eq. (3.54) can be evaluated, so that the first-order variation of the functional I_L is given by

$$\Delta I_L = -\frac{\Delta(\cot \delta_L)}{k}. \tag{3.61}$$

For the exact wave function with exact phase shift, $I_L = 0$, and Eq. (3.61) reduces to the Rubinow variational principle [8]:

$$[\cot \delta_L]_{IK} = \cot \delta_L + \frac{I_L(\cot \delta_L)}{k}. \tag{3.62}$$

As the variational function $\cot \delta_L$ of the Rubinow variational principle is the inverse of the variational function $\tan \delta_L$ of the Kohn variational principle, the Rubinow variational principle (3.62) is usually known as the *inverse Kohn variational principle*.

Next let us consider the trial function (3.46) with the asymptotic property

$$\psi_L^t(r) \to \sin\left(kr - \frac{L\pi}{2} + \delta_L\right), \tag{3.63}$$

where δ_L is the approximate phase shift. The asymptotic property (3.63) is essentially the same as (1.137). If we take $\delta_L = \delta_L^e + \Delta\delta_L$, where δ_L^e is the exact phase shift, the asymptotic behavior (3.63) can be expanded for small $\Delta\delta_L$ and one obtains

$$\Delta\psi_L(r) \to \cos\left(kr - \frac{L\pi}{2} + \delta_L\right)\Delta\delta_L. \tag{3.64}$$

Using these asymptotic properties, the limit in the first-order variation of the functional I_L given by Eq. (3.54) can be evaluated, so that this first-order variation is given by

$$\Delta I_L - k\,\Delta\delta_L = 0. \tag{3.65}$$

For the exact phase shift, $I_L = 0$. Hence, Eq. (3.65) reduces to

$$[\delta_L]_H = \delta_L + \frac{I_{Lt}}{k}. \tag{3.66}$$

This is the *Hulthén variational principle* [7,8]. In the original work, Hulthén imposed a further condition that $I_{Lt} = 0$ for the approximate wave function and approximate phase shift, so that

$$[\delta_L]_H = \delta_L; \qquad I_{Lt} = 0. \tag{3.67}$$

We have written an explicit Kato identity (3.58) for the Kohn variational principle. However, we could write similar identities for the Hulthén and inverse Kohn variational principles. The second-order correction term in these identities are not of any particular sign, and this term is neglected in the variational principles. So the errors in these variational calculations are not of any definite sign. Hence, the Kohn, Rubinow, and Hulthén variational principles do not provide any bound on the phase shift. Recalling that as $k \to 0$, $k\cot\delta_0 \to -a^{-1}$, where a is the S-wave scattering length, Eq. (3.62) at zero energy gives a variational principle for a. This variational principle also does not provide any bound. Rosenberg, et al. suggested a calculational scheme for the scattering length which provides bounds on the result [14].

3.3 VARIATIONAL BASIS-SET METHODS

3.3.1 Kohn Method

The variational principles of Section 3.2 can be used to develop practical calculational schemes for the solution of scattering problems. Schwartz and Nesbet [9,12] suggested simple basis-set calculational schemes for Kohn variational principles, which we describe in the following. They expanded the unknown trial functions of these methods in appropriate basis sets and used the variational principles to reduce the scattering problem to the solution of a set of algebraic equations.

For the Kohn variational principle (3.56), it is convenient to work with the trial function

$$\psi_t(r) = [v^S(r) + S(r)] + \lambda[v^C(r) + C(r)], \tag{3.68}$$

where $\lambda = \tan\delta$ is the tangent of the approximate phase shift, and the explicit dependencies of the functions on angular momentum L have been suppressed in Eq. (3.68), and unless there is a chance of confusion they will be suppressed in the following. The functions $v^S(r)$ and $v^C(r)$ are $\mathcal{L}^2$ and are taken to satisfy Eq. (1.35) or (1.38). The functions $S(r)$ and $C(r)$ are known functions with the following asymptotic behaviors:

$$\lim_{r\to\infty} S(r) \to \sin\left(kr - \frac{L\pi}{2}\right), \tag{3.69}$$

$$\lim_{r\to\infty} C(r) \to \cos\left(kr - \frac{L\pi}{2}\right). \tag{3.70}$$

Then the trial wave function of Eq. (3.68) has the asymptotic behavior (1.136), as required in the Kohn variational principle. The functions v^S and v^C are expanded in terms of the set of $\mathcal{L}^2$ functions χ_i, $i = 1, \ldots, n$:

$$v^S(r) = \sum_{i=1}^{n} b_i^S \chi_i(r), \tag{3.71}$$

$$v^C(r) = \sum_{i=1}^{n} b_i^C \chi_i(r), \tag{3.72}$$

where the expansion coefficients b_i^S and b_i^C are to be determined from the stationary property of Eq. (3.56).

The oscillating functions $S(r)$ and $C(r)$ are usually taken as

$$C(r) \equiv f_1(r) = [1 - \exp(-\beta r)] \cos\left(kr - \frac{L\pi}{2}\right), \tag{3.73}$$

$$S(r) \equiv f_2(r) = \sin\left(kr - \frac{L\pi}{2}\right), \tag{3.74}$$

where β is an arbitrary parameter.

Two common choices for the $\mathcal{L}^2$ functions $\chi_i(r)$ are

$$\chi_i(r) \equiv f_i(r) = r^{(i+L)} \exp(-\alpha r), \qquad i = 1, 2, ..., n. \tag{3.75}$$

or

$$\chi_i(r) \equiv f_i(r) = r^{(L+1)} \exp[-i\alpha r], \qquad i = 1, 2, ..., n. \tag{3.76}$$

In Eqs. (3.75) and (3.76), α is an arbitrary parameter. The parameters β and α of Eqs. (3.73), (3.75), and (3.76) are to be chosen by trial, to obtain the best convergence.

Use of the trial function (3.68) in the Kohn variational principle (3.56) leads to

$$[\lambda]_K = \lambda - \frac{I(\lambda)}{k}, \tag{3.77}$$

where

$$I(\lambda) = B_0 + \lambda B_1 + \lambda^2 B_2 + W_0 + \lambda W_1 + \lambda^2 W_2, \tag{3.78}$$

with

$$W_0 = \sum_{i,j=1}^{n} b_i^S b_j^S \mathcal{M}_{ij} + 2\sum_{i=1}^{n} b_i^S R_i^S, \tag{3.79}$$

$$W_1 = 2\sum_{i,j=1}^{n} b_i^S b_j^C \mathcal{M}_{ij} + 2\sum_{i=1}^{n} b_i^S R_i^C + 2\sum_{i=1}^{n} b_i^C R_i^S, \tag{3.80}$$

$$W_2 = \sum_{i,j=1}^{n} b_i^C b_j^C \mathcal{M}_{ij} + 2\sum_{i=1}^{n} b_i^C R_i^C, \tag{3.81}$$

$$\mathcal{M}_{ij} = \mathcal{M}_{ji} = \int_0^\infty \chi_i(r)(H - E)\chi_j(r)\, dr, \tag{3.82}$$

$$R_i^S = \int_0^\infty \chi_i(r)(H - E)S(r)\, dr = \int_0^\infty S(r)(H - E)\chi_i(r)dr, \tag{3.83}$$

$$R_i^C = \int_0^\infty \chi_i(r)(H - E)C(r)\, dr = \int_0^\infty C(r)(H - E)\chi_i(r)dr, \tag{3.84}$$

$$B_0 = \int_0^\infty S(r)(H - E)S(r)\, dr, \tag{3.85}$$

$$B_1 = \int_0^\infty S(r)(H - E)C(r)\, dr + \int_0^\infty C(r)(H - E)S(r)\, dr, \tag{3.86}$$

$$B_2 = \int_0^\infty C(r)(H - E)C(r)\, dr. \tag{3.87}$$

The elements (integrals) $\mathcal{M}_{ij}$, where two $\mathcal{L}^2$ functions appear, are called bound–bound elements (integrals). The quantities R_i^S and R_i^C, where $\mathcal{L}^2$ and unrenormalizable functions appear, are referred to as bound–free, and the quantities B_i, where two unrenormalizable functions appear, are referred to as free–free. The free–free elements involve infinite integrals over oscillating functions and are the most difficult to evaluate numerically.

The expression (3.77) is stationary with respect to independent variations of the trial phase shift and trial wave function, for example,

$$\frac{\partial[\lambda]_K}{\partial b_i^C} = \frac{\partial[\lambda]_K}{\partial b_i^S} = \frac{\partial[\lambda]_K}{\partial \lambda} = 0. \tag{3.88}$$

The stationary properties with respect to variations of b_i^C and b_i^S for all λ lead to

$$\sum_{j=1}^{n} \mathcal{M}_{ij} b_j^S = -R_i^S, \tag{3.89}$$

$$\sum_{j=1}^{n} \mathcal{M}_{ij} b_j^C = -R_i^C. \tag{3.90}$$

As variations of the coefficients b_i^C and b_i^S and λ are independent in the derivative with respect to b_i^C and b_i^S, coefficients of each power of λ are separately zero. Equations (3.89) and (3.90) involve the same matrix as in the standard Rayleigh–Ritz variational method for the bound-state problem defined by Eq. (3.52). These equations can be solved for b_i^S and b_i^C unless the matrix $\mathcal{M}$ becomes accidentally singular at some energy. We shall see later that the appearance of this accidental singularity is not fatal for the variational calculation of the phase shift, and a finite phase shift exists in this case. However, when this happens, numerical care might be needed. When b_i^S and b_i^C are known, the $\mathcal{L}^2$ functions v^S and v^C are determined from Eqs. (3.71) and (3.72). Then one needs to find the optimum λ in expression (3.77) consistent with the stationary property.

The stationary property of Eqs. (3.77) and (3.78) with respect to variations of λ is still to be implemented. If we use Eqs. (3.89) and (3.90), the functional of Eq. (3.78) becomes

$$I(\lambda) = M_{00} + (M_{01} + M_{10})\lambda + M_{11}\lambda^2, \tag{3.91}$$

where

$$M_{00} = \langle S|(H - E)|v^S + S\rangle, \tag{3.92}$$

$$M_{01} = \langle S|(H - E)|v^C + C\rangle, \tag{3.93}$$

$$M_{10} = \langle C|(H - E)|v^S + S\rangle, \tag{3.94}$$

$$M_{11} = \langle C|(H - E)|v^C + C\rangle. \tag{3.95}$$

In Eqs. (3.92) to (3.95) and in the following, we are using notations such as

$$\langle S|(H - E)|C\rangle = \int_0^\infty S(r)(H - E)C(r)\, dr. \tag{3.96}$$

The solution of Eqs. (3.89) and (3.90) can be written as

$$b_j^S = -\sum_{i=1}^{n} (\mathcal{D})_{ji}^{(n)} R_i^S, \tag{3.97}$$

$$b_j^C = -\sum_{i=1}^{n} (\mathcal{D})_{ji}^{(n)} R_i^C, \tag{3.98}$$

respectively, where

$$(\mathcal{D}^{-1})_{ij} = \mathcal{M}_{ij}. \tag{3.99}$$

If we use Eqs. (3.71), (3.72), (3.97), and (3.98), Eqs. (3.92), (3.93), (3.94), and (3.95) can be rewritten as

$$M_{00} = \langle S|(H-E)|S\rangle - \sum_{i,j=1}^{n} R_i^S (\mathcal{D})_{ij}^{(n)} R_j^S, \tag{3.100}$$

$$M_{01} = \langle S|(H-E)|C\rangle - \sum_{i,j=1}^{n} R_i^S (\mathcal{D})_{ij}^{(n)} R_j^C, \tag{3.101}$$

$$M_{10} = \langle C|(H-E)|S\rangle - \sum_{i,j=1}^{n} R_i^C (\mathcal{D})_{ij}^{(n)} R_j^S, \tag{3.102}$$

$$M_{11} = \langle C|(H-E)|C\rangle - \sum_{i,j=1}^{n} R_i^C (\mathcal{D})_{ij}^{(n)} R_j^C. \tag{3.103}$$

Equations (3.100) to (3.103) provide a calculational scheme for the matrix elements M_{ij}. This is done in two steps. First the bound–bound problem is solved by inverting the $n \times n$ matrix $\mathcal{M}_{ij}$, then certain bound–free and free–free matrix elements are calculated in order to evaluate the matrix elements $M_{ij}, i, j = 0, 1$. The variational result λ for the tangent of the phase shift is then expressed in terms of these matrix elements.

Noting that v^C and v^S are $\mathcal{L}^2$ functions, we have

$$M_{01} = M_{10} + k. \tag{3.104}$$

As the kinetic energy operator is not Hermitian in the space of non-$\mathcal{L}^2$ functions, we have $M_{01} \neq M_{10}$. Equation (3.104) follows from the following identity:

$$\int_0^\infty S(r)\left(-\frac{d^2}{dr^2}\right)C(r)\,dr = k + \int_0^\infty C(r)\left(-\frac{d^2}{dr^2}\right)S(r)\,dr. \tag{3.105}$$

This equation can be established by integration by parts and using the boundary conditions satisfied by the functions $S(r)$ and $C(r)$. The stationary property of Eqs. (3.77) and (3.91) with respect to variations of λ leads to

$$1 - \frac{M_{01} + M_{10}}{k} - \frac{2\lambda M_{11}}{k} = 0. \tag{3.106}$$

Equations (3.104) and (3.106) yield the following optimum or zero-order estimate of λ in the Kohn method:

$$\lambda_I = -\frac{M_{10}}{M_{11}}. \tag{3.107}$$

When we use Eq. (3.107) in Eqs. (3.77) and (3.91), we get the Kohn variational result:

$$[\lambda]_K = \lambda_I - \frac{I(\lambda_I)}{k} \tag{3.108}$$

$$= -\frac{1}{k}\left(M_{00} - M_{10}\frac{1}{M_{11}}M_{10}\right) \tag{3.109}$$

$$= -\frac{M_{10}}{M_{11}} - \frac{1}{kM_{11}}\det M, \tag{3.110}$$

where

$$\det M = M_{00}M_{11} - M_{01}M_{10}. \tag{3.111}$$

The equivalence between Eqs. (3.109) and (3.110) follows with the use of Eqs. (3.104) and (3.111). If trial λ represents the exact solution, so does λ_I; consequently, $\det M = 0$. However, $\det M = 0$ does not necessarily guarantee that the exact solution has been obtained. Equations (3.108) to (3.110) are the fundamental equations for an algebraic basis-set calculational scheme based on the Kohn variational principle.

To summarize the Kohn variational basis-set method, first one has to solve the inhomogeneous bound–bound problem (3.89) and (3.90). This determines the $\mathcal{L}^2$ functions v^S and v^C via Eqs. (3.71) and (3.72). The basis functions χ_i of these equations are supposed to contain a variational parameter. Then the four matrix elements $M_{ij}, i, j = 0, 1$, are calculated via Eqs. (3.100) to (3.103). The phase shift is finally obtained from these elements with the use of Eqs. (3.107), (3.109), and (3.110). The ideal value of the parameter in the $\mathcal{L}^2$ functions is determined by experimentation so as to yield well-converged result with a small basis set.

3.3.2 Inverse Kohn Method

A basis-set calculational scheme for the Rubinow or inverse Kohn variational principle (3.62) can be obtained in a similar fashion. For obtaining such a scheme it is convenient to employ the following trial function:

$$\psi_t(r) = \lambda^{-1}[v^S(r) + S(r)] + [v^C(r) + C(r)], \tag{3.112}$$

where $\lambda = \tan\delta$ and the functions have the same meaning as in Eq. (3.68). If we use this trial function, with v^S and v^C given by Eqs. (3.71) and (3.72), respectively, in the inverse Kohn variational principle (3.62), we get

$$[\lambda^{-1}]_{IK} = \lambda^{-1} + \frac{I(\lambda)}{k}, \tag{3.113}$$

where

$$I(\lambda) = B_2 + \lambda^{-1}B_1 + \lambda^{-2}B_0 + W_2 + \lambda^{-1}W_1 + \lambda^{-2}W_0. \tag{3.114}$$

The parameters of this expression are the same as those in Eq. (3.78). Expression (3.113) is stationary with respect to independent variations of b_i^C, b_i^S, and λ. The stationary properties with respect to variations of b_i^C and b_i^S again lead to Eqs. (3.89) and (3.90). If we use Eqs. (3.89) and (3.90), the functional $I(\lambda)$ of Eq. (3.114) is given by

$$I(\lambda) = M_{11} + (M_{01} + M_{10})\lambda^{-1} + M_{00}\lambda^{-2}. \tag{3.115}$$

The elements M_{ij} are defined by Eqs. (3.100) to (3.103). The stationary property of (3.113) and (3.115) with respect to variations of λ^{-1} leads to

$$1 + \frac{M_{01} + M_{10}}{k} + \frac{2M_{00}\lambda^{-1}}{k} = 0. \tag{3.116}$$

Equations (3.116) and (3.104) lead to the following optimum or zero-order estimate λ of the inverse Kohn variational method:

$$\lambda_{II} = -\frac{M_{00}}{M_{01}}. \tag{3.117}$$

When Eq. (3.117) is used in Eqs. (3.113) and (3.115), we get the following inverse Kohn variational result:

$$[\lambda^{-1}]_{IK} = \lambda_{II}^{-1} + \frac{I(\lambda_{II})}{k} \tag{3.118}$$

$$= \frac{1}{k}\left(M_{11} - M_{01}\frac{1}{M_{00}}M_{01}\right) \tag{3.119}$$

$$= -\frac{M_{01}}{M_{00}} + \frac{1}{kM_{00}}\det M. \tag{3.120}$$

The equivalence between Eqs. (3.119) and (3.120) follows with the use of Eqs. (3.104) and (3.111). Equations (3.118) to (3.120) provide a basis-set calculational scheme based on the Rubinow or inverse Kohn variational principle.

The Kohn and inverse Kohn variational results (3.110) and (3.120) can be expressed as

$$[\lambda]_K = \lambda_I - \frac{\det M}{kM_{11}}, \tag{3.121}$$

$$[\lambda^{-1}]_{IK} = \lambda_{II}^{-1} + \frac{\det M}{kM_{00}}. \tag{3.122}$$

The corrections on the optimum results λ_I and λ_{II} in these cases involve $\det M$, which would be small with a good trial wave function and phase shift. For an exact solution $\det M = 0$ and

$$[\lambda]_K = ([\lambda^{-1}]_{IK})^{-1} = \lambda_I = \lambda_{II}. \tag{3.123}$$

For a small $\det M$ a power series expansion of $([\lambda^{-1}]_{IK})^{-1}$ of Eq. (3.120) in powers of $\det M$ reveals that

$$[\lambda]_K - ([\lambda^{-1}]_{IK})^{-1} = \frac{(\det M)^2}{M_{01}^2}\left(\frac{1}{kM_{11}} + \frac{M_{00}}{k^2 M_{01}}\right) + \cdots, \tag{3.124}$$

so that tangents of the phase shift of the Kohn and inverse Kohn variational methods are equal to terms of the order of $\det M$.

3.3.3 Hulthén Method

The Hulthén variational principle can also be used to develop a basis set calculational scheme. In this case the trial function is taken to have the general form

$$\psi_t(r) = \kappa_0[v^S(r) + S(r)] + \kappa_1[v^C(r) + C(r)], \tag{3.125}$$

where the functions have the same meaning as in Eq. (3.68). Equation (3.125) is a generalization of the forms (3.68) and (3.112). For the Kohn variational method $\kappa_0 = 1$ and $\kappa_1 = \lambda = \tan\delta$; and for the inverse Kohn variational method $\kappa_1 = 1$ and $\kappa_0 = \lambda^{-1} = \cot\delta$. In the Hulthén method we use $\kappa_0 = \cos\delta$ and $\kappa_1 = \sin\delta$. This trial function, with v^S and v^C given by Eqs. (3.71) and (3.72), is then used to calculate the functional I of Eq. (3.44). This functional is now given by

$$I(\lambda) = \cos^2\delta[M_{00} + (M_{01} + M_{10})\lambda + M_{11}\lambda^2], \tag{3.126}$$

$$= \sin^2\delta[M_{00}\lambda^{-2} + (M_{01} + M_{10})\lambda^{-1} + M_{11}]. \tag{3.127}$$

If we use the definition $\lambda = \tan\delta$, equivalence between (3.126) and (3.127) follows. The phase shifts in Eqs. (3.126) and (3.127) are approximate and implementation of the Hulthén variational principle, $I = 0$ of Eq. (3.67), can be done via either of these equations.

With $I(\lambda)$ given by (3.126), the phase shift of the Hulthén variational method (3.67) is given by

$$I(\lambda_H) \equiv M_{00} + (M_{01} + M_{10})\lambda_H + M_{11}\lambda_H^2 = 0, \tag{3.128}$$

so that

$$[\lambda]_H = -\frac{M_{10}}{M_{11}} - \frac{k}{2M_{11}} + \frac{\sqrt{k^2 - 4\det M}}{2M_{11}}. \tag{3.129}$$

There could be a $\pm$ sign in front of the square root in Eq. (3.129). In the exact solution limit, $\det M = 0$, and $[\lambda]_H$ should reduce to $[\lambda]_K$ of Eq. (3.123). This condition rules out the negative sign in the square-root term in Eq. (3.129), and the positive sign is chosen. Also, if $4\det M > E$, the solutions of Eq. (3.129) are complex and the method fails. This means that the trial solution is "so much" away from the exact result that the variational method fails to improve it.

With $I(\lambda)$ given by (3.127), the phase shift of the Hulthén variational method (3.67) is given, equivalently, by

$$I(\lambda_H) \equiv M_{11} + (M_{01} + M_{10})\lambda_H^{-1} + M_{00}\lambda_H^{-2} = 0, \tag{3.130}$$

so that

$$[\lambda^{-1}]_H = -\frac{M_{01}}{M_{00}} + \frac{k}{2M_{00}} - \frac{\sqrt{k^2 - 4\det M}}{2M_{00}}. \tag{3.131}$$

Only the physically acceptable sign in front of the square root has been shown in Eq. (3.131).

The Kohn variational result (3.110) is the linear term in the power series expansion of the Hulthén result (3.129) in powers of $(\det M)$, so that

$$[\lambda]_H = [\lambda]_K - \frac{(\det M)^2}{k^3 M_{11}} + \cdots. \tag{3.132}$$

Similarly, the inverse Kohn result (3.120) is the linear term in the power series expansion of the Hulthén result (3.131) in powers of $\det M$. In Eq. (3.124) we have seen that the tangents of the Kohn and inverse Kohn variational phase shifts agree with each other to terms on the order of $\det M$. Hence, for small $\det M$, the Kohn, inverse Kohn, and Hulthén results agree to terms on the order of $\det M$.

Equations (3.109) and (3.110), (3.119) and (3.120), and (3.129) and (3.131) provide calculational schemes of the Kohn, inverse Kohn, and Hulthén variational principles, respectively, in terms of the matrix elements M_{ij}, which are calculated using Eqs. (3.100) to (3.103). The essential calculational scheme of these three methods are the same. First solve for the $\mathcal{L}^2$ functions via Eqs. (3.89) and (3.90). Then calculate the four elements M_{ij}, $i, j = 0, 1$, via Eqs. (3.100) to (3.103). The phase shifts of the three methods are obtained from these M_{ij}'s. The calculation of the free–free matrix elements B_0, B_1, and B_2 is numerically the most troublesome part of these methods. These elements involve infinite integrals over oscillatory functions at infinity.

3.3.4 Variational *R*-Matrix Method

Rayleigh Ritz and Kohn methods provide two applications of the variational functional (3.44). There are other applications of this functional in solving the scattering problem. It can be used to find the phase shift via the R-matrix method. For potential scattering, the R matrix is defined by [15]

$$\frac{1}{R} = \left[\frac{r}{\psi(r)}\frac{d\psi(r)}{dr}\right]_{r=r_0}, \tag{3.133}$$

where r_0 is presumably larger than the range of the interaction potential, so that the solution of the Schrödinger equation at this point is a linear combination of free-particle solutions. Hence, at $r = r_0$, the asymptotic region implicit in Eq. (1.136) has been attained and the R matrix is given by

$$R = (kr_0)^{-1}\tan\left(kr_0 - \frac{L\pi}{2} + \delta_L\right). \tag{3.134}$$

As L, k, and r_0 are assumed known, knowledge of the R matrix leads to the scattering phase shifts δ_L.

Using variational functional (3.44), we describe below a method [4] for calculating the R matrix. The entire configuration space is first divided into two parts: $r > r_0$ and $r \leq r_0$. In Eq. (3.44) we introduce a trial wave function, $\psi_t(r)$, which is assumed to be exact for $r > r_0$. Consequently, troublesome infinite integrals over the oscillating functions as encountered in Kohn methods can be avoided. With this trial function, the functional (3.44) is written as

$$[I] = \int_0^{r_0} \psi_t(r)(H - E)\psi_t(r)\,dr \tag{3.135}$$

$$= \int_0^{r_0}\left[\left(\frac{d\psi_t(r)}{dr}\right)^2 + \left(V(r) + \frac{L(L+1)}{r^2} - k^2\right)[\psi_t(r)]^2\right]dr - \kappa[\psi_t(r_0)]^2, \tag{3.136}$$

where

$$\kappa = \frac{1}{Rr_0} = \left[\frac{1}{\psi_t(r)}\frac{\psi_t(r)}{dr}\right]_{r=r_0}. \tag{3.137}$$

An integration by parts over the kinetic energy term has been performed to obtain Eq. (3.136). As the trial wave function is exact for $r > r_0$, a truncation of the integral in the functional (3.135) has taken place.

Let us first study the condition of the stationary property of the functional (3.135) for variations $\psi_t(r) \to \psi(r) + \Delta\psi(r)$. The first-order variation of the functional I of (3.135) is given by

$$\Delta I = \int_0^{r_0} \Delta\psi(r)(H - E)\psi(r)\, dr + \int_0^{r_0} \psi(r)(H - E)\Delta\psi(r) dr \quad (3.138)$$

$$= 2\int_0^{r_0} \Delta\psi(r)(H - E)\psi(r)\, dr + \left[\Delta\psi(r)\frac{d\psi(r)}{dr} - \psi(r)\frac{d\Delta\psi(r)}{dr}\right]_{r=r_0}. \quad (3.139)$$

In order to derive (3.139), an integration by parts has been performed in the last term of Eq. (3.138). Equation (3.139) should be compared with Eq. (3.48), where there is no restriction on the range of the integral of the functional (3.44). The first term on the right-hand side of Eq. (3.139) is zero by using the Schrödinger equation. The functional of Eq. (3.136) is stationary if the last term on the right-hand side of Eq. (3.139) is also zero. This happens when

$$\left[\frac{1}{\Delta\psi(r)}\frac{d\,\Delta\psi(r)}{dr}\right]_{r=r_0} = \left[\frac{1}{\psi(r)}\frac{d\psi(r)}{dr}\right]_{r=r_0}. \quad (3.140)$$

From Eqs. (3.137) and (3.140) we find that for the functional I of Eq. (3.136) to be stationary, the trial wave function should have the same logarithmic derivative at $r = r_0$ as the exact wave function. However, it is difficult to implement this condition.

Next, $\psi_t(r)$ is expanded in a linearly independent basis set

$$\psi_t(r) = \sum_{i=1}^{n} b_i\chi_i(r), \quad (3.141)$$

where χ_i are the basis functions and b_i are the expansion coefficients. This expansion is then substituted into the variational expression (3.136) and the resultant expression is stationary with respect to small variations of the coefficients b_i. Using this stationary property, we obtain

$$\sum_{j=1}^{n} \mathcal{M}_{ij}b_j = \kappa\chi_i(r_0)\psi_t(r_0), \quad (3.142)$$

where

$$\mathcal{M}_{ij} = \int_0^{r_0} \left[\frac{d\chi_i(r)}{dr} \frac{d\chi_j(r)}{dr} + \left(V(r) + \frac{L(L+1)}{r^2} - k^2 \right) \chi_i(r) \chi_j(r) \right] dr. \tag{3.143}$$

Equation (3.142) can be solved for the coefficients b_i to yield

$$b_i = \kappa \sum_{j=1}^{n} [\mathcal{M}^{-1}]_{ij} \chi_j(r_0) \psi_t(r_0). \tag{3.144}$$

From Eqs. (3.141) and (3.144) we obtain

$$\psi_t(r_0) \equiv \sum_{i=1}^{n} b_i \chi_i(r_0) = \kappa \sum_{i,j=1}^{n} \chi_i(r_0) [\mathcal{M}^{-1}]_{ij} \chi_j(r_0) \psi_t(r_0). \tag{3.145}$$

Equation (3.145) gives the condition of existence of a solution to Eq. (3.142). Canceling $\psi_t(r_0)$ from both sides of this equation, we obtain for the R matrix

$$R = \frac{1}{\kappa r_0} = \frac{1}{r_0} \sum_{i,j=1}^{n} \chi_i(r_0) [\mathcal{M}^{-1}]_{ij} \chi_j(r_0). \tag{3.146}$$

Equation (3.146) is the desired expression for the R matrix. From a knowledge of this R matrix, the phase shifts can be calculated using Eq. (3.134).

In the derivation of the R matrix above, we have imposed no condition on the linearly independent basis functions χ_i. We have seen that for the functional of (3.136) to be stationary, Eq. (3.140) must be satisfied. This means that the trial wave function should have the same logarithmic derivative as the exact wave function at $r = r_0$. The logarithmic derivative of the exact wave function is not known. However, we can choose a fixed approximate logarithmic derivative for all the expansion functions:

$$\kappa_t = \left[\frac{1}{\chi(r)} \frac{d\chi(r)}{dr} \right]_{r=r_0}. \tag{3.147}$$

Then the resultant R matrix of (3.146) leads to a logarithmic derivative κ of the wave function at $r = r_0$, which is different from the imposed logarithmic derivative κ_t of Eq. (3.147). This will result in a different slope of wave function than that implied by Eq. (3.147). This discontinuity of slope of the wave function at $r = r_0$ leads to a slow convergence of the R-matrix variational basis-set method. The situation can be improved by considering alternative methods [16].

3.4 HARRIS METHODS

Kohn variational methods are arbitrary in the sense that they are a few of a presumably infinite set of prescriptions leading to phase shifts whose errors are of second order. Harris and Michels suggested alternative nonvariational basis-set methods [10,11] that have some advantages over the Kohn, inverse Kohn, and Hulthén methods.

Instead of considering the trial wave functions with correct asymptotic behaviors, such as in Eq. (1.136), the positive-energy Schrödinger equation is first solved, in partial wave L with normalizable $\mathcal{L}^2$ wave function, just as in the negative-energy Rayleigh–Ritz variational method. The Schrödinger equation permits mathematically correct $\mathcal{L}^2$ solutions at positive energies. These solutions are eliminated by the physical boundary conditions. The variational function for a specific L is taken of the form (3.50)

$$v_\nu(r) = \sum_{i=1}^{n} b_i^{(\nu)} \chi_i(r), \tag{3.148}$$

where the $\mathcal{L}^2$ functions χ_i are supposed to contain an arbitrary parameter. The function (3.148) represents a mathematical solution to the problem. The Rayleigh–Ritz variational equation (3.51) is then solved at a discrete positive energy E_ν, where one is interested to calculate the scattering phase shift. First, the free parameter in the $\mathcal{L}^2$ functions is varied so that one of the energy eigenvalues of the homogeneous algebraic equation (3.51), E_ν, becomes the energy at which the phase shift is to be calculated. There are n such discrete eigenvalues E_ν. At each of these discrete energies the bound–bound matrix $\mathcal{M}_{ij}, i, j = 1, ..., n$, of Eq. (3.52) is singular. Solution of the eigenfunction-eigenvalue problem (3.51) yields the coefficients $b_i^{(\nu)}$ of Eq. (3.148).

The physical scattering wave function is then taken to be

$$\psi_t(r) = v_\nu(r) + S(r) + \lambda C(r), \tag{3.149}$$

where the $\mathcal{L}^2$ functions $v_\nu(r)$ is given by (3.148) and where the νth eigenvalue E_ν is the energy under consideration. The non-$\mathcal{L}^2$ functions $S(r)$ and $C(r)$ have the asymptotic form as in Eqs. (3.69) and (3.70), respectively. At the discrete energy E_ν, and in the space spanned by the $\mathcal{L}^2$ functions, χ_i, the Hamiltonian has two mathematically correct eigenfunctions $\psi_t(r)$ and $v_\nu(r)$. Of these, only $\psi_t(r)$ satisfies the proper boundary conditions and is physically acceptable. These two eigenfunctions should, however, be orthogonal to each other. The phase shift is calculated from the orthogonality condition

$$\langle v_\nu|(H - E_\nu)|\psi_t\rangle = 0, \tag{3.150}$$

or

$$\langle v_\nu|(H - E_\nu)|S + \lambda C\rangle = 0, \tag{3.151}$$

so that

$$\lambda_{\mathcal{H}} = -\frac{\langle v_\nu|(H - E_\nu)|S\rangle}{\langle v_\nu|(H - E_\nu)|C\rangle}. \tag{3.152}$$

Here we have used the property $\langle v_\nu|(H - E_\nu)|v_\nu\rangle = 0$. Equation (3.152) yields phase shifts of the Harris method [10].

Use of the alternative function

$$\psi_t(r) = v_\nu(r) + \lambda^{-1}S(r) + C(r), \tag{3.153}$$

in place of (3.149) also leads to the same result (3.152). In the Harris method one needs to evaluate only the bound–free integrals. The Harris method has the advantage over Kohn methods that it avoids the troublesome free–free integrals.

However, there is one trouble with the Harris method. If, by this method, one is to calculate the phase shift using a finite $\mathcal{L}^2$ basis set (3.148) at two neighboring energies, a discontinuous result could be obtained because of the discrete and discontinuous nature of the Rayleigh–Ritz eigenfunction–eigenvalue problem. This is not of great concern, as the converged phase shift becomes continuous in the exact solution limit of infinite basis functions. Also, as the number of $\mathcal{L}^2$ functions is increased, the variational parameter has to be varied to keep the test energy as one of the eigenvalues E_ν. Hence, it is not possible to use any extrapolation procedure to predict the converged result from calculations with a finite number of $\mathcal{L}^2$ functions χ_i, $i = 1, 2, ..., n$.

It is instructive to relate the Harris result (3.152) with the results of the Kohn, inverse Kohn, and Hulthén variational methods [12]. In this discussion the Kohn, inverse Kohn, and Hulthén methods apply but with a singular $\mathcal{M}_{ij}$ of Eq. (3.82). The Harris method corresponds to a singular $\mathcal{M}_{ij}$. Multiplying both sides of Eq. (3.89) by $b_i^{(\nu)}$ and summing over i and j, we get

$$\langle v_\nu|(H - E)|v^S\rangle = -\langle v_\nu|(H - E)|S\rangle, \tag{3.154}$$

where v_ν is the $\mathcal{L}^2$ function (3.148) of the Harris method and where v^S is the same as in the Kohn method in Eqs. (3.68) and (3.71). In the space spanned by the $\mathcal{L}^2$ functions χ_i, from Eq. (3.51) we find that v_ν is an eigenfunction of H with an eigenvalue E_ν. Hence,

$$\langle v_\nu|v^S\rangle = -\frac{\langle v_\nu|(H - E)|S\rangle}{(E_\nu - E)}. \tag{3.155}$$

Similarly, from Eq. (3.90), one obtains

$$\langle v_\nu|v^C\rangle = -\frac{\langle v_\nu|(H - E)|C\rangle}{(E_\nu - E)}. \tag{3.156}$$

The functions v_ν are assumed to be orthonormal, hence, M_{00}, given by Eq. (3.92), can be rewritten as

$$M_{00} = M_{SS} + \sum_\nu \langle S|(H - E)|v_\nu\rangle\langle v_\nu|v^S\rangle. \tag{3.157}$$

Using Eq. (3.155), Eq. (3.157) becomes

$$M_{00} = M_{SS} + \sum_\nu M_{S\nu}(E - E_\nu)^{-1}M_{\nu S}, \tag{3.158}$$

where $M_{SS} = \langle S|(H - E)|S\rangle$ and $M_{S\nu} = M_{\nu S} = \langle v_\nu|(H - E)|S\rangle$. Analogously, from Eqs. (3.93), (3.94), and (3.95) we get

$$M_{01} = M_{SC} + \sum_\nu M_{S\nu}(E - E_\nu)^{-1}M_{\nu C}, \tag{3.159}$$

$$M_{10} = M_{CS} + \sum_{\nu} M_{C\nu}(E - E_{\nu})^{-1} M_{\nu S}, \tag{3.160}$$

$$M_{11} = M_{CC} + \sum_{\nu} M_{C\nu}(E - E_{\nu})^{-1} M_{\nu C}, \tag{3.161}$$

where $M_{SC} = \langle S|(H - E)|C\rangle$, $M_{CS} = \langle C|(H - E)|S\rangle$, $M_{CC} = \langle C|(H - E)|C\rangle$, and $M_{C\nu} = M_{\nu C} = \langle v_{\nu}|(H - E)|C\rangle$.

The analysis presented above provides alternative calculational schemes for different Kohn algebraic variational methods. The Kohn, inverse Kohn, and Hulthén variational methods require the quantities M_{ij}, $i, j = 0, 1$. This can be done via Eqs. (3.158), (3.159), (3.160), and (3.161) provided that the functions v_{ν} are precalculated by solving an eigenfunction–eigenvalue problem. The calculational scheme of Section 3.3 requires solution of inhomogeneous bound–bound problems (3.89) and (3.90) at each energy where the phase shift is needed. The present calculational scheme, on the other hand, requires the solution of a single homogeneous bound–bound problem (3.51), which provides several energies where the scattering problem can be solved.

As E approaches an eigenvalue E_{ν} of Eq. (3.51), the elements M_{00}, M_{01}, M_{10}, and M_{11} diverge. However, one has from Eqs. (3.158), (3.159), (3.160), and (3.161),

$$M_{00} \to \frac{M_{S\nu} M_{\nu S}}{E - E_{\nu}}, \tag{3.162}$$

$$M_{01} \to \frac{M_{S\nu} M_{\nu C}}{E - E_{\nu}}, \tag{3.163}$$

$$M_{10} \to \frac{M_{C\nu} M_{\nu S}}{E - E_{\nu}}, \tag{3.164}$$

$$M_{11} \to \frac{M_{C\nu} M_{\nu C}}{E - E_{\nu}}, \tag{3.165}$$

as $E \to E_{\nu}$. In this notation the result for the Harris method of Eq. (3.152) at $E = E_{\nu}$ is given by

$$\lambda_{\mathcal{H}} = -\frac{M_{\nu S}}{M_{\nu C}}. \tag{3.166}$$

Using Eqs. (3.162) to (3.165) we find that as $E \to E_\nu$, the zero-order estimates of the Kohn and inverse-Kohn variational methods given by Eqs. (3.107) and (3.117) can be rewritten as

$$\lambda_I = \lambda_{II} = -\frac{M_{\nu S}}{M_{\nu C}}. \tag{3.167}$$

Next we relate these results with the result of the Hulthén variational phase shift given by (3.129). The determinant $\det M$ of Eq. (3.111) is quadratic in M_{ij}, and each M_{ij} has a simple pole in energy at $E = E_\nu$. However, one of these poles cancels and the determinant develops a simple pole in energy. From Eqs. (3.162) to (3.165) we find that, as $E \to E_\nu$, this determinant has the behavior

$$\det M \to \frac{D_\nu}{E - E_\nu}, \tag{3.168}$$

where D_ν is smoothly varying. Using Eqs. (3.162) to (3.165) and noting that both M_{11} and $\det M$ have simple poles in energy at $E = E_\nu$, the Hulthén variational result (3.129) at this energy becomes

$$[\lambda]_H = -\frac{M_{\nu S}}{M_{\nu C}}. \tag{3.169}$$

From Eqs. (3.166), (3.167), and (3.169) we find that the Hulthén, Harris, and zero-order estimates of the Kohn and inverse Kohn variational results are all identical. The Kohn and inverse Kohn results given by Eqs. (3.110) and (3.120) are variational improvements on this zero-order result and are different. Formally, a serious difficulty with the Harris method is that unlike in the case of Kohn and inverse Kohn methods, there is no variational principle to improve on the Harris result. There is one difficulty with the Hulthén method also. As $\det M$ passes through ∞ at $E = E_\nu$, the square root term in Eq. (3.129) changes from real to imaginary. So in the vicinity of $E = E_\nu$ the Hulthén method does not behave smoothly.

3.4.1 Minimum-Norm Kohn and Inverse Kohn Methods

Later, Harris and Michels [11] suggested another method, which has certain advantages over other commonly used approaches. For the exact wave function one has $(H - E)\psi = 0$. Harris and Michels suggested minimization of the norm of $(H - E)\psi_t$, where ψ_t is a trial wave function, a method known as the minimum-norm method. After this minimization, the result can be used in the Kohn or inverse-Kohn variational principle. This approach is also very useful for multi-channel problems.

In the single-channel version of this approach, the approximate wave function is given by Eq. (3.125). In the present approach, as in the Kohn and inverse Kohn variational methods, the bound–bound problem is first solved via Eqs. (3.89) and (3.90). Once this is done, the functions v^S and v^C of Eq. (3.125) are known via Eqs. (3.71) and (3.72). Instead of determining the phase shift variationally as in the Kohn methods, it is obtained by minimizing certain norms.

In the space of oscillating functions S and C, the projection of $(H - E)\psi_t \neq 0$. With ψ_t given by Eq. (3.125), this projection is given by

$$\sum_{j=0,1} M_{ij}\kappa_j \neq 0, \tag{3.170}$$

$i = 0, 1$, where M_{ij} are defined by Eqs. (3.92) to (3.95). For the exact solution, the right-hand side of Eq. (3.170) is zero. The best set of solutions, given the trial function (3.125), can be obtained by minimizing the norm

$$\left(\sum_{l=0,1} M_{il}\kappa_l\right)^T \left(\sum_{j=0,1} M_{ij}\kappa_j\right) \tag{3.171}$$

of the left-hand side of Eq. (3.170), where ^{T} denotes the transpose. This procedure determines κ_0 and κ_1. The quantity $\lambda_I = \kappa_1/\kappa_0$ provides an estimate of $\tan\delta$ in the minimum-norm method. This estimate can be used in the Kohn variational result (3.108) for further ad hoc improvement. This procedure is the minimum-norm Kohn method. Similarly, $\lambda_{II}^{-1} = \kappa_0/\kappa_1$ provides an estimate for $\cot\delta$ in the minimum-norm method which can be used in the inverse Kohn variational result (3.118) for further ad hoc improvement. This procedure is known as the minimum-norm inverse Kohn method.

The minimum-norm method does not provide a unique prescription for defining the quantities κ_0 and κ_1. Also, use of the results of the minimum-norm method in the Kohn and inverse Kohn variational principles are quite ad hoc, as determination of λ in the minimum-norm method is not performed from the stationary property of a functional. Nevertheless, the present determination of λ is intuitively reasonable. In the multichannel case, we consider several other approaches based on the present idea of minimizing the norm.

3.5 ANOMALOUS BEHAVIOR

Schwartz [9] made a careful study of the physically interesting single-channel electron-hydrogen scattering using the Kohn variational method. His calculation for the S wave phase shift not only provided numerical solution of an important physical problem by the Kohn method, but presented some numerical problems with this approach. In the numerical calculation, several sets of $\mathcal{L}^2$ functions χ_i were employed. These functions had a free variational parameter that was supposed to be varied to obtain ideal convergence with an increase in the number of functions. It was found in both spin singlet and triplet states that results for the S-wave phase shift for fixed energy and a fixed number of $\mathcal{L}^2$ functions could vary rapidly with a variation in the free parameter. This phase shift could also vary rapidly for a fixed value of the parameter with a variation in energy. In fact, the calculated tangent of the phase shift $\lambda = \tan \delta$ could be singular at any energy for some value of the parameter. When this happens the phase shift increases (resonance) or decreases (antiresonance) by π as the energy is varied, and the cross section derived from this phase shift exhibits a resonancelike behavior. Such behavior could occur for both the zero-order phase shift (3.107) end the variational result (3.108). Exact solution of the Schrödinger or Lippmann–Schwinger equation with the same potential does not exhibit these resonances or anti-resonances. Hence, these resonances are spurious and with an increase of the number of $\mathcal{L}^2$ functions or with a change in the variational parameter, such resonances disappear from this specific energy but may reappear at some other energy. A similar resonance can also appear in the inverse Kohn variational method. These behaviors are referred to as anomalous behaviors or anomalies.

If one makes a study of the function λ versus the free variational parameter, it is possible to identify the range of numerical values of the parameter for which λ remains reasonably stable at a fixed energy for a fixed number of basis functions. Then the calculation should be performed with a parameter in this range. For calculations performed at another energy or with a different number of basis functions, the same procedure has to be repeated.

In addition to the above-mentioned anomalies, there could also be true resonance in a scattering problem. They are common in S-wave positron–atom and electron–atom scattering [18]. The spurious resonances encountered above are artifacts of the variational methods. True resonance will tend to remain at the same energy as the number of $\mathcal{L}^2$ functions is increased or the variational parameter varied. Hence, the spurious resonances could be identified and avoided by varying the variational parameter or by increasing the number of functions in a presumably converged calculation. However, this makes the calculational scheme unnecessarily tedious and lengthy, as each such resonance has to be tested to be false or true by changing the number of $\mathcal{L}^2$ functions or the free parameter.

Since Schwartz identified these anomalies, their origin has been a subject of investigation. A detailed account of these investigations has been given by Truhlar et al. [19]. In the beginning it was thought that the anomaly at a specific energy appears when the bound–bound matrix $\mathcal{M}$ of Eqs. (3.89) or (3.90) accidentally becomes singular. However, we have seen in the discussion related to the Harris method that at the accidental singularity of $\mathcal{M}$, both the Kohn and inverse–Kohn variational methods yield well-defined results with smooth behavior. Later, Nesbet [12] pointed out that the anomaly of the Kohn method is related to the appearance of an accidental zero of the matrix element M_{11} of Eq. (3.95). This can be realized from the expression (3.110) for the Kohn variational result. From Eq. (3.210) it follows that similar anomalies will also appear in the inverse–Kohn method as the denominator M_{00} of Eq. (3.92) vanishes accidentally. Also, it is realized that the particular electron-scattering problem studied by Schwartz has nothing to do with the appearance of the anomalies; similar anomalies could appear in the numerical solution of any other scattering problem using the Kohn variational method.

Such anomalies may appear as the results of the Kohn or inverse Kohn method with finite basis sets are not bounds on the phase shifts at arbitrary energies. In general, λ may vary between $+\infty$ and $-\infty$. These anomalies not only appear in the unconverged approximate solution but may also appear in the limit when these variational principles yield the almost converged solution. In the exact solution limit, $\det M = 0$ must be satisfied. This means that in application of the Kohn variational method, if $M_{11} = 0$, at a specific energy, from Eqs. (3.107), (3.110), and (3.111) we find that $M_{10} = 0$, too. Hence, $\det M = 0$ and the Kohn variational $[\lambda]_K$ is finite. Similarly, if $M_{00} = 0$ at some energy, $M_{01} = 0$ also. For an almost converged solution,

$\det M \simeq 0$ and M_{11} and M_{10} will not vanish at a specific energy but at nearby energies. Hence, $[\lambda]_K$ of Eq. (3.110) will be infinity when $M_{11} = 0$ and will be almost zero when $M_{10} = 0$. This will lead to a resonancelike behavior of the Kohn variational phase shift with a rapid variation of $[\lambda]_K$ with energy. Similarly, anomalies may appear in the almost converged phase shift of the inverse Kohn variational method. However, Nuttall [20] showed that as the number of $\mathcal{L}^2$ functions are increased, the anomalies have such a small width that they are unobservable in practice and the phase shifts actually converge to the correct value.

3.5.1 Anomaly-Free Methods

It is quite plausible that the element M_{00} or M_{11} of Eqs. (3.92) and (3.95), respectively, could vanish accidentally, so that both Kohn and inverse Kohn methods may exhibit these anomalies. However, the elements M_{00} and M_{11} cannot vanish simultaneously at the same energy in a variational calculation with identical trial functions. This means that Kohn and inverse Kohn variational methods do not exhibit anomalies under identical conditions. Using this point, Nesbet [12] suggested the anomaly-free Kohn and inverse Kohn methods. He advocated using the Kohn method when $|M_{11}|/|M_{00}| > 1$ and the inverse Kohn method when $|M_{11}|/|M_{00}| < 1$. This simple prescription avoids all difficulties with anomalies. However, there are a few cautionary remarks. First, as the Kohn and inverse Kohn variational methods lead to different results for a model problem, the Nesbet [12] prescription to avoid anomaly will lead to discontinuous results for phase shifts as energy (or the variational parameter) is varied. Second, a physical resonance may appear in a Kohn calculation in the same fashion as an anomaly, $M_{11} = 0, M_{00} \neq 0$. In this case the use of Nesbet criteria will erroneously eliminate the physical resonance and hence lead to an inferior result. A similar problem appears in the inverse Kohn variational method when $M_{00} = 0$ yields the physically correct result. The use of Nesbet criteria will again lead to an inferior result.

Rudge [21] suggested a procedure for tackling the anomalies. By varying the arbitrary parameter in the $\mathcal{L}^2$ functions, he always imposed the condition $\det M = 0$. Then, by Eqs. (3.123) and (3.132), the phase shifts calculated by the Kohn, inverse Kohn, and Hulthén methods are identical to each other. The anomalies of the Kohn and inverse Kohn methods correspond to zeros of M_{11} and M_{00}, respectively. Then if $M_{00} = 0$, one must have $M_{01} = 0$ also, so that the Kohn and inverse Kohn phase shifts calculated from (3.110) and (3.120), respectively, could be identical. Similarly, if $M_{11} = 0$, then $M_{10} = 0$, too, as the phase shifts obtained by the Kohn and inverse Kohn methods should agree with each other. However, one could have $M_{11} = M_{01} = 0$ also, which by Eqs. (3.110) and (3.120) should denote resonance in both the Kohn and inverse Kohn methods and hence should be a true resonance. The last nontrivial possibility, $M_{00} = M_{10} = 0$, should imply, by Eqs. (3.107) and (3.117), that the phase shift is zero. Thus the Rudge method not only avoids a spurious resonance but can also identify a true resonance.

3.6 MULTICHANNEL SCATTERING

So far we have discussed applications of different on-shell variational principles to the study of potential scattering. In most physical scattering processes, many scattering channels are open. Such a realistic scattering process can be described by multichannel scattering equations. In this section we generalize the variational principles of potential scattering to the case of two-cluster multichannel scattering [12]. Multicluster channels are mathematically difficult to treat and are not considered here.

There are different ways of formulating two-cluster multichannel scattering. We consider the two-cluster multichannel close-coupling (CC) equations developed in Section 1.6.2 [12]. These two-cluster multichannel Schrödinger equations can be written as

$$(H - E)|\psi\rangle \equiv \sum_{\beta=1}^{N}(H_{\alpha\beta} - E\delta_{\alpha\beta})|\psi_\beta\rangle = 0, \qquad (3.172)$$

where the indices α, β, τ, ρ, and so on, denote the different channels that run from 1 to N. It is assumed that a partial-wave projection has already been carried out and an on-shell wave number k_α is associated with a channel, α. The idea of channels can be extended to the case where both the target and projectile are composite, and where rearrangement can occur.

3.6.1 Kohn Methods

The wave function $|\psi_\alpha\rangle$ for scattering initiated in channel α has the asymptotic form

$$\lim_{r_\alpha \to \infty} \psi_\alpha(r_\alpha) \sim \kappa_{0\alpha} S_\alpha(r_\alpha) + \kappa_{1\alpha} C_\alpha(r_\alpha). \tag{3.173}$$

Here $S_\alpha(r_\alpha)$ and $C_\alpha(r_\alpha)$ are appropriate sine and cosine functions in the asymptotic region

$$\lim_{r_\alpha \to \infty} S_\alpha(r_\alpha) \sim \sin\left(k_\alpha r_\alpha - \frac{L_\alpha \pi}{2}\right), \tag{3.174}$$

$$\lim_{r_\alpha \to \infty} C_\alpha(r_\alpha) \sim \cos\left(k_\alpha r_\alpha - \frac{L_\alpha \pi}{2}\right), \tag{3.175}$$

where r_α is the relative separation between clusters in channel α, k_α is the on-shell wave number, and L_α is the angular momentum variable. For small r_α, the wave function should vanish as $r_\alpha^{L_\alpha+1}$. These conditions should be compared with Eqs. (1.35) and (1.136) of potential scattering. The coefficients κ determine the probability of transition from one channel to another and are in general complex in the presence of other open channels. In potential scattering they are related to phase shifts.

A numerical attempt to solve Eq. (3.172) with a K-matrix boundary condition will lead to N degenerate solutions, $|\psi^\tau\rangle$, at a particular energy, labeled by the index τ, which runs from 1 to N. If the coefficients of one of the solutions $|\psi^\tau\rangle$ of Eqs. (3.172) satisfy

$$\kappa_{0\alpha}^\tau = \delta_{\alpha\tau}, \qquad \alpha = 1, ..., N, \tag{3.176}$$

for each $\tau = 1, ..., N$, the computed values of $\kappa_{1\alpha}^\tau \equiv \lambda_{\tau\alpha}$ are related to the dimensionless K-matrix elements of Eq. (1.206) by

$$K_{\tau\alpha} = \lambda_{\tau\alpha}. \tag{3.177}$$

This definition of λ is useful for writing the multichannel Kohn variational principle.

An alternative set of definitions leads directly to the matrix K^{-1} suitable for writing the multichannel inverse Kohn variational principle. If the coefficients of one of the solutions $|\psi^\tau\rangle$ of Eqs. (3.172) satisfy

$$\kappa_{1\alpha}^\tau = \delta_{\alpha\tau}, \qquad \alpha = 1, ..., N \tag{3.178}$$

for each τ, the elements of the dimensionless K^{-1} matrix are given in terms of the computed values of $\kappa_{0\alpha}^{\tau} \equiv \bar{\lambda}_{\tau\alpha}$ by

$$(K^{-1})_{\tau\alpha} = \bar{\lambda}_{\tau\alpha}. \tag{3.179}$$

The quantity $\bar{\lambda}$ will be used in the formulation of the multichannel inverse Kohn variational principle.

The algebraic formulation and numerical analysis proceeds in a way analogous to that in potential scattering. We consider the trial function

$$\begin{aligned}\psi_{\alpha}^{\tau}(r_{\alpha}) &= \sum_{\gamma=1}^{N}[\kappa_{0\gamma}^{\tau}\psi_{0\alpha}^{\gamma}(r_{\alpha}) + \kappa_{1\gamma}^{\tau}\psi_{1\alpha}^{\gamma}(r_{\alpha})]\\ &\equiv \sum_{\gamma=1}^{N}\{\kappa_{0\gamma}^{\tau}[v_{S\alpha}^{\gamma}(r_{\alpha}) + \delta_{\gamma\alpha}S_{\alpha}(r_{\alpha})] + \kappa_{1\gamma}^{\tau}[v_{C\alpha}^{\gamma}(r_{\alpha}) + \delta_{\gamma\alpha}C_{\alpha}(r_{\alpha})]\},\end{aligned} \tag{3.180}$$

where $\tau = 1, ..., N$ labels the N solutions of the coupled set of equations (3.172) and α denotes the initial physical channel. In Eq. (3.180), $v_{S\alpha}^{\gamma}$ and $v_{C\alpha}^{\gamma}$ are $\mathcal{L}^2$ functions and the oscillating functions S_{α} and C_{α} are included in appropriate physical channels. Equation (3.180) is the appropriate multichannel generalization of potential scattering function (3.125) [12].

The functions $v_{S\alpha}^{\gamma}$ and $v_{C\alpha}^{\gamma}$ are next expanded in a set of $\mathcal{L}^2$ basis functions as in single-channel expansions (3.71) and (3.72):

$$v_{S\alpha}^{\gamma}(r_{\alpha}) = \sum_{i=1}^{n} b_{iS}^{\gamma\alpha}\chi_{i}^{\alpha}(r_{\alpha}), \tag{3.181}$$

$$v_{C\alpha}^{\gamma}(r_{\alpha}) = \sum_{i=1}^{n} b_{iC}^{\gamma\alpha}\chi_{i}^{\alpha}(r_{\alpha}). \tag{3.182}$$

The additional explicit index α on the functions allows us to employ different basis functions for different channels and also permits us to develop a consistent matrix notation in channel and function space.

In the multichannel case, the generalization of the variational functional I of Eq. (3.44) is given by

$$I_{\tau\rho} = \sum_{\alpha,\beta=1}^{N} \langle \psi_\alpha^\tau | (H_{\alpha\beta} - E\delta_{\alpha\beta}) | \psi_\beta^\rho \rangle \tag{3.183}$$

$$= \sum_{\alpha,\beta,\gamma,\sigma=1}^{N} [\kappa_{0\gamma}^\tau \kappa_{0\sigma}^\rho (\delta_{\gamma\alpha} B_0^{\alpha\beta} \delta_{\sigma\beta} + W_{0\gamma\sigma}^{\alpha\beta})$$

$$+ \kappa_{0\gamma}^\tau \kappa_{1\sigma}^\rho (\delta_{\gamma\alpha} B_1^{\alpha\beta} \delta_{\sigma\beta} + W_{1\gamma\sigma}^{\alpha\beta}) \tag{3.184}$$

$$+ \kappa_{1\gamma}^\tau \kappa_{0\sigma}^\rho (\delta_{\gamma\alpha} B_3^{\alpha\beta} \delta_{\sigma\beta} + W_{3\gamma\sigma}^{\alpha\beta}) + \kappa_{1\gamma}^\tau \kappa_{1\sigma}^\rho (\delta_{\gamma\alpha} B_2^{\alpha\beta} \delta_{\sigma\beta} + W_{2\gamma\sigma}^{\alpha\beta})],$$

where

$$W_{0\gamma\sigma}^{\alpha\beta} = \sum_{i,j=1}^{n} b_{iS}^{\gamma\alpha} \mathcal{M}_{ij}^{\alpha\beta} b_{jS}^{\sigma\beta} + \sum_{i=1}^{n} b_{iS}^{\gamma\alpha} R_{iS}^{\alpha\beta} \delta_{\sigma\beta} + \sum_{i=1}^{n} b_{iS}^{\sigma\beta} R_{iS}^{\beta\alpha} \delta_{\alpha\gamma}, \tag{3.185}$$

$$W_{1\gamma\sigma}^{\alpha\beta} = \sum_{i,j=1}^{n} b_{iS}^{\gamma\alpha} \mathcal{M}_{ij}^{\alpha\beta} b_{jC}^{\sigma\beta} + \sum_{i=1}^{n} b_{iS}^{\gamma\alpha} R_{iC}^{\alpha\beta} \delta_{\sigma\beta} + \sum_{i=1}^{n} b_{iC}^{\sigma\beta} R_{iS}^{\beta\alpha} \delta_{\alpha\gamma}, \tag{3.186}$$

$$W_{3\gamma\sigma}^{\alpha\beta} = \sum_{i,j=1}^{n} b_{iC}^{\gamma\alpha} \mathcal{M}_{ij}^{\alpha\beta} b_{jS}^{\sigma\beta} + \sum_{i=1}^{n} b_{iS}^{\sigma\beta} R_{iC}^{\beta\alpha} \delta_{\alpha\gamma} + \sum_{i=1}^{n} b_{iC}^{\gamma\alpha} R_{iS}^{\alpha\beta} \delta_{\beta\sigma}, \tag{3.187}$$

$$W_{2\gamma\sigma}^{\alpha\beta} = \sum_{i,j=1}^{n} b_{iC}^{\gamma\alpha} \mathcal{M}_{ij}^{\alpha\beta} b_{jC}^{\sigma\beta} + \sum_{i=1}^{n} b_{iC}^{\gamma\alpha} R_{iC}^{\alpha\beta} \delta_{\beta\sigma} + \sum_{i=1}^{n} b_{iC}^{\sigma\beta} R_{iC}^{\beta\alpha} \delta_{\alpha\gamma}, \tag{3.188}$$

$$\mathcal{M}_{ij}^{\alpha\beta} = \mathcal{M}_{ji}^{\beta\alpha} = \langle \chi_i^\alpha | (H_{\alpha\beta} - E\delta_{\alpha\beta}) | \chi_j^\beta \rangle, \tag{3.189}$$

$$R_{iS}^{\tau\beta} = \langle \chi_i^\tau | (H_{\tau\beta} - E\delta_{\tau\beta}) | S_\beta \rangle, \tag{3.190}$$

$$R_{iC}^{\tau\beta} = \langle \chi_i^\tau | (H_{\tau\beta} - E\delta_{\tau\beta}) | C_\beta \rangle, \tag{3.191}$$

$$B_0^{\alpha\beta} = \langle S_\alpha | (H_{\alpha\beta} - E\delta_{\alpha\beta}) | S_\beta \rangle, \tag{3.192}$$

$$B_1^{\alpha\beta} = \langle S_\alpha | (H_{\alpha\beta} - E\delta_{\alpha\beta}) | C_\beta \rangle, \tag{3.193}$$

$$B_2^{\alpha\beta} = \langle C_\alpha | (H_{\alpha\beta} - E\delta_{\alpha\beta}) | C_\beta \rangle, \tag{3.194}$$

$$B_3^{\alpha\beta} = \langle C_\alpha | (H_{\alpha\beta} - E\delta_{\alpha\beta}) | S_\beta \rangle. \tag{3.195}$$

The multichannel variational principles are written in terms of the functional (3.184). The multichannel generalization of the Kohn variational principle is given by

$$[\lambda_{\rho\tau}]_K = \lambda_{\rho\tau} - \frac{I_{\tau\rho}(\lambda_{\rho\tau})}{k_\tau}, \tag{3.196}$$

with $I_{\tau\rho}$ given by Eq. (3.184). As in the single-channel case, this functional should be stationary with respect to variations of expansion coefficients $b_{iS}^{\gamma\alpha}$ and $b_{iC}^{\gamma\alpha}$ as well as quantities $\lambda_{\rho\tau}$. The stationary property with respect to variations of coefficients $b_{iS}^{\gamma\alpha}$ and $b_{iC}^{\gamma\alpha}$ for any $\lambda_{\rho\tau}$ leads to the following equations satisfied by the coefficients of the $\mathcal{L}^2$ functions of Eqs. (3.181) and (3.182):

$$\sum_{\beta=1}^{N}\sum_{j=1}^{n} \mathcal{M}_{ij}^{\alpha\beta} b_{jS}^{\rho\beta} = -R_{iS}^{\alpha\rho}, \tag{3.197}$$

$$\sum_{\beta=1}^{N}\sum_{j=1}^{n} \mathcal{M}_{ij}^{\alpha\beta} b_{jC}^{\rho\beta} = -R_{iC}^{\alpha\rho}. \tag{3.198}$$

Equations (3.197) and (3.198) should be compared with single-channel equations (3.89) and (3.90), respectively. The matrix $\mathcal{M}$ is of dimension $2N + \mathcal{N}$; $2N$ is the total number of oscillating functions used in N channels, and $\mathcal{N}$ is the total number of $\mathcal{L}^2$ functions used in N channels. If an equal number n of $\mathcal{L}^2$ functions are used in each of the N channels, then $\mathcal{N} = nN$. Unless the matrix $\mathcal{M}$ is accidentally singular, the coefficients $b_{jS}^{\rho\beta}$ and $b_{jC}^{\rho\beta}$ can be determined, from Eqs. (3.197) and (3.198) so that the $\mathcal{L}^2$ functions $v_{S\alpha}^{\gamma}$ and $v_{C\alpha}^{\gamma}$ of Eqs. (3.180), (3.181), and (3.182) are known.

If we use Eqs. (3.197) and (3.198), the variational functional $I_{\tau\rho}$ of Eq. (3.184) can be simplified and rewritten as

$$I_{\tau\rho} = \sum_{i,j=0}^{1}\sum_{\gamma,\sigma=1}^{N} \kappa_{i\gamma}^{\tau} M_{ij}^{\gamma\sigma} \kappa_{j\sigma}^{\rho}, \tag{3.199}$$

where

$$M_{00}^{\gamma\sigma} = \sum_{\beta=1}^{N} \langle S_\gamma | (H_{\gamma\beta} - E\delta_{\gamma\beta}) | \psi_{0\beta}^{\sigma} \rangle, \tag{3.200}$$

$$M_{01}^{\gamma\sigma} = \sum_{\beta=1}^{N} \langle S_\gamma | (H_{\gamma\beta} - E\delta_{\gamma\beta}) | \psi_{1\beta}^{\sigma} \rangle, \tag{3.201}$$

$$M_{10}^{\gamma\sigma} = \sum_{\beta=1}^{N} \langle C_\gamma | (H_{\gamma\beta} - E\delta_{\gamma\beta}) | \psi_{0\beta}^{\sigma} \rangle, \tag{3.202}$$

$$M_{11}^{\gamma\sigma} = \sum_{\beta=1}^{N} \langle C_\gamma | (H_{\gamma\beta} - E\delta_{\gamma\beta}) | \psi_{1\beta}^{\sigma} \rangle. \tag{3.203}$$

As the functions $\psi_{i\beta}^{\sigma}$, $i = 0, 1$, are known, the matrix elements $M_{ij}^{\gamma\sigma}$, $i, j = 0, 1$, can be calculated. In the matrix elements $M_{ij}^{\gamma\sigma}$ there are free–free matrix elements involving the kinetic energy operator. In these elements, if the direction of the derivative in the kinetic energy operator is changed, there could be an extra contribution, as in Eq. (3.104). Consequently, these matrix elements satisfy

$$M_{01}^{\alpha\alpha} = M_{10}^{\alpha\alpha} + k_\alpha, \tag{3.204}$$

$$M_{ij}^{\alpha\beta} = M_{ji}^{\beta\alpha}, \qquad \alpha \neq \beta \tag{3.205}$$

for $\alpha, \beta = 1, \ldots, N$, $i, j = 0, 1$.

In the case of the Kohn variational principle (3.196), the coefficients κ of Eq. (3.180) are taken to be $\kappa_{1\gamma}^{\tau} = \lambda_{\tau\gamma}$, $\kappa_{0\gamma}^{\tau} = \delta_{\tau\gamma}$, so that the multichannel functional I is given by

$$I_{\tau\rho} = M_{00}^{\tau\rho} + \sum_{\sigma=1}^{N} M_{01}^{\tau\sigma} \lambda_{\rho\sigma} + \sum_{\gamma=1}^{N} \lambda_{\tau\gamma} M_{10}^{\gamma\rho} + \sum_{\gamma,\sigma=1}^{N} \lambda_{\tau\gamma} M_{11}^{\gamma\sigma} \lambda_{\rho\sigma}. \tag{3.206}$$

With this choice of the coefficients κ, the K matrix is defined by Eq. (3.177). The stationary property of Eqs. (3.196) and (3.206), with respect to variations of $\lambda_{\rho\tau}$, yields the following zeroth-order $\lambda_{\rho\tau}$ in the multichannel Kohn method:

$$\sum_{\sigma=1}^{N} M_{11}^{\gamma\sigma} \lambda_{\rho\sigma}^{I} = -M_{10}^{\gamma\rho}. \tag{3.207}$$

For each channel index ρ, this is a set of N inhomogeneous equations for N unknowns $\lambda_{\rho\sigma}^{I}$. This system of equations has the well-defined solution

$$\lambda^I_{\rho\sigma} = -\sum_{\gamma=1}^{N}(M_{11}^{-1})^{\sigma\gamma}M_{10}^{\gamma\rho}, \tag{3.208}$$

unless, accidentally, det $M_{11} = 0$. The use of Eq. (3.208) in Eq. (3.196) leads to the following improved Kohn formula for the multichannel problem:

$$[\lambda_{\rho\tau}]_K = \lambda^I_{\rho\tau} - \frac{1}{k_\tau}\left(M_{00}^{\tau\rho} + \sum_{\beta=1}^{N} M_{01}^{\tau\beta}\lambda^I_{\rho\beta}\right) \tag{3.209}$$

$$= -\sum_{\alpha=1}^{N}(M_{11}^{-1})^{\tau\alpha}M_{10}^{\alpha\rho}$$

$$-\frac{1}{k_\tau}\left(M_{00}^{\tau\rho} - \sum_{\beta=1}^{N} M_{01}^{\tau\beta}\sum_{\alpha=1}^{N}(M_{11}^{-1})^{\beta\alpha}M_{10}^{\alpha\rho}\right) \tag{3.210}$$

$$= -\frac{1}{k_\tau}\left(M_{00}^{\tau\rho} - \sum_{\beta,\alpha=1}^{N} M_{10}^{\beta\tau}(M_{11}^{-1})^{\beta\alpha}M_{10}^{\alpha\rho}\right) \tag{3.211}$$

Equations (3.204) and (3.205) has been used in deriving Eq. (3.211).

The multichannel generalization of the inverse Kohn variational principle is given by

$$[\bar{\lambda}_{\rho\tau}]_{IK} = \bar{\lambda}_{\rho\tau} + \frac{I_{\tau\rho}(\bar{\lambda}_{\rho\tau})}{k_\tau} \tag{3.212}$$

Expression (3.212) is stationary with respect to variations of the expansion coefficients $b_{iS}^{\gamma\alpha}$ and $b_{iC}^{\gamma\alpha}$ as well as quantities $\bar{\lambda}_{\rho\tau}$. The stationary property with respect to variations of $b_{iS}^{\gamma\alpha}$ and $b_{iC}^{\gamma\alpha}$ for any $\bar{\lambda}_{\rho\tau}$ leads to Eqs. (3.197) and (3.198). In this case, the coefficients κ of Eq. (3.180) are taken to be $\kappa_{0\alpha}^{\tau} = \bar{\lambda}_{\tau\alpha}$, $\kappa_{1\alpha}^{\tau} = \delta_{\tau\alpha}$, so that the multichannel functional I is given by

$$I_{\tau\rho} = M_{11}^{\tau\rho} + \sum_{\beta=1}^{N} M_{10}^{\tau\beta}\bar{\lambda}_{\rho\beta} + \sum_{\alpha=1}^{N}\bar{\lambda}_{\tau\alpha}M_{01}^{\alpha\rho} + \sum_{\alpha,\beta=1}^{N}\bar{\lambda}_{\tau\alpha}M_{00}^{\alpha\beta}\bar{\lambda}_{\rho\beta}. \tag{3.213}$$

With this choice of the coefficients κ the K matrix is defined by Eq. (3.179). The stationary property of Eqs. (3.212) and (3.213), with respect to variations of $\bar{\lambda}$, leads to the following zeroth-order $\bar{\lambda}_{\rho\alpha}$ in the multichannel inverse Kohn method:

$$\sum_{\alpha=1}^{N} M_{00}^{\tau\alpha} \bar{\lambda}_{\rho\alpha}^{II} = -M_{01}^{\tau\rho}. \tag{3.214}$$

This system of equations, for N unknowns $\bar{\lambda}_{\rho\alpha}^{II}$, has the zeroth-order multichannel inverse Kohn result

$$\bar{\lambda}_{\rho\alpha}^{II} = -\sum_{\gamma=1}^{N} (M_{00}^{-1})^{\alpha\gamma} M_{01}^{\gamma\rho}, \tag{3.215}$$

unless accidentally, $\det M_{00} = 0$. When substituted into Eqs. (3.212) and (3.213), the zeroth-order result (3.215) leads to the following improved inverse Kohn formula for the multichannel problem:

$$[\bar{\lambda}_{\rho\alpha}]_{IK} = \bar{\lambda}_{\rho\alpha}^{II} + \frac{1}{k_\alpha}\left(M_{11}^{\alpha\rho} + \sum_{\beta=1}^{N} M_{10}^{\alpha\beta} \bar{\lambda}_{\rho\beta}^{II} \right), \tag{3.216}$$

$$= -\sum_{\tau=1}^{N} (M_{00}^{-1})^{\alpha\tau} M_{01}^{\tau\rho} + \frac{1}{k_\alpha}\left(M_{11}^{\alpha\rho} - \sum_{\beta=1}^{N} M_{10}^{\alpha\beta} \sum_{\tau=1}^{N} (M_{00}^{-1})^{\beta\tau} M_{01}^{\tau\rho} \right) \tag{3.217}$$

$$= \frac{1}{k_\alpha}\left(M_{11}^{\alpha\rho} - \sum_{\beta,\tau=1}^{N} M_{01}^{\beta\alpha} (M_{00}^{-1})^{\beta\tau} M_{01}^{\tau\rho} \right). \tag{3.218}$$

In deriving Eq. (3.218), Eqs. (3.204) and (3.205) have been used. It is realized that for the converged solutions of Eqs. (3.211) and (3.218), the matrices λ and $\bar{\lambda}$ are related by $\bar{\lambda} = \lambda^{-1}$.

In the present generalization, most of the functions of the single-channel formulation have been replaced by matrices in the channel space. In the single-channel case, anomalies appear in the Kohn and inverse Kohn variational methods due to the vanishing of M_{11} and M_{00}, respectively. In the present multichannel formulation, the functions M_{11} and M_{00} are matrices and the anomalies correspond to $\det M_{11} = 0$ and $\det M_{00} = 0$, respectively. One can use Nesbet's anomaly-free approach as in the single-channel case. He suggested using the Kohn method when $|\det M_{11}|/|\det M_{00}| > 1$ and the inverse

Kohn method when $|\det M_{11}|/|\det M_{00}| < 1$. As in the single-channel case, this procedure may lead to a discontinuous result as energy or the variational parameter of the basis set is changed. The Hulthén variational principle has not been generalized to the multichannel case because of the difficulty in finding the solution of $I_{\rho\tau} = 0$ in this case.

3.6.2 Variational *R*-Matrix Method

The variational R-matrix method of Section 3.3.4 can also be generalized to the multichannel case. In this case there are different channels, and in each two-cluster channel α, there is a variable r_α denoting separation between clusters. The multichannel R matrix is defined by

$$\psi_\alpha^\rho(r_0) = r_0 \sum_{\tau=1}^{N} R_{\rho\tau} \left[\frac{d\psi_\tau^\alpha(r)}{dr}\right]_{r=r_0}, \tag{3.219}$$

where $r = r_0$ is outside the range of interaction. In definition (3.219) we have used the same r_0 for all channels. This condition simplifies the algebra and can in principle be removed. The method is based on the stationary property of the functional (3.183), which is a multichannel generalization of the functional of Eq. (3.44). In Eq. (3.183) we introduce a trial wave function that is assumed to be exact for $r > r_0$. With this trial function, the integral in Eq. (3.183) is from 0 to r_0, and consequently, troublesome infinite integrals over the oscillating functions as encountered in the Kohn method can be avoided. The functional (3.183) is written as

$$I_{\tau\rho} = \sum_{\alpha,\beta=1}^{N} \int_0^{r_0} \psi_\alpha^\tau(r)(H_{\alpha\beta} - E\delta_{\alpha\beta})\psi_\beta^\rho(r)\, dr, \tag{3.220}$$

where the operator $H_{\alpha\beta} = (H_0)_{\alpha\alpha}\delta_{\alpha\beta} + V_{\alpha\beta}$, and H_0 is the kinetic energy. Then by an integration by parts, as in the single-channel case, this functional can be written as

$$I_{\tau\rho} = \sum_{\alpha,\beta=1}^{N} \int_0^{r_0} \left[\left(\frac{d\psi_\alpha^\tau(r)}{dr}\delta_{\alpha\beta}\frac{d\psi_\beta^\rho(r)}{dr}\right) + \psi_\alpha^\tau(r)\left[V_{\alpha\beta}(r) - E\delta_{\alpha\beta}\right]\psi_\beta^\rho(r)\right] dr$$

$$- \sum_{\alpha,\beta=1}^{N} \psi_\alpha^\tau(r_0)\delta_{\alpha\beta}\left[\frac{d\psi_\beta^\rho(r)}{dr}\right]_{r=r_0}. \tag{3.221}$$

If we use definition (3.219) for the multichannel R matrix, Eq. (3.221) can be rewritten as

$$I_{\tau\rho} = \sum_{\alpha,\beta=1}^{N} \int_0^{r_0} \left[\left(\frac{d\psi_\alpha^\tau(r)}{dr} \delta_{\alpha\beta} \frac{d\psi_\beta^\rho(r)}{dr} \right) + \psi_\alpha^\tau(r) \big(V_{\alpha\beta}(r) - E\delta_{\alpha\beta} \big) \psi_\beta^\rho(r) \right] dr$$

$$- \sum_{\alpha,\beta=1}^{N} \frac{1}{r_0} \psi_\alpha^\tau(r_0)(R^{-1})_{\alpha\beta} \psi_\beta^\rho(r_0). \tag{3.222}$$

In operator form Eq. (3.222) becomes

$$I = \psi^T A \psi - \frac{1}{r_0} \psi^T(r_0) R^{-1} \psi(r_0), \tag{3.223}$$

where the explicit matrix representation of the operator A appears in the first term on the right-hand side of Eq. (3.222).

As in the single-channel case, we make an expansion of the trial function in a linearly independent set of basis functions χ_i,

$$\psi_\alpha^\tau(r) = \sum_{i=1}^{n} b_i^{\tau\alpha} \chi_i(r). \tag{3.224}$$

In this expansion the same basis functions have been used in all channels. However, as in the case of Kohn methods discussed above, one can also use a different set of functions in different channels. If we substitute Eq. (3.224) in Eq. (3.223), the functional $I_{\tau\rho}$ becomes

$$I_{\tau\rho} = \sum_{\alpha,\beta=1}^{N} \sum_{i,j=1}^{n} b_i^{\tau\alpha} \left[A_{ij}^{\alpha\beta} - \frac{1}{r_0} \chi_i(r_0)(R^{-1})_{\alpha\beta} \chi_j(r_0) \right] b_j^{\rho\beta}, \tag{3.225}$$

where $A_{ij}^{\alpha\beta}$ is the matrix element of the operator A.

As in the single-channel case, the functional $I_{\tau\rho}$ is supposed to be stationary with respect to variations of the trial functions or with respect to the coefficients $b_j^{\rho\beta}$. This condition leads to the equation

$$\sum_{\beta=1}^{N} \sum_{j=1}^{n} \left[A_{ij}^{\alpha\beta} - \frac{1}{r_0} \chi_i(r_0)(R^{-1})_{\alpha\beta} \chi_j(r_0) \right] b_j^{\rho\beta} = 0. \tag{3.226}$$

If we use Eq. (3.224), the solution of Eq. (3.226) can be written as

$$b_i^{\tau\alpha} = \sum_{\beta=1}^{N}\sum_{j=1}^{n}\frac{1}{r_0}(A^{-1})_{ij}^{\tau\beta}\chi_j(r_0)\sum_{\gamma=1}^{\infty}(R^{-1})_{\beta\gamma}\psi_\gamma^\alpha(r_0). \qquad (3.227)$$

Equations (3.224) and (3.227) lead to

$$\psi_\alpha^\tau(r_0) = \sum_{i,j=1}^{n}\sum_{\beta=1}^{N}\frac{1}{r_0}\chi_i(r_0)(A^{-1})_{ij}^{\tau\beta}\chi_j(r_0)\sum_{\gamma=1}^{\infty}(R^{-1})_{\beta\gamma}\psi_\gamma^\alpha(r_0). \qquad (3.228)$$

The multichannel equations of this section have a simple matrix structure in channel and function space. Although explicit forms of these equations are needed for performing a numerical calculation, operator forms are often convenient for algebraic manipulation. For example, in operator form Eq. (3.228) can be written as

$$\psi(r_0) = \frac{1}{r_0}\chi^T(r_0)A^{-1}\chi(r_0)R^{-1}\psi(r_0), \qquad (3.229)$$

with the following solution for the R matrix:

$$R = \frac{1}{r_0}\chi^T(r_0)A^{-1}\chi(r_0), \qquad (3.230)$$

which has the following explicit representation:

$$R_{\tau\rho} = \frac{1}{r_0}\sum_{i,j=1}^{n}\chi_i(r_0)(A^{-1})_{ij}^{\tau\rho}\chi_j(r_0). \qquad (3.231)$$

By construction, the operator A and hence the R matrix of the present formulation are symmetric in channel space.

As in the single-channel case, the trial function of the R-matrix method should have the correct logarithmic derivative at the boundary, and it is difficult to satisfy this condition. Usually, an approximate logarithmic derivative is chosen for the basis functions. The resultant R matrix leads to a different logarithmic derivative at $r = r_0$. This is unavoidable and makes the numerical convergence of the multichannel R-matrix method slow [22].

3.7 OPTIMIZED MINIMUM-NORM AND ANOMALY-FREE METHODS

Using the idea of Harris and Michels of Section 3.4.1 [11], Nesbet and Oberoi [17] suggested a minimum-norm method for multichannel processes, which has been generalized to yield the optimized minimum-norm and optimized anomaly-free methods for multichannel scattering. In the following, we often use compact operator notation instead of exhibiting the explicit channel indices [11,17,18]. For the exact wave function, the functional I of Eq. (3.199) is zero, and we have

$$M\kappa \equiv \sum_{\beta=1}^{N}\sum_{j=0}^{1} M_{ij}^{\alpha\beta}\kappa_{j\beta}^{\rho} = 0. \tag{3.232}$$

This equation is a system of $2N$ homogeneous equations for $2N$ unknown coefficients $\kappa_{j\beta}^{\rho}, j = 0, 1$ and $\beta = 1, ..., N$ for a fixed ρ. However, by the usual asymptotic boundary condition, not all these coefficients are independent. In Section 3.6.1 we have employed two common choices: (1) $\kappa_{0\beta}^{\rho} = \delta_{\rho\beta}$ and $\kappa_{1\beta}^{\rho} = \lambda_{\rho\beta}$ or (2) $\kappa_{1\beta}^{\rho} = \delta_{\rho\beta}$ and $\kappa_{0\beta}^{\rho} = \bar{\lambda}_{\rho\beta}$. In either case there are only N independent coefficients and the remaining N coefficients are known. Hence, with the exact wave function, Eq. (3.232) should produce N independent solutions for the quantities $\kappa_{j\beta}^{\rho}$. This implies that for the exact solution, the $2N \times 2N$ matrix M has only N nontrivial eigenvalues. In an approximate calculation, this problem does not appear, as Eq. (3.232) is not satisfied. Hence, a reasonable approach for finding the optimal solution, with a trial basis set, is to minimize the norm

$$(M\kappa)^T(M\kappa) = \kappa^T(M^TM)\kappa, \tag{3.233}$$

as in the single-channel case. This norm is minimized with a trial κ_t. As in the multichannel Kohn method, a trial matrix λ^I can be obtained from this κ_t in the form

$$\lambda^I = \kappa_{t1}\kappa_{t0}^{-1}. \tag{3.234}$$

To obtain an improved result, this estimate could be used in the Kohn variational formula (3.209). Alternatively, as in the multichannel inverse Kohn method, from this κ_t a trial matrix λ^{II} can be obtained in the form

$$\bar{\lambda}^{II} = \kappa_{t0}\kappa_{t1}^{-1}. \tag{3.235}$$

The use of this estimate in the inverse Kohn variational formula (3.216) should yield an improved result.

In general, it is difficult to minimize norm (3.233). As the norm of a matrix is independent of representation, one can find it in any representation. The optimized minimum-norm method works in a different representation and has advantages. The diagonal representation is advantageous for calculating the norm. However, it is difficult to find the diagonal representation of the nonsymmetric matrix M.

The 2×2 matrix M with four submatrices can be written as

$$M = \begin{pmatrix} M_{00} & M_{01} \\ M_{10} & M_{11} \end{pmatrix}, \tag{3.236}$$

where the $M_{ij}, i, j = 0, 1$, defined by Eqs. (3.200) to (3.203), are $N \times N$ matrices. The matrix M is not symmetric and satisfies

$$M - M^T = k\begin{pmatrix} 0 & I \\ -I & 0 \end{pmatrix}, \tag{3.237}$$

where I is the $N \times N$ unit matrix. Equation (3.237) is equivalent to Eqs. (3.204) and (3.205).

Let us consider the difficulties with an arbitrary norm-preserving orthogonal transformation in this $2N$-dimensional linear space of the form

$$U = \{u_\alpha u_\beta\} \equiv \begin{pmatrix} \alpha_0 & \beta_0 \\ \alpha_1 & \beta_1 \end{pmatrix}, \tag{3.238}$$

where u_α and u_β are both $2N \times N$ matrices. Each of these matrices represent N orthonormal column vectors, so that the condition of orthogonality of matrix U yields

$$u_\alpha^T u_\alpha = u_\beta^T u_\beta = I, \qquad u_\alpha^T u_\beta = u_\beta^T u_\alpha = 0. \tag{3.239}$$

The transformed matrix M' can be written as

$$M' \equiv U^T M U = \begin{pmatrix} M'_{00} & M'_{01} \\ M'_{10} & M'_{11} \end{pmatrix} = \begin{pmatrix} u_\alpha^T M u_\alpha & u_\alpha^T M u_\beta \\ u_\beta^T M u_\alpha & u_\beta^T M u_\beta \end{pmatrix}. \tag{3.240}$$

Unfortunately, since M is not symmetric, neither M' nor any of its submatrices can be diagonal. Moreover, Eq. (3.237) is not valid for the transformed matrices. Equation (3.237) is crucial in deriving the Kohn variational principle (3.211) in the multichannel case. Hence, any transformation of the form (3.238) can not be used to minimize the norm and also preserve the Kohn variational principle (3.211). The optimized minimum-norm method provides a way to minimize the norm (3.233) and also maintain the stationary property.

3.7.1 Optimized Minimum-Norm Method

In the limit of an exact solution, the matrix λ_I of Eq. (3.234) is a real symmetric matrix in channel space. The optimized minimum-norm method uses the symmetric part of this matrix to define an orthogonal transformation that preserves Eq. (3.237) and the Kohn variational principle [17].

A real symmetric λ matrix, called $\tan\delta$, is constructed via

$$\lambda_I = \tan\delta \equiv \frac{1}{2}\{\kappa_{t1}\kappa_{t0}^{-1} + (\kappa_{t1}\kappa_{t0}^{-1})^T\} \tag{3.241}$$

The real symmetric matrices $\cos\delta$ and $\sin\delta$ are constructed by diagonalizing the matrix $\tan\delta$ and then recombining the appropriate functions of the eigenvalues with the corresponding eigenvectors. In the single-channel case, δ is the phase shift. The transformation matrix U is now taken as the orthogonal matrix,

$$U = \begin{pmatrix} \cos\delta & -\sin\delta \\ \sin\delta & \cos\delta \end{pmatrix}, \tag{3.242}$$

with the following properties satisfied by the submatrices:

$$(\cos\delta)^2 + (\sin\delta)^2 = I, \tag{3.243}$$

$$\cos\delta\sin\delta = \sin\delta\cos\delta. \tag{3.244}$$

The transformed matrix M' is defined explicitly by

$$M' \equiv \begin{pmatrix} M'_{00} & M'_{01} \\ M'_{10} & M'_{11} \end{pmatrix}$$

$$= \begin{pmatrix} \cos\delta & \sin\delta \\ -\sin\delta & \cos\delta \end{pmatrix} \begin{pmatrix} M_{00} & M_{01} \\ M_{10} & M_{11} \end{pmatrix} \begin{pmatrix} \cos\delta & -\sin\delta \\ \sin\delta & \cos\delta \end{pmatrix} \tag{3.245}$$

If we use Eqs. (3.243) and (3.244), by straightforward algebra one can show that submatrices M'_{00} and M'_{11} are symmetrical [17]. Also, if Eq. (3.237) holds for the original matrix M, it also holds for the transformed matrix M'. Equation (3.237) is essential in deriving the Kohn and inverse Kohn variational principles (3.211) and (3.218). As Eq. (3.237) is valid in the transformed system, the Kohn and inverse Kohn variational principles are also valid and are explicitly given, respectively, by

$$[\lambda'_{\rho\alpha}]_K = -\frac{1}{k_\alpha}\left(M'^{\alpha\rho}_{00} - \sum_{\beta,\tau=1}^{N} M'^{\beta\alpha}_{10} (M'^{-1}_{11})^{\beta\tau} M'^{\tau\rho}_{10} \right) \tag{3.246}$$

$$[\bar{\lambda}'_{\rho\alpha}]_{IK} = \frac{1}{k_\alpha}\left(M'^{\alpha\rho}_{11} - \sum_{\beta,\tau=1}^{N} M'^{\beta\alpha}_{01} (M'^{-1}_{00})^{\beta\tau} M'^{\tau\rho}_{01} \right) \tag{3.247}$$

First, let us consider the Kohn form (3.246). In the primed coordinate system, the asymptotic form of the wave function is specified by $\kappa'_0 = I$, $\kappa'_1 = \lambda'$. In the unprimed system we have

$$\kappa_0 = \cos\delta - (\sin\delta)(\lambda'), \tag{3.248}$$

$$\kappa_1 = \sin\delta + (\cos\delta)(\lambda'). \tag{3.249}$$

Then using Eq. (3.234) the untransformed Kohn formula becomes

$$[\lambda]_K = [\sin\delta + (\cos\delta)(\lambda')][\cos\delta - (\sin\delta)(\lambda')]^{-1}. \tag{3.250}$$

This is the expression for λ in the optimized minimum-norm Kohn method.

Similarly, the inverse Kohn condition is specified by $\kappa'_1 = I$,

$\kappa'_0 = \bar{\lambda}'$. Then the untransformed inverse Kohn formula becomes

$$[\bar{\lambda}]_{IK} = [(\cos\delta)(\bar{\lambda}') - \sin\delta][(\sin\delta)(\bar{\lambda}') + \cos\delta]^{-1}. \tag{3.251}$$

This is the expression for λ in the optimized minimum-norm inverse Kohn method.

The optimum-norm method using the Kohn and inverse Kohn approaches fails when the matrices M'_{11} and M'_{00} are singular, respectively; or when $\det M'_{11} = 0$ and $\det M'_{00} = 0$, respectively. As in the Kohn and inverse Kohn methods, without minimum-norm transformation, one could choose between Kohn and inverse Kohn methods using the ratio $|\det M'_{00}|/|\det M'_{11}|$. This procedure, due to Nesbet, avoids anomalies in the transformed system but gives discontinuous results as energy E or the variational parameters are varied. This difficulty could be removed in the optimized anomaly-free method.

3.7.2 Optimized Anomaly-Free Method

Let us again consider the matrix M. As M is not symmetric, this matrix usually cannot be diagonalized by an orthogonal transformation. When all the eigenvalues of this matrix are real, this matrix could be reduced to a real upper triangular form by an orthogonal transformation [17]. After transformation, the M matrix will have 2×2 diagonal blocks and the remaining portion below the principal diagonal will be zero.

Let us assume that the matrix M has real eigenvalues only and is transformed to a real upper triangular form. Consequently,

$$M'_{10} \sim u_\beta^T M u_\alpha = 0, \tag{3.252}$$

where the transformation U is defined as $U = \{u_\alpha u_\beta\}$. The details of the triangularization procedure of the M matrix are given by Nesbet and Oberoi [17]. If this transformation maintains Eq. (3.237), the Kohn variational principle (3.246) remains valid after transformation. If Eq. (3.252) is satisfied, the last term in the Kohn variational principle (3.246) vanishes and no anomalies can appear in this approach. In general, after transformation, Eq. (3.237) is not valid. However, one can develop an anomaly-free method based on Eq. (3.252).

The variational functional I of Eq. (3.199) remains unchanged under this transformation provided that we also effect the same transformation on the coefficients of this equation. The transformed inverse Kohn functional of Eq. (3.213) can be written as

$$I_{\tau\rho} = M'^{\tau\rho}_{11} + + \sum_{\alpha=1}^{N} \bar{\lambda}'_{\tau\alpha} M'^{\alpha\rho}_{01} + \sum_{\alpha,\beta=1}^{N} \bar{\lambda}'_{\tau\alpha} M'^{\alpha\beta}_{00} \bar{\lambda}'_{\rho\beta}, \tag{3.253}$$

where we have used Eq. (3.252). The stationary property of Eq. (3.212) in the transformed system now leads to an equation similar to (3.215):

$$\bar{\lambda}'_{\rho\alpha} = -\sum_{\tau=1}^{N} (M'^{-1}_{00})^{\alpha\tau} M'^{\tau\rho}_{01}, \tag{3.254}$$

which can be inverted to yield

$$\lambda'_{\rho\alpha} \equiv (\bar{\lambda}'^{-1})_{\rho\alpha} = -\sum_{\tau=1}^{N} (M'^{-1}_{01})^{\alpha\tau} M'^{\tau\rho}_{00}. \tag{3.255}$$

The original matrix M is not symmetric and has 2×2 block structure. After transformation, M'_{00} is diagonal. However, as M'_{10} is identically zero, the submatrix M'_{01} is expected to be nonsingular, so that $\det M'_{01} \neq 0$. Hence, the anomalous behavior of the result related to a singular M'_{01} can be avoided. Once λ' is known in the transformed system via Eq. (3.255), one can consider the inverse transformation to find λ in the original system. The quantity $\bar{\lambda}'$ of Eq. (3.255) is the K matrix via Eq. (3.177). If the transformation matrix U is given by Eq. (3.238), in the unprimed system we have

$$\kappa_0 = \alpha_0 + (\beta_0)(\lambda'), \tag{3.256}$$

$$\kappa_1 = \alpha_1 + (\beta_1)(\lambda'). \tag{3.257}$$

In the unprimed system λ is now given by

$$\lambda \equiv \frac{\kappa_1}{\kappa_0} = [\alpha_1 + \beta_1\lambda'][\alpha_0 + \beta_0\lambda']^{-1}. \tag{3.258}$$

This is the result for λ in the optimized anomaly-free method. This matrix λ is not manifestly symmetric. This undesirable feature is not present in other methods described previously, where λ is symmetric. Moreover, if M has complex eigenvalues, the λ matrix of Eq. (3.258) could be complex, which requires special care [17].

3.8 VARIATIONAL LEAST SQUARES METHOD

The minimum-norm method of Harris and Michels has been used by several authors to develop the variational least squares method [18,19,23]. As in the minimum-norm method, this method develops an expression for the zero-order K-matrix elements which can be improved upon by using a Kohn variational principle.

For the multichannel problem, the effective Schrödinger equation is given by Eq. (3.172). However, for an approximate trial function ψ_t, $(H - E)\psi_t \neq 0$. The quantity $(H - E)\psi_t$ represents a measure of the error in the trial wave function. The variational least squares method minimizes $|(H - E)\psi_t|^2$ in such a way that a zeroth-order estimate of λ is found, which is, subsequently, improved by the Kohn or inverse Kohn variational principles.

In this method one uses the following trial function

$$\psi^{\mathcal{T}}_{\beta}(r_\beta) = v^{\mathcal{T}}_{\beta}(r_\beta) + \kappa^{\mathcal{T}}_{0\beta} S_\beta(r_\beta) + \kappa^{\mathcal{T}}_{1\beta} C_\beta(r_\beta). \tag{3.259}$$

This expression is consistent with the general form (3.180) of the multichannel wave function. The function $v^{\mathcal{T}}_{\beta}$ is then expanded in terms of a set of $\mathcal{L}^2$ functions:

$$v^{\mathcal{T}}_{\beta}(r_\beta) = \sum_{i=1}^{n} b_i^{\mathcal{T}\beta} \chi_i^{\mathcal{T}}(r_\beta). \tag{3.260}$$

The total number of expansion functions – $\mathcal{L}^2$ or not – used in all channels is $2N + \mathcal{N}$, where $\mathcal{N} \equiv nN$ is the total number of $\mathcal{L}^2$ functions in all channels and N the number of channels. The three projections of the Schrödinger equation on sin, cos, and $\mathcal{L}^2$ functions can be written as

$$\sum_{\alpha,\beta=1}^{N} \langle S_\alpha|(H_{\alpha\beta} - E_{\alpha\beta})|\psi_\beta^\tau\rangle = 0, \tag{3.261}$$

$$\sum_{\alpha,\beta=1}^{N} \langle C_\alpha|(H_{\alpha\beta} - E_{\alpha\beta})|\psi_\beta^\tau\rangle = 0, \tag{3.262}$$

$$\sum_{\alpha,\beta=1}^{N} \langle \chi_i^\alpha|(H_{\alpha\beta} - E_{\alpha\beta})|\psi_\beta^\tau\rangle = 0. \tag{3.263}$$

In compact operator form these equations can be written as

$$\begin{pmatrix} \mathcal{M}_{SS} & \mathcal{M}_{SC} & \mathcal{M}_{Sb} \\ \mathcal{M}_{CS} & \mathcal{M}_{CC} & \mathcal{M}_{Cb} \\ \mathcal{M}_{bS} & \mathcal{M}_{bC} & \mathcal{M}_{bb} \end{pmatrix} \begin{pmatrix} \kappa_0 \\ \kappa_1 \\ b \end{pmatrix} = 0. \tag{3.264}$$

The label b stands for $\mathcal{N}$ $\mathcal{L}^2$ functions in all channels, and the label S (C) stands for the N sin (cos) functions in all channels. Explicitly,

$$\mathcal{M}_{bb} \equiv \mathcal{M}_{ij}^{\tau\beta} = \langle \chi_i^\tau|(H_{\tau\beta} - E\delta_{\tau\beta})|\chi_j^\beta\rangle, \tag{3.265}$$

$$\mathcal{M}_{SS} \equiv B_0^{\tau\beta} = \langle S_\tau|(H_{\tau\beta} - E\delta_{\tau\beta})|S_\beta\rangle, \tag{3.266}$$

$$\mathcal{M}_{bS} \equiv R_{iS}^{\tau\beta} = \langle \chi_i^\tau|(H_{\tau\beta} - E\delta_{\tau\beta})|S_\beta\rangle, \tag{3.267}$$

and so on [see Eqs. (3.189) to (3.195)]. Equation (3.264) can be written in the form of the matrix equation

$$\mathcal{M}\mathcal{X} = 0, \tag{3.268}$$

where $\mathcal{M}$ is a square matrix of dimension $2N + \mathcal{N}$ and $\mathcal{X}$ is a column vector.

For a finite basis set, Eq. (3.268) cannot be satisfied exactly and there will be a small error matrix $\mathcal{D}$ on the right-hand side of this equation, so that one should have

$$\mathcal{M}\mathcal{X} = \mathcal{D}. \tag{3.269}$$

In the variational least squares method, it is required that $\mathcal{D}^T\mathcal{D}$ be stationary, or

$$\delta(\mathcal{D}^T\mathcal{D}) = 0, \tag{3.270}$$

so that

$$\delta(\mathcal{X}^T\mathcal{M}^T\mathcal{M}\mathcal{X}) = [(\delta\mathcal{X})^T\mathcal{M}^T\mathcal{M}\mathcal{X}] + [(\delta\mathcal{X})^T\mathcal{M}^T\mathcal{M}\mathcal{X}]^T = 0. \tag{3.271}$$

In order that Eq. (3.271) is valid for arbitrary variations $\delta\mathcal{X}$, we should have

$$(\mathcal{M}^T\mathcal{M})\mathcal{X} = 0. \tag{3.272}$$

Condition (3.272) should lead to the optimal solution. This equation should be solved subject to $\kappa_0 = I$ and $\kappa_1 = \lambda$. Then Eq. (3.272) can be written as

$$\begin{pmatrix} (\mathcal{M}^T\mathcal{M})_{SC} & (\mathcal{M}^T\mathcal{M})_{Sb} \\ (\mathcal{M}^T\mathcal{M})_{CC} & (\mathcal{M}^T\mathcal{M})_{Cb} \\ (\mathcal{M}^T\mathcal{M})_{bC} & (\mathcal{M}^T\mathcal{M})_{bb} \end{pmatrix} \begin{pmatrix} \lambda \\ b \end{pmatrix} = -\begin{pmatrix} (\mathcal{M}^T\mathcal{M})_{SS} \\ (\mathcal{M}^T\mathcal{M})_{CS} \\ (\mathcal{M}^T\mathcal{M})_{bS} \end{pmatrix}. \tag{3.273}$$

As we have defined N of the unknowns via $\kappa_0 = I$, Eq. (3.273) constitutes $2N + \mathcal{N}$ equations for $N + \mathcal{N}$ unknowns. So this set of equations cannot be solved. One way out is to discard N of these equations arbitrarily. Wladawski, and Abdallah and Truhlar, suggested discarding the first row of this equation arbitrarily [18,23]. Consequently, one is left with

$$\begin{pmatrix} (\mathcal{M}^T\mathcal{M})_{CC} & (\mathcal{M}^T\mathcal{M})_{Cb} \\ (\mathcal{M}^T\mathcal{M})_{bC} & (\mathcal{M}^T\mathcal{M})_{bb} \end{pmatrix} \begin{pmatrix} \lambda \\ b \end{pmatrix} = -\begin{pmatrix} (\mathcal{M}^T\mathcal{M})_{CS} \\ (\mathcal{M}^T\mathcal{M})_{bS} \end{pmatrix}. \tag{3.274}$$

Equation (3.274) constitute $N + \mathcal{N}$ equations for $N + \mathcal{N}$ unknowns, which can be solved and the matrix λ determined. The matrix λ is the required zeroth-order solution. This λ can be used in the Kohn variational principle (3.209) for further improvement.

Another possibility is to take $\kappa_0 = \bar{\lambda}$ and $\kappa_1 = I$, as in the inverse Kohn variational principle, and find $\bar{\lambda}$. Once the zero-order $\bar{\lambda} = \lambda^{-1}$ is obtained, an improved result could be found by using these results in the inverse-Kohn method.

Anomalies of the Kohn or inverse Kohn methods arise from the singularities of matrices M_{11} and M_{00}. The variational least squares procedure is supposed to be free of such difficulties. The only matrix inversion in this method is inversion of the matrix in Eq. (3.274). Anomalies of this method, if any, should arise when the determinant of this matrix is zero. This smaller matrix is denoted by $(\mathcal{M}^T\mathcal{M})'$. It is demonstrated below that the matrix $(\mathcal{M}^T\mathcal{M})'$ is singular when the matrix $\mathcal{M}$ itself is singular. As the matrix $\mathcal{M}$ is assumed to be nonsingular, this possibility is never realized.

The matrix $\mathcal{M}^T\mathcal{M}$ is positive definite for any vector $\mathcal{X}$:

$$\mathcal{X}^T\mathcal{M}^T\mathcal{M}\mathcal{X} > 0. \tag{3.275}$$

Now consider the vector

$$\mathcal{X} = \begin{pmatrix} 0 \\ \lambda \\ b \end{pmatrix}. \tag{3.276}$$

With this vector

$$\mathcal{X}^T\mathcal{M}^T\mathcal{M}\mathcal{X} \equiv \mathcal{Y}^T(\mathcal{M}^T\mathcal{M})'\mathcal{Y} > 0, \tag{3.277}$$

where

$$\mathcal{Y} = \begin{pmatrix} \lambda \\ b \end{pmatrix}. \tag{3.278}$$

Hence, in the 2×2 matrix space of Eq. (3.274), the matrix $(\mathcal{M}^T\mathcal{M})'$ is positive definite as long as the matrix $\mathcal{M}$ is nonsingular. Vanishing of the determinant of $\mathcal{M}$ is a necessary (but not sufficient) condition for the existence of the singularity of $(\mathcal{M}^T\mathcal{M})'$. consequently, the present method is not supposed to lead to anomalies, as the determinant of $\mathcal{M}$ is taken to be nonzero.

3.9 NOTES AND REFERENCES

[1] For a general review of the topics covered in this chapter, see Gerjuoy et al. (1983), Nesbet (1975, 1978, 1980), Callaway (1978a), and Truhlar et al. (1974).

[2] Schwinger (1947a,b).

[3] See, for example, Hochstadt (1973) and Vorobyev (1965), for a discussion of Fredhohm integral equations and different methods of solution.

[4] Kohn (1948).

[5] See, for example, Newton (1982).

[6] Takatsuka and McKoy (1981a,b).

[7] Hulthén (1944).

[8] Hulthén (1948), Rubinow (1955).

[9] Schwartz (1961a,b).

[10] Harris (1967).

[11] Harris (1969a,b), Harris and Michels (1971).

[12] Nesbet (1968, 1969a,b, 1980).

[13] Kato (1950).

[14] Rosenberg et al. (1960).

[15] Wigner and Eisenbud (1947), Lane and Thomas (1958), Burke and Robb (1975), Nesbet (1980).

[16] Buttle (1967), Fano and Lee (1973), Nesbet (1980).

[17] Nesbet and Oberoi (1972).

[18] For a review on variational principles and their applications in atomic scattering, see Callaway (1978a).

[19] For a review of different variational principles, see Truhlar et al. (1974).

[20] Nuttall (1969).

[21] Rudge (1973, 1975).

[22] Pioneering work on the multichannel variational *R*-matrix method was done by Jackson (1951). Similar studies have been undertaken by Lane and Robson (1969) and Oberoi and Nesbet (1973a), and a review of these studies is given by Nesbet (1980).

[23] Bardsley et al. (1972), Wladawsky (1973), Abdallah and Truhlar (1974).

CHAPTER 4

VARIATIONAL PRINCIPLES FOR OFF-SHELL AMPLITUDES

4.1 INTRODUCTION

The variational principles for on-shell K-matrix elements discussed in Chapter 3 are related to phase shifts, cross sections, and to on-shell S- and t-matrix elements. However, the K and t matrices are operators with an infinite number of matrix elements for infinite possibilities in incoming and outgoing momentum variables, such as $K(p', p, k^2)$, where k is the on-shell wave number and in general $p' \neq k \neq p$. Physical observables for the two-particle system are related to the on-shell quantity $K(k, k, k^2)$. The off-shell elements $K(p', p, k^2)$ correspond to unphysical or virtual transitions, which occur in the presence of other particles. For example, nucleon–nucleon off-shell scattering is realized when the nucleon–nucleon system is embedded in a multinucleon system: for example, in a trinucleon system or in nuclear matter. The nonconservation of energy in the intermediate states allows the possibility of off-shell scattering.

In this chapter we discuss variational principles based on the operator nature of the K or t matrices and not for their on-shell elements only. Consequently, in addition to yielding the physical on-shell quantities, these variational principles yield the off-shell quantities, which could be useful in solving multiparticle problems in which the present system is a subsystem. The present variational principles are for the off-shell partial-wave quantities $t_L(p', p, k^2)$ or

$K_L(p', p, k^2)$. However, if one is not interested in off-shell scattering, the present variational principles yield the on-shell observables, as in Chapter 3, provided that one considers the special case $p' = p = k$.

All these variational principles are suitable for numerical application with a finite basis set. When this procedure is adopted, as in Chapter 3, the numerical task is reduced to simple linear algebra: inversion of a matrix of relatively small dimension and matrix-vector multiplication. Then, numerically, the present variational principles lead to algebraic variational basis-set methods for scattering problems.

Apart from permitting the above-mentioned off-shell extension, most of the variational principles of the present chapter involve the free outgoing-wave or standing-wave Green's function explicitly, which builds in the correct outgoing-wave or standing-wave scattering boundary condition. These Green's functions build in the appropriate asymptotic boundary condition. When standing-wave boundary conditions are used, the variational principle leads to the K matrix, and when outgoing-wave boundary conditions are employed, it yields the t matrix. This approach can also be used to derive variational principles for the nonsingular Γ matrix.

The first of these variational principles was suggested by Schwinger in 1947 [1] for the t or K matrix. Then, in 1966, Newton [2] suggested a variational principle for $t - V$ or $K - V$. In the usual form of the Schwinger (Newton) variational principle, the unknown function is a non-$\mathcal{L}^2$ ($\mathcal{L}^2$) function. Hence, for obtaining optimal convergence, one should use non-$\mathcal{L}^2$ ($\mathcal{L}^2$) functions in the case of the Schwinger (Newton) variational principle. However, as the free Green's function appears in these variational principles, there is no need to include non-$\mathcal{L}^2$ functions in these variational principles. Numerically, it is easier to deal with $\mathcal{L}^2$ functions and one does not have to perform integrations over non-$\mathcal{L}^2$ oscillating functions as in Kohn-type variational principles. This is a great advantage over the methods of Chapter 3, where some non-$\mathcal{L}^2$ functions must be included.

It is possible to extend this progression and write a variational principle for $t - V - VG_0^{(+)}V$ [3], which deals with the first two terms of the Born–Neumann series exactly, and the presumably small difference between the exact t matrix and the truncated Born–Neumann series is treated variationally. This procedure can be continued in treating more terms of the Born–Neumann series exactly and the difference terms variationally. These variational principles can be written in two equivalent forms. The first form involves a trilinear combination of different terms, and the second form, often known as the fractional form, is an appropriate ratio of different terms.

The utility and power of the Schwinger variational principle were fully appreciated only some 25 years later, when a practical finite-rank form for performing numerical calculations using the Schwinger variational principle was suggested and used in several numerical calculations in 1974 by Sloan and the author [4]. Similar finite-rank form for the Newton variational principle was suggested and studied by Sloan and Brady [5]. Since then numerous applications of these variational principles have been made in nuclear, atomic, molecular, and chemical physics.

Most of the variational principles described in this chapter involve a (multiple) integral involving the free Green's function, which could be cumbersome in practice. To avoid such integrals and the anomalies of the usual Kohn variational principle of Chapter 3, Miller and Jansen op de Haar suggested in 1987 [6] the complex Kohn variational principle for the t- and S-matrix elements, which uses complex basis functions in the Kohn variational principle compatible with the outgoing-wave boundary condition for the t matrix. This last variational principle requires only the matrix elements of the Hamiltonian operator. However, unlike the usual Kohn variational principle, the complex Kohn variational principle is easily extended to the calculation of off-shell scattering elements. The use of real basis functions compatible with the standing-wave boundary condition in the formulation of Miller and Jansen op de Haar leads to a practical variational method for calculating off-shell K-matrix elements. In both cases, however, one needs to evaluate integrals over non-$\mathcal{L}^2$ oscillating functions as in the on-shell Kohn methods of Chapter 3.

Unlike the Kohn variational principles of Chapter 3, the variational principles of this chapter were believed to possess no anomaly. However, it has recently been demonstrated that for all variational principles presented in this chapter, anomalies may occur in these variational principles under exceptional choice of potential and expansion functions [7–10]. Unlike in the Kohn variational principles of Chapter 3, these anomalies are found to be rare and hence should not be of great concern in actual numerical application.

The variational principles of this chapter are written for the t-matrix elements, one of the exceptions being the variational principle for the Γ matrix and the real Kohn variational principle. However, most of the variational principle of this chapter can be reformulated for the K-matrix elements by replacing the outgoing-wave free Green's function in these variational principles by the standing-wave free Green's function. The complex Kohn variational principle for the t matrix can be modified to yield the real K matrix by changing a complex basis function of this approach to an appropriate real basis function.

In Section 4.2 we develop the Schwinger variational principle and derive several algebraic basis-set methods based on them. A discussion on the choice of the basis functions and on the application to the nuclear problems are presented. In Section 4.3 a description of the Newton variational principle and related basis-set methods are given. In Section 4.4 we describe an alternative approximate iterative method for using the Schwinger and the Newton variational principles in practical calculation with certain advantages over the usual algebraic basis-set methods. In Section 4.5 variational principles are derived for the higher-order terms of the Born–Neumann series. In Section 4.6 we describe variational basis-set methods for the t and K matrices based on a form of the Kohn variational principle. In Section 4.7 we describe a class of variational principles employing distorted waves, which facilitate the numerical task. In Section 4.8 we describe a technical scheme for calculating the different matrix elements encountered in these variational methods. A general scheme for deriving these variational principles is given in Section 4.9. In Section 4.10 we derive variational principles for the Γ matrix and develop algebraic basis-set methods based on them. Finally, in Section 4.11 we discuss anomalous behaviors of these basis-set methods. An anomaly-free Kohn method is also suggested in this section.

4.2 SCHWINGER VARIATIONAL PRINCIPLE

4.2.1 Formulation

The algebraic basis-set methods developed from the Kohn variational principles of Chapter 3 requires the evaluation of integrals over non-$\mathcal{L}^2$ functions. These integrals are often difficult to evaluate numerically. Basis-set methods based on most of the variational principles of this chapter can use only $\mathcal{L}^2$ functions. The Schwinger variational principle is the first variational principle of this type [1], which we describe below. It was originally suggested for calculating the scattering phase shifts. However, it can be readily modified for calculating the off-shell t- or K-matrix elements and we present the modified version in the following.

The Schwinger variational principle for the t-matrix element is based on the identity

$$\begin{aligned}\langle p'|t|p\rangle &= \langle p'|V|\psi_p^{(+)A}\rangle + \langle\psi_{p'}^{(-)A}|V|p\rangle \\ &- \langle\psi_{p'}^{(-)A}|(V - VG_0^{(+)}V)|\psi_p^{(+)A}\rangle \\ &+ \langle\psi_{p'}^{(-)A} - \psi_{p'}^{(-)}|(V - VG_0^{(+)}V)|\psi_p^{(+)A} - \psi_p^{(+)}\rangle,\end{aligned} \tag{4.1}$$

where exact wave functions $|\psi_p^{(+)}\rangle$ and $\langle\psi_{p'}^{(-)}|$ satisfy

$$|\psi_p^{(+)}\rangle = |p\rangle + G_0^{(+)}V|\psi_p^{(+)}\rangle, \tag{4.2}$$

$$\langle\psi_{p'}^{(-)}| = \langle p'| + \langle\psi_{p'}^{(-)}|VG_0^{(+)}. \tag{4.3}$$

Here $|\psi_p^{(+)A}\rangle$ and $\langle\psi_{p'}^{(-)A}|$ are supposed to be approximations to exact scattering wave functions $|\psi_p^{(+)}\rangle$ and $\langle\psi_{p'}^{(-)}|$, respectively. Equation (4.1) is an identity independent of $|\psi_p^{(+)A}\rangle$ and $\langle\psi_{p'}^{(-)A}|$. However, if $|\psi_p^{(+)A}\rangle$ and $\langle\psi_{p'}^{(-)A}|$ are good approximations to the exact wave functions $|\psi_p^{(+)}\rangle$ and $\langle\psi_{p'}^{(-)}|$, respectively, the last term in Eq. (4.1) is second order in deviation. Equations (4.1) to (4.2) and most of the equations of this chapter are valid for the partial-wave observables. In these equations, the energy dependence of the t matrix and the Green's function and the angular momentum label have been suppressed and Dirac bra-ket notation has been used for convenience:

$$\langle p'|t|p\rangle \equiv t_L(p', p, k^2). \tag{4.4}$$

Keeping only first-order terms in deviation in Eq. (4.1) and suppressing the index A from approximate wave functions, we obtain the Schwinger variational principle,

$$[\langle p'|t|p\rangle] = \langle p'|V|\psi_p^{(+)}\rangle + \langle\psi_{p'}^{(-)}|V|p\rangle - \langle\psi_{p'}^{(-)}|(V - VG_0^{(+)}V)|\psi_p^{(+)}\rangle. \tag{4.5}$$

Approximate trial wave functions are to be used in variational principle (4.5). Equations (4.1) and (4.5) should be compared with Eqs. (3.1) and (3.2), respectively. There are three terms in Eq. (4.5). If exact wave functions are used in this equation, each of these three terms becomes equal to the exact t matrix and Eq. (4.5) is an identity. All the variational principles of this chapter are based on similar trilinear forms. The present construction of the variational principle should be compared to that of Eq. (3.11) in relation to a standard Fredholm equation. Much of the discussion related to Eq. (3.11) also holds in this case. In particular, as in Eq. (3.26), one has the following equivalent fractional form for the Schwinger variational principle:

$$[\langle p'|t|p\rangle] = \frac{\langle p'|V|\psi_p^{(+)}\rangle\langle\psi_{p'}^{(-)}|V|p\rangle}{\langle\psi_{p'}^{(-)}|(V - VG_0^{(+)}V)|\psi_p^{(+)}\rangle}.$$

This t matrix is obtained if one solves the partial-wave Lippmann–Schwinger equation (1.121) with the following potential:

$$\langle p'|V|p\rangle = \frac{\langle p'|V|\psi_p^{(+)}\rangle\langle\psi_{p'}^{(-)}|V|p\rangle}{\langle\psi_{p'}^{(-)}|V|\psi_p^{(+)}\rangle}. \tag{4.7}$$

For real trial functions, $\mathcal{L}^2$ or not, the potential of Eq. (4.7) is real and Hermitian and hence the Schwinger variational forms (4.5) and (4.6) satisfy the conditions of unitarity and time-reversal symmetry. The separable form (4.6) for the t matrix is particularly useful if one is interested in solving the multiparticle problem.

In this section the t-matrix version of the Schwinger variational principle is considered. The K-matrix version of this variational principle is straightforwardly obtained by replacing the outgoing-wave Green's function by the standing-wave Green's function. As both these possibilities correspond to the same underlying separable potential (4.7), the t- and K-matrix Schwinger variational principles based on Eqs. (4.5) and (4.6), respectively, lead to the same phase shift.

The following alternative form of the Schwinger variational principle can be obtained if the trial wave functions of Eq. (4.5) are rewritten in terms of the t-matrix elements:

$$\begin{aligned}[\langle p'|t|p\rangle] = {} & \langle p'|V(I + G_0^{(+)}t_1)|p\rangle + \langle p'|(I + t_2G_0^{(+)})V|p\rangle \\ & - \langle p'|(I + t_2G_0^{(+)})(V - VG_0^{(+)}V)(I + G_0^{(+)}t_1)|p\rangle.\end{aligned} \tag{4.8}$$

In Eq. (4.8), t_1 and t_2 are two independent trial estimates for the exact t matrix. The stationary property of Eq. (4.8) can be established if we use the Lippmann–Schwinger equation for the t matrix. We shall see in Section 4.4 how the fractional form of the stationary expression (4.8) can be used iteratively in an approximate calculational scheme.

In Eqs. (4.5) and (4.6) there is a multiple integral over the Green's function. The trouble associated with this multiple integral could be reduced, for example, by considering the alternative form [11]

$$[\langle p'|t|p\rangle] = \langle p'|\chi_p^{(+)}\rangle + \langle\chi_{p'}^{(-)}|p\rangle - \langle\chi_{p'}^{(-)}|(V^{-1} - G_0^{(+)})|\chi_p^{(+)}\rangle, \quad (4.9)$$

where we have introduced the form factors

$$|\chi_p^{(+)}\rangle = V|\psi_p^{(+)}\rangle, \quad (4.10)$$

$$\langle\chi_{p'}^{(-)}| = \langle\psi_{p'}^{(-)}|V. \quad (4.11)$$

The Schwinger variational expression (4.9) is stationary to first order with respect to variations of trial functions $|\chi_p^{(+)}\rangle$ and $\langle\chi_{p'}^{(-)}|$. In Eq. (4.9) one has a simpler integral over the free Green's function than in Eq. (4.5). However, there is also an integral over the inverse potential operator V^{-1} which deserves special care.

One can also rewrite the Schwinger variational principle (4.5) or (4.9) in the following form:

$$[\langle p'|t|p\rangle] = \langle p'|V|\psi_p^{(+)}\rangle + \langle\chi_{p'}^{(-)}|p\rangle - \langle\chi_{p'}^{(-)}|(I - G_0^{(+)}V)|\psi_p^{(+)}\rangle, \quad (4.12)$$

which is stationary to first order, with respect to variations of trial functions $|\psi_p^{(+)}\rangle$ and $\langle\chi_{p'}^{(-)}|$. In form (4.12) there is a simpler integral over the Green's function than in Eqs. (4.5) and (4.6), and no integral over the inverse operator V^{-1}.

Expressions (4.9) and (4.12) can also be written in the equivalent fractional forms:

$$[\langle p'|t|p\rangle] = \frac{\langle p'|\chi_p^{(+)}\rangle\langle\chi_{p'}^{(-)}|p\rangle}{\langle\chi_{p'}^{(-)}|(V^{-1} - G_0^{(+)})|\chi_p^{(+)}\rangle}, \quad (4.13)$$

$$[\langle p'|t|p\rangle] = \frac{\langle p'|V|\psi_p^{(+)}\rangle\langle\chi_{p'}^{(-)}|p\rangle}{\langle\chi_{p'}^{(-)}|(I - G_0^{(+)}V)|\psi_p^{(+)}\rangle}, \quad (4.14)$$

respectively. Variational principles (4.12) and (4.14) are not time-reversal symmetric. However, this is not a bar for numerical applications. Corresponding to variational principle (4.14), one has two transposed variational principles

$$[\langle p'|t|p\rangle] = \langle\psi_{p'}^{(-)}|V|p\rangle + \langle p'|\chi_p^{(+)}\rangle - \langle\psi_{p'}^{(-)}|(I - VG_0^{(+)})|\chi_p^{(+)}\rangle, \quad (4.15)$$

and

$$[\langle p'|t|p\rangle] = \frac{\langle p'|\chi_p^{(+)}\rangle\langle\psi_{p'}^{(-)}|V|p\rangle}{\langle\psi_{p'}^{(-)}|(I - VG_0^{(+)})|\chi_p^{(+)}\rangle}, \quad (4.16)$$

respectively.

A calculational scheme using the foregoing variational principles is obtained by taking the trial functions as linear combinations of a given set of basis functions. In the ideal situation, these basis functions should be orthonormal and linearly independent. However, these requirements are not usually realized in numerical calculation. The coefficients in the expansion of the trial function in terms of the basis set are determined by considering the stationary property.

A practical calculational scheme can be developed from (4.5) with the following choice of trial functions:

$$|\psi_p^{(+)}\rangle = \sum_{i=1}^{n} a_i(p)|f_i\rangle, \quad (4.17)$$

$$\langle\psi_{p'}^{(-)}| = \sum_{j=1}^{n}\langle f_j|b_j(p'), \quad (4.18)$$

where a_i and b_j are coefficients to be determined by the variational requirement and $|f_i\rangle$, $i = 1, ..., n$, form the basis set. We have used the same set of states for both trial functions. This is not essential, but it has the virtue of maintaining the result time-reversal symmetric.

To determine the coefficients a_i and b_j, we substitute Eqs. (4.17) and (4.18) into the Schwinger variational principle (4.5) and require that the resulting expression be stationary with respect to independent variations of a_i and b_j. It then follows that

$$b_j(p') = \sum_{i=1}^{n}\langle p'|V|f_i\rangle D_{ij}^{(n)}, \quad (4.19)$$

$$a_i(p) = \sum_{j=1}^{n} D_{ij}^{(n)}\langle f_j|V|p\rangle, \quad (4.20)$$

where

$$(D^{-1})_{ji} = \langle f_j|(V - VG_0^{(+)}V)|f_i\rangle, \qquad i,j = 1, \dots, n. \tag{4.21}$$

Here $D^{(n)}$ is an $n \times n$ matrix, constructed by inverting the presumably nonsingular D^{-1} matrix of Eq. (4.21). The n dependence of the D matrix should be noted: for example, $D_{11}^{(1)} \neq D_{11}^{(2)}$. With these expansion coefficients the trial functions of Eqs. (4.17) and (4.18) become

$$|\psi_p^{(+)}\rangle = \sum_{i,j=1}^{n} |f_i\rangle D_{ij}^{(n)} \langle f_j|V|p\rangle, \tag{4.22}$$

$$\langle\psi_{p'}^{(-)}| = \sum_{i,j=1}^{n} \langle p'|V|f_i\rangle D_{ij}^{(n)} \langle f_j|. \tag{4.23}$$

When these results are substituted into Eq. (4.5), we obtain the Schwinger t matrix in the rank-n degenerate form:

$$[\langle p'|t_n|p\rangle] = \sum_{i,j=1}^{n} \langle p'|V|f_i\rangle D_{ij}^{(n)} \langle f_j|V|p\rangle. \tag{4.24}$$

For $n = 1$, Eq. (4.24) reduces to the rank-1 fractional form (4.6) for the Schwinger variational principle, and hence it should be considered as the multirank generalization of the rank-1 result (4.6). The result (4.24) can now be recognized to be the one obtained with the well-known method of moments for the solution of Fredholm integral equations [i.e., Eq. (3.22)].

The rank-n t matrix of Eq. (4.24) is obtained if one solves Lippmann–Schwinger equation (1.121) with the following rank-n degenerate potential:

$$\langle p'|V_n|p\rangle = \sum_{i,j=1}^{n} \langle p'|V|f_i\rangle B_{ij}^{(n)} \langle f_j|V|p\rangle, \tag{4.25}$$

where the $n \times n$ B matrix is defined through its inverse elements

$$(B^{-1})_{ji} = \langle f_j|V|f_i\rangle, \qquad i,j = 1, \dots, n. \tag{4.26}$$

For real expansion functions $|f_i\rangle$ and real Hermitian potential V, the rank-n potential $\langle p'|V_n|p\rangle$ is also real and Hermitian. Hence, the variational t matrix (4.24) satisfies unitarity automatically and the resultant phase shifts are real. The rank-n potential V_n of Eq. (4.25) has the property of being exact when it operates on any linear combination of the n functions $|f_i\rangle, i = 1, ..., n$, since

$$\langle f_j|V_n = \langle f_j|V, \qquad j = 1, ..., n, \tag{4.27}$$

$$V_n|f_i\rangle = V|f_i\rangle, \qquad i = 1, ..., n. \tag{4.28}$$

Assuming that the set of functions $|f_i\rangle$ becomes complete as $n \to \infty$, in the same limit $V_n \to V$ and $t_n \to t$. Hence, as more and more functions are introduced in the basis set, a more accurate t matrix should result.

A calculational scheme could be developed from variational principle (4.12) with the following choice of trial functions:

$$|\psi_p^{(+)}\rangle = \sum_{i=1}^{n} a_i(p)|f_i\rangle, \tag{4.29}$$

$$\langle\chi_{p'}^{(-)}| = \sum_{j=1}^{n} \langle v_j|b_j(p'), \tag{4.30}$$

where a_i and $b_i, i = 1, ..., n$, are expansion coefficients and $|f_i\rangle$ and $\langle v_i|, i = 1, ..., n$, form the basis sets. As variational principle (4.12) is not symmetric, there is no need to take the same basis set for two trial functions. Next we substitute functions (4.29) and (4.30) into variational principle (4.12) and require that the resultant expression be stationary with respect to independent variations of a_i and b_j. Then we get

$$b_j(p') = \sum_{i=1}^{n} \langle p'|V|f_i\rangle C_{ij}^{(n)}, \tag{4.31}$$

$$a_i(p) = \sum_{j=1}^{n} C_{ij}^{(n)}\langle v_j|p\rangle, \tag{4.32}$$

where

$$(C^{-1})_{ji} = \langle v_j|(I - G_0^{(+)}V)|f_i\rangle, \qquad i, j = 1, ..., n. \tag{4.33}$$

When these expansion coefficients are substituted into Eqs. (4.29), (4.30), and (4.12), we obtain the following rank-n variational t matrix:

$$[\langle p'|t_n|p\rangle] = \sum_{i,j=1}^{n} \langle p'|V|f_i\rangle C_{ij}^{(n)} \langle v_j|p\rangle. \tag{4.34}$$

Equation (4.34) should be compared with Eq. (3.22) obtained by the method of moments for a general Fredholm equation. This result could have been derived from Eq. (4.24) by taking $\langle v_j| = \langle f_j|V$, because variational principles (4.5) and (4.12) are also related by such a mapping: $\langle \chi_{p'}^{(-)}| = \langle \psi_{p'}^{(-)}|V$.

One could derive several other forms of the Schwinger variational method by taking appropriate forms of trial functions. We consider two examples below. First, by taking $|f_i\rangle = G_0^{(+)}|g_i\rangle$ and a similar equation for its transpose in Eq. (4.24), we obtain

$$[\langle p'|t_n|p\rangle] = \sum_{i,j=1}^{n} \langle p'|VG_0^{(+)}|g_i\rangle J_{ij}^{(n)} \langle g_j|G_0^{(+)}V|p\rangle, \tag{4.35}$$

with

$$(J^{-1})_{ji} = \langle g_j|(G_0^{(+)}VG_0^{(+)} - G_0^{(+)}VG_0^{(+)}VG_0^{(+)})|g_i\rangle, \qquad i,j = 1, \ldots, n. \tag{4.36}$$

In this case, the implicit rank-n separable potential is not Hermitian at positive energies, so this t_n does not satisfy unitarity. A Hermitian V_n is obtained if we employ the principal-value Green's function in Eq. (4.35). Expansion (4.35) is numerically more complicated because of the appearance of multiple integrals over Green's functions. Nevertheless, this expansion has proved to be useful in situations where V has complicated singularity structures [12]. For example, the three-particle scattering equation can be written in the Lippmann–Schwinger form with an energy-dependent V which is complex at positive energies. Because of the explicit presence of $VG_0^{(+)}$ in the form factor, the expansion (4.35) is better able to reproduce the singularity structure of the t matrix in a low-rank calculation. Successful application of this variational principle has been made to the three-nucleon problem [12].

Finally, if we take $V|f_i\rangle = |v_i\rangle$ and a similar equation for its transpose in Eq. (4.24), we obtain the alternative form for the

Schwinger variational method:

$$[\langle p'|t_n|p\rangle] = \sum_{i,j=1}^{n} \langle p'|v_i\rangle U_{ij}^{(n)} \langle v_j|p\rangle, \tag{4.37}$$

with

$$(U^{-1})_{ji} = \langle v_j|(V^{-1} - G_0^{(+)})|v_i\rangle, \qquad i, j = 1, ..., n. \tag{4.38}$$

This result is time-reversal symmetric and unitary. Expansion (4.37) could also be derived from the trilinear form (4.9) by making an expansion of the form factors χ of Eqs. (4.10) and (4.11) in terms of the basis functions v_i, $i = 1, ..., n$. As these form factors are dressed by the potential, they are $\mathcal{L}^2$.

4.2.2 Choice of Basis Functions

With this discussion about several forms of the Schwinger variational principle, we now discuss how to choose the basis functions. A proper choice of basis functions is fundamental for the success of a variational principle in numerical application. The trial functions of a variational method should be chosen so as to satisfy certain asymptotic properties of the actual function they approximate. The choice should also reflect the properties of the potential under consideration: short-range or not, energy-dependent or not, Hermitian or not, and so on.

The trial functions $|\psi_p^{(+)}\rangle$ and $\langle\psi_{p'}^{(-)}|$ of the Schwinger variational principles (4.5) or (4.6) can be taken to be $\mathcal{L}^2$. However, these scattering wave functions are nonnormalizable, energy-dependent, and have oscillating behavior at ∞. If this asymptotic behavior is included in the trial function, one can expect rapid convergence. The following trial function incorporates proper asymptotic boundary conditions (1.35) and (1.36) or (1.136) for L-wave scattering:

$$\psi(r) = c_1 S(r) + c_2 C(r) + \sum_{i=1}^{n} c_i r^{(i+L)} \exp(-\alpha r), \tag{4.39}$$

where $C(r)$ and $S(r)$ are defined by Eqs. (3.73) and (3.74), respectively [13]. In actual calculation, this corresponds to taking $f_1(r) = S(r)$,

$f_2(r) = C(r)$, and the following $\mathcal{L}^2$ functions:

$$f_i(r) = r^{(L+i-2)} \exp[-\alpha r], \qquad i = 3, 4, ..., n. \tag{4.40}$$

The alternative choice for the $\mathcal{L}^2$ functions,

$$f_i(r) = r^{(L+1)} \exp[-(i-2)\alpha r], \qquad i = 3, 4, ..., n, \tag{4.41}$$

can also be used in numerical calculation in place of Eq. (4.40).

The inclusion of proper asymptotic boundary condition in the Schwinger variational method based on Eq. (4.5) or (4.6) is only desirable for obtaining rapid convergence and is not necessary. Hence, calculations based on only $\mathcal{L}^2$ basis functions $f_i(r)$ of Eq. (3.75) or (3.76) have produced good convergence. The inclusion of nonnormalizable functions (3.73) and (3.74) in the basis set should improve convergence. In Eq. (4.24) the trial functions are always multiplied by the potential. Thus, for a very short-ranged potential, the long-range asymptotic behavior of the trial function is suppressed, and only the short-range behavior of the trial function is important. Therefore, for very short-range potentials, there is no real advantage in using oscillating trial functions.

For short-ranged potentials, the exact functions of Eqs. (4.10) and (4.11) are $\mathcal{L}^2$. Hence, in Eq. (4.37), the trial functions should be chosen to be $\mathcal{L}^2$. There are two functions in variational principles (4.12) to (4.16). The exact function ψ is nonnormalizable, whereas χ is $\mathcal{L}^2$. For a numerical calculation, both these functions can be taken to be $\mathcal{L}^2$. However, for good convergence, approximate ψ should have oscillating behavior at ∞, whereas approximate χ should be taken to be $\mathcal{L}^2$.

The Schwinger variational method (4.37) has also been used successfully in the numerical calculation for local potentials in configuration space [11]. However, in this form there is an integral over the inverse potential operator V^{-1}, which may not be defined for all momentum and configuration space variables. This may not be of concern. What one needs in this approach is not the operator V^{-1}, but the matrix elements $\langle v_j|V^{-1}|v_i\rangle$, which could be finite for appropriate trial functions even when the operator V^{-1} diverges. For example, for a Yukawa potential $V(r) \sim \exp(-\mu r)/r$, and consequently, $V^{-1}(r) \sim r\exp(\mu r)$, so that $V^{-1}(r)$ diverges as $r \to \infty$. This divergence can be tackled by the following $\mathcal{L}^2$ functions: $v_j(r) \equiv \langle r|v_j\rangle = \langle v_j|r\rangle = \exp(-j\alpha r), j = 1, 2, ..., n$ with $\alpha > \mu/2$. Then the matrix element $\langle v_j|V^{-1}|v_i\rangle$ is finite. In this case the basis set $v_j(r)$ is supposed to be a good representation of the function $\langle r|V|\psi_k^{(+)}\rangle$, which is $\mathcal{L}^2$ for short-range potentials. The use of $\mathcal{L}^2$ basis functions (3.75) or (3.76) in Eq. (4.37) should produce good convergence. Equation (4.37) has rarely been used in numerical calculations. As this variational method involves simpler integrals than the usual Schwinger variational method (4.24), it should receive more attention in the future. However, most of the commonly used potentials in nuclear and atomic physics are nonlocal and have a repulsive core at short distances and a long-range attraction. For such potentials V^{-1} is singular at a finite r, and special care in the choice of the basis functions is needed in using (4.37) than was necessary in the simple application with Yukawa or exponential potentials.

4.2.3 Application to Nuclear Problems

The Schwinger variational method has been used extensively in few-nucleon problems for expressing the resultant t matrix in finite-rank forms. The use of a finite-rank two-nucleon t matrix in the three-nucleon problem reduces the latter to an effective two-body problem [14,15]. The use of finite-rank two- and three-nucleon t matrices in the four-nucleon problem also reduces the latter to an effective two-body problem. In the remaining of this section we describe the choices of basis functions used for expressing the few-nucleon t matrix in finite-rank or separable forms. The two-nucleon t matrix has a pole at energy corresponding to a two-nucleon bound or virtual state. The momenta and energy dependencies of the low-energy two-nucleon t-matrix elements are usually dominated by the presence of this pole. Once this pole is reproduced in an approximation to the two-nucleon t matrix, this approximation becomes a good one at all energies. Similar considerations apply to the three-nucleon t matrix also. However, in problems of atomic physics, one usually has several bound and/or virtual states. In these problems a low-rank separable expansion has not provided a faithful reproduction of the actual t matrix. We describe below several choices for the expansion functions that have been used successfully in nuclear few-body scattering problems [14,15].

Unitary Pole Expansion (UPE). One way of making the basis functions linearly independent and orthonormal, as one should in a numerical calculation, is to take them to be the eigenfunctions of an auxiliary eigenfunction—eigenvalue problem. The UPE corresponds to the following energy-independent choice for basis functions [16]:

$$V|f_i\rangle = |\psi_i(-B)\rangle, \tag{4.42}$$

$$\langle f_j|V = \langle\bar{\psi}_j(-B)|, \tag{4.43}$$

to be used in Eq. (4.24). Here, $|\psi_i(-B)\rangle$ and $\langle\bar{\psi}_j(-B)|$ are defined by

$$|\psi_i(-B)\rangle = \lambda_i(-B)VG_0^{(+)}(-B)|\psi_i(-B)\rangle, \tag{4.44}$$

$$\langle\bar{\psi}_j(-B)| = \lambda_j(-B)\langle\bar{\psi}_j(-B)|G_0^{(+)}(-B)V, \tag{4.45}$$

and are normalized according to

$$\langle\bar{\psi}_j(-B)|G_0^{(+)}(-B)|\psi_i(-B)\rangle = \delta_{ij}, \qquad i,j = 1, ..., n. \tag{4.46}$$

Typically, B is the ground-state binding energy. With this choice of the expansion functions the UPE for the t matrix can explicitly be written as

$$[t_n] = \sum_{i,j=1}^{n} |\psi_i(-B)\rangle J_{ij}^{(n)}\langle\bar{\psi}_j(-B)|, \tag{4.47}$$

where

$$[J^{-1}]_{ji} = \lambda_j(-B)\delta_{ji} - \langle\bar{\psi}_j(-B)|G_0^{(+)}(E)|\psi_i(-B)\rangle. \tag{4.48}$$

The function $|\psi_i(-B)\rangle$ with eigenvalue $\lambda_i(-B) = 1$ corresponds to the bound state so that $|\psi_i(-B)\rangle$ is the bound-state wave function at energy $E = -B$. If this function is included in UPE (4.47), the UPE t matrix builds in the correct bound-state pole and the residue. If there is no bound state, B could be taken to be zero. With the choice $B = 0$, the UPE builds in the correct t matrix at zero energy. The UPE uses energy-independent functions, satisfies unitarity, and has been frequently used in the solution of the two-, three-, and four-nucleon problems.

Hilbert–Schmidt Expansion (HSE). The UPE is the special case of a general expansion scheme called HSE, which also uses the eigenfunctions of a related eigenfunction–eigenvalue problem. The HSE corresponds to the energy-dependent choice for the basis functions [17]

$$V|f_i\rangle = |\psi_i(E)\rangle, \tag{4.49}$$

$$\langle f_j|V = \langle\bar{\psi}_j(E)|, \tag{4.50}$$

to be used in Eq. (4.24). Here $|\psi_i(E)\rangle$ and $\langle\bar{\psi}_j(E)|$ are defined by

$$|\psi_i(E)\rangle = \lambda_i(E)VG_0^{(+)}(E)|\psi_i(E)\rangle, \tag{4.51}$$

$$\langle\bar{\psi}_j(E)| = \lambda_j(E)\langle\bar{\psi}_j(E)|G_0^{(+)}(E)V, \tag{4.52}$$

and are normalized according to

$$\langle\bar{\psi}_j(E)|G_0^{(+)}(E)|\psi_i(E)\rangle = \delta_{ij}, \qquad i,j = 1,...,n. \tag{4.53}$$

With this choice of expansion functions, the HSE for the t matrix can be written as

$$[t_n] = \sum_{i=1}^{n} \frac{|\psi_i(E)\rangle\langle\bar{\psi}_i(E)|}{\lambda_i(E) - 1} \tag{4.54}$$

At $E = -B$, the HSE [17] and UPE [16] are identical. The HSE uses energy-dependent trial functions to be found by solving a eigenfunction–eigenvalue problem at each energy under consideration, and hence is numerically more complicated than the UPE from a practical point of view. The HSE has been used in the solution of the few-nucleon problem [15].

Bateman Expansion. The Bateman expansion employs in Eq. (4.24) the following set of basis functions:

$$|f_i\rangle = |p_i\rangle, \qquad i = 1,...,n, \tag{4.55}$$

where $|p_i\rangle$ are the momentum eigenfunctions of the free Hamiltonian and satisfy $H_0|p_i\rangle = E_i|p_i\rangle, E_i = p_i^2$. With this choice, the finite-rank potential of Eq. (4.25) becomes exact at the n preselected momenta $p_i, i = 1,...,n$:

$$V_n|p_i\rangle = V|p_i\rangle; \qquad \langle p_i|V_n = \langle p_i|V. \tag{4.56}$$

If these momenta are cleverly chosen, the finite-rank potential should be a good approximation to the exact potential, and a similar quality is expected of the finite-rank t matrix of Eq. (4.24). This choice is simple to use, satisfies unitarity, has acceptable convergence properties, and has been employed in solving few-nucleon problems.

Ernst-Shakin-Thaler Expansion (ESTE). Here one takes the $|f_i\rangle$'s of Eq. (4.24) to be the eigenstates (bound or continuum) of the full Hamiltonian [18]. For a system with a single bound state, typically, one takes the first expansion function $|f_1\rangle = |B\rangle$, the ground-state wave function, and the remaining expansion functions to be scattering wave function at several energies, E_i: $|f_i\rangle = |\psi^{(+)}(E_i)\rangle$. The resultant t matrix has the correct bound-state pole position and residue. In this expansion scheme one has the interesting properties

$$t(E_i)|k_i\rangle = t_n(E_i)|k_i\rangle; \qquad \langle k_i|t(E_i) = \langle k_i|t_n(E_i) \qquad E_i = k_i^2. \quad (4.57)$$

The ESTE t matrix is exact half-on-shell at these preselected energies, satisfies unitarity, and converges rapidly. This method has been used successfully for solving the realistic three-nucleon problem employing modern meson-theoretic potentials [15].

Analytic Separable Expansions (ASE). This method uses a set of analytic or quasianalytic basis functions $|f_i\rangle$'s in Eq. (4.24) and has certain advantages [4]. If these functions are cleverly chosen, the method produces rapid convergence. One does not have to evaluate the expansion functions numerically as in the UPE, HSE, and ESTE. The use of such simple analytic functions facilitates the numerical task further for the solution of three-particle scattering problem by the method of contour rotation. In application of the ESTE, Plessas and Haidenbauer [19] approximate the expansion functions by simple analytic form factors. Thus, they combine the virtues of both the ASE and ESTE.

Separable Expansion for Energy-Dependent Potentials. The foregoing discussion for the implementation of the Schwinger variational principle or the method of separable expansion is limited to energy-independent Hermitian potentials. In many situations of physical interest, the potential operator depends on the parametric energy, is complex, and is not Hermitian. This happens when some energetically open channels are suppressed in the process of deriving an effective interaction between composite objects. Some examples are meson-theoretic nucleon–nucleon potentials with mesonic degrees of freedom suppressed above the meson production threshold, or neutron–deuteron scattering above the deuteron breakup threshold.

In such cases it is not clear if it is possible to implement useful methods, such as UPE or ESTE, with energy-independent form factors [20]. However, it would be extremely useful if separable expansion with energy-independent form factors could be derived for energy-dependent complex potentials. This can actually be implemented for a class of energy-dependent potentials. Generalizations of the UPE and EST expansion techniques have been made to generate separable expansions with energy-independent form factors for energy-dependent potentials in the three-nucleon system [21].

4.3 NEWTON VARIATIONAL PRINCIPLE

Another class of variational principles for the t or the K matrix involving the Green's function can be established as in Chapter 3.1. One of them, the Newton variational principle, is based on the operator identity [2,5]

$$t = V + VG_0^{(+)}t_1 + t_2G_0^{(+)}V - t_2(G_0^{(+)} - G_0^{(+)}VG_0^{(+)})t_1 + (t_2 - t)(G_0^{(+)} - G_0^{(+)}VG_0^{(+)})(t_1 - t), \tag{4.58}$$

valid for arbitrary operators t_1 and t_2. We shall be interested in taking t_1 and t_2 to be independent approximations to the t matrix. Then the last term in Eq. (4.58) is of second order in deviations $t_1 - t$ and $t_2 - t$. The Newton variational principle for the t matrix is obtained by omitting the last term in Eq. (4.58) and rewriting it as

$$[t - V] = VG_0^{(+)}t_1 + t_2G_0^{(+)}V - t_2(G_0^{(+)} - G_0^{(+)}VG_0^{(+)})t_1. \tag{4.59}$$

The quantity V has been taken to the left-hand side in Eq. (4.59) to give this equation the same trilinear form as other variational principles. If exact t matrices are used in place of t_1 and t_2, the three terms on the right-hand side are each equal to $t - V$ and Eq. (4.59) is an identity. This is why the Newton variational principle is a variational principle for $t - V$ rather than for t. This variational principle is of special advantage when $t - V$ is small compared to t or V or when the Born approximation is a good approximation to the t matrix.

The momentum-space matrix element of Eq. (4.59) is given by [2]

$$[\langle p'|t|p\rangle] = \langle p'|V|p\rangle + \langle p'|VG_0^{(+)}|\chi_p^{(+)}\rangle + \langle\chi_{p'}^{(-)}|G_0^{(+)}V|p\rangle$$
$$- \langle\chi_{p'}^{(-)}|(G_0^{(+)} - G_0^{(+)}VG_0^{(+)})|\chi_p^{(+)}\rangle, \qquad (4.60)$$

where for simplicity we have used $t_1 = t_2 = t_A$. In Eq. (4.60), amplitude densities $|\chi_p^{(+)}\rangle \equiv t_A|p\rangle$ and $\langle\chi_{p'}^{(-)}| \equiv \langle p'|t_A$ are the trial functions. The exact functions are defined in Eqs. (4.10) and (4.11). For short-range potentials, the exact functions are $\mathcal{L}^2$. Hence, the trial functions of the Newton variational principle should also be $\mathcal{L}^2$. This is an advantage of the Newton variational principle over the Schwinger and Kohn variational principles in numerical application. In Kohn variational methods, the basis functions must be nonnormalizable, consistent with the appropriate scattering boundary conditions. For obtaining good convergence with the Schwinger variational method, non-$\mathcal{L}^2$ oscillating functions should also be included in the basis set. Numerically, it is tedious to perform integrals over such functions.

Variational principle (4.60) possesses integrals over single and dual Green's functions. Such integrals are difficult to deal with in the configuration space and are usually handled in the momentum space. One can dispense with integrals over the Green's function by considering the new trial functions

$$|\eta_p^{(+)}\rangle \equiv G_0^{(+)}V|\psi_p^{(+)}\rangle \equiv G_0^{(+)}|\chi_p^{(+)}\rangle = |\psi_p^{(+)}\rangle - |\phi_p\rangle, \qquad (4.61)$$

$$\langle\eta_{p'}^{(-)}| \equiv \langle\psi_{p'}^{(-)}|VG_0^{(+)} \equiv \langle\chi_{p'}^{(-)}|G_0^{(+)} = \langle\psi_p^{(-)}| - \langle\phi_p|. \qquad (4.62)$$

Then variational principle (4.60) becomes

$$[\langle p'|t|p\rangle] = \langle p'|V|p\rangle + \langle p'|V|\eta_p^{(+)}\rangle + \langle\eta_{p'}^{(-)}|V|p\rangle$$
$$- \langle\eta_{p'}^{(-)}|(E - H_0 - V)|\eta_p^{(+)}\rangle. \qquad (4.63)$$

The functions $|\eta_p^{(+)}\rangle$ and $\langle\eta_{p'}^{(-)}|$ are, respectively, the scattered waves for outgoing- and incoming-wave boundary conditions. Variational principle (4.63) does not possess any Green's function and hence requires oscillating trial functions satisfying outgoing-wave asymptotic boundary conditions. This variational principle will be identified as the complex Kohn variational principle for the t matrix. If standing-wave scattered waves are used in Eq. (4.63), the standing-wave real Kohn variational principle for the K matrix will be obtained.

Variational principles (4.60) and (4.63), respectively, can also be written in the following equivalent fractional forms:

$$[\langle p'|t|p\rangle] = \langle p'|V|p\rangle + \frac{\langle p'|VG_0^{(+)}|\chi_p^{(+)}\rangle\langle\chi_{p'}^{(-)}|G_0^{(+)}V|p\rangle}{\langle\chi_{p'}^{(-)}|(G_0^{(+)} - G_0^{(+)}VG_0^{(+)})|\chi_p^{(+)}\rangle}, \tag{4.64}$$

$$[\langle p'|t|p\rangle] = \langle p'|V|p\rangle + \frac{\langle p'|V|\eta_p^{(+)}\rangle\langle\eta_{p'}^{(-)}|V|p\rangle}{\langle\eta_{p'}^{(-)}|(E - H_0 - V)|\eta_p^{(+)}\rangle}. \tag{4.65}$$

To transform variational principle (4.60) into finite-rank form, we expand the trial function (4.30) and its conjugate into $\mathcal{L}^2$ basis sets:

$$|\chi_p^{(+)}\rangle = \sum_{i=1}^{n} a_i(p)|f_i\rangle, \tag{4.66}$$

$$\langle\chi_{p'}^{(-)}| = \sum_{j=1}^{n} b_j(p')\langle f_j|. \tag{4.67}$$

Next we substitute these expansions in Eq. (4.60) and require that the resultant expression be stationary with respect to independent variations of a_i and b_j. Then we obtain

$$b_j(p') = \sum_{i=1}^{n}\langle p'|VG_0^{(+)}|f_i\rangle P_{ij}^{(n)}, \tag{4.68}$$

$$a_i(p) = \sum_{j=1}^{n} P_{ij}^{(n)}\langle f_j|G_0^{(+)}V|p\rangle, \tag{4.69}$$

where

$$(P^{-1})_{ji} = \langle f_j|(G_0^{(+)} - G_0^{(+)}VG_0^{(+)})|f_i\rangle, \qquad i, j = 1, \ldots, n. \tag{4.70}$$

If we use expansion coefficients (4.68) and (4.69), the expansion functions defined by Eqs. (4.66) and (4.67) become

$$|\chi_p^{(+)}\rangle = \sum_{i,j=1}^{n} |f_i\rangle P_{ij}^{(n)} \langle f_j|G_0^{(+)}V|p\rangle, \tag{4.71}$$

$$\langle \chi_{p'}^{(-)}| = \sum_{i,j=1}^{n} \langle p'|VG_0^{(+)}|f_i\rangle P_{ij}^{(n)} \langle f_j|. \tag{4.72}$$

Finally, we substitute Eqs. (4.71) and (4.72) into Eq. (4.60) to obtain

$$[\langle p'|t_n|p\rangle] = \langle p'|V|p\rangle + \sum_{i,j=1}^{\infty} \langle p'|VG_0^{(+)}|f_i\rangle P_{ij}^{(n)} \langle f_j|G_0^{(+)}V|p\rangle. \tag{4.73}$$

This result is equivalent to one of the degenerate-kernel schemes mentioned in relation to the discussion of Fredholm integral equations [i.e., Eqs. (3.37), (3.38), and (3.35)]. For a numerical evaluation of the t matrix using Eqs. (4.73) and (4.70), it is of advantage to work in momentum space and to calculate and store the elements $\langle p'|VG_0^{(+)}|f_i\rangle$. Then the calculation of the P^{-1} matrix of (4.70) involves a single integral and the evaluation of a double integral over the Green's function can be avoided.

The versions of the Newton variational principle discussed above are time-reversal symmetric. Equivalent nonsymmetric forms of the Newton variational principle can be formulated. In such nonsymmetric forms, the matrix to be inverted is simpler and involves a single Green's function. The asymmetric form of Newton variational principle (4.60) can be written as

$$[\langle p'|t|p\rangle] = \langle p'|V|p\rangle + \langle p'|V|\eta_p^{(+)}\rangle + \langle \chi_{p'}^{(-)}|G_0^{(+)}V|p\rangle$$
$$- \langle \chi_{p'}^{(-)}|(I - G_0^{(+)}V)|\eta_p^{(+)}\rangle, \tag{4.74}$$

where $|\eta_p^{(+)}\rangle = G_0^{(+)}|\chi_p^{(+)}\rangle$. The function $|\eta_p^{(+)}\rangle$ is not $\mathcal{L}^2$, whereas the function $\langle \chi_{p'}^{(-)}|$ is $\mathcal{L}^2$. In a numerical implementation both these trial functions can be taken as $\mathcal{L}^2$ in (4.74). But for obtaining rapid convergence the appropriate oscillating boundary condition should be included in trial $|\eta_p^{(+)}\rangle$.

Variational principle (4.74) can also be written in the following degenerate kernel form in the usual fashion by expanding the trial functions in appropriate basis sets:

$$[\langle p'|t|p\rangle] = \langle p'|V|p\rangle + \sum_{i,j=1}^{n} \langle p'|V|f_i\rangle C_{ij}^{(n)} \langle f_j|G_0^{(+)}V|p\rangle, \qquad (4.75)$$

where C^{-1} is defined by Eq. (4.33) with $v_j = f_j$. As variational principle (4.74) is nonsymmetric, two different basis sets have been used in the expansion. A transposed form can also be used in place of (4.75).

We have considered several forms of the Newton variational principle. These variational principles for the t matrix—time-reversal symmetric or not—do not satisfy unitarity. Even for real Hermitian potentials they do not produce real phase shifts for real trial functions. Distinct K-matrix versions of these variational principles are obtained by replacing the outgoing-wave Green's function by the standing-wave Green's function. If K-matrix versions of the Newton variational principle are considered, unitarity is automatically guaranteed and the consequent phase shifts are real. The magnitude of the imaginary parts of the phase shifts of a real Hermitian potential in a numerical application of the t-matrix Newton variational method should yield a measure on the rate of convergence.

4.4 ITERATIVE VARIATIONAL SCHEMES

In the usual implementation of variational principles, some unknown trial function is expanded in a set of functions involving some arbitrary parameter(s). This scheme reduces the variational principle to a set of algebraic equations. The success of such a method depends on an appropriate choice of the basis functions and the parameter(s) involved. However, in practice, it is seldom known how to make the ideal choice in a realistic problem. This is responsible for the slow convergence obtained in many applications of variational methods.

We consider an iterative method that can improve the convergence properties [22]. The method makes use of a variational principle for the t (K)-matrix elements, where the trial functions are also t (K)-matrix elements. We illustrate the method with two variational principles for the t matrix. However, a similar version exists for the K matrix with the outgoing-wave free Green's function replaced by the standing-wave free Green's function. A very simple trial function (e.g., $t = 0$) can be used in the variational principle. The resultant t-

matrix elements are then substituted in the variational principle again for calculating an improved t matrix. This procedure is repeated until a numerical result of the desired precision is obtained. In this approach there are no basis functions with free parameters, as in other applications of variational methods. This scheme leads to quick convergence and high-precision results for many potentials.

For the on-shell t-matrix elements, the following variational principles can be used in iterative study [22]:

$$[\langle k|t|k\rangle] = \frac{\langle k|V(I+G_0^{(+)}t)|k\rangle\langle k|(I+tG_0^{(+)})V|k\rangle}{\langle k|(I+tG_0^{(+)})(V-VG_0^{(+)}V)(I+G_0^{(+)}t)|k\rangle}, \tag{4.76}$$

$$[\langle k|t|k\rangle] = \langle k|V|k\rangle + \frac{\langle k|VG_0^{(+)}t|k\rangle\langle k|tG_0^{(+)}V|k\rangle}{\langle k|t(G_0^{(+)}-G_0^{(+)}VG_0^{(+)})t|k\rangle}. \tag{4.77}$$

Equation (4.76) is the fractional form of the alternative Schwinger variational principle (4.8), and Eq. (4.77) is the fractional form of the Newton variational principle (4.59). The half-shell t operators on the right-hand side of Eqs. (4.76) and (4.77) are trial inputs for calculating the on-shell t-matrix elements. If trial t matrices with small errors on the order of Δ are used in these variational principles, the resulting on-shell t matrices have smaller errors, on the order of Δ^2. Equations (4.76) and (4.77) can be used as the starting point of an iterative calculational scheme, which proceeds by substituting successively the improved t matrix on the right-hand side of these equations.

The t matrices on the left-hand side of Eqs. (4.76) and (4.77) are on-shell t-matrix elements, whereas the t operators that appear on the right-hand side are half-shell. For the iterative scheme to function, estimates of half-shell t matrices are needed. Variational principles can be written for the half-shell t-matrix elements in terms of the off-shell t-matrix elements. For example, a generalization of Eq. (4.76) to half-shell t-matrix elements is given by

$$[\langle p|t|k\rangle] = \frac{\langle p|V(I+G_0^{(+)}t)|k\rangle\langle p|(I+tG_0^{(+)})V|k\rangle}{\langle p|(I+tG_0^{(+)})(V-VG_0^{(+)}V)(I+G_0^{(+)}t)|k\rangle}. \tag{4.78}$$

This variational principle is quite involved in requiring the off-shell t-matrix elements as input for calculating the half-shell t-matrix elements variationally.

Instead of using the variational half-shell t-matrix elements of (4.78), the following nonvariational half-shell extension of stationary expressions (4.76) and (4.77) can be used:

$$\langle p|t|k\rangle = \frac{\langle p|V(I+G_0^{(+)}t)|k\rangle\langle k|(I+tG_0^{(+)})V|k\rangle}{\langle k|(I+tG_0^{(+)})(V-VG_0^{(+)}V)(I+G_0^{(+)}t)|k\rangle}, \tag{4.79}$$

$$\langle p|t|k\rangle = \langle p|V|k\rangle + \frac{\langle p|VG_0^{(+)}t|k\rangle\langle k|tG_0^{(+)}V|k\rangle}{\langle k|t(G_0^{(+)}-G_0^{(+)}VG_0^{(+)})t|k\rangle}. \tag{4.80}$$

Expressions (4.79) and (4.80) are stationary for the on-shell t-matrix elements. However, the nonvariational half-shell t matrices obtained from these expressions are expected to be good approximations to the exact elements and can be used in an iterative study.

A useful iterative scheme for scattering problems can be developed from Eqs. (4.79) and (4.80). The iterative method could be started with any initial guess for the t matrix. For example, the lowest (zeroth)-order t matrix, t_0, can be generated if on the right-hand side of Eqs. (4.79) and (4.80) one uses $t=0$. The first-order t matrix, t_1, is then generated if on the right-hand side of these equations one uses $t=t_0$, and so on. This procedure should be repeated until a result of the desired precision is obtained. This nonvariational iterative scheme is expected to converge unless the potential is too strong. In numerical application high-precision results have been obtained for several local potentials with a bound state.

Usually, the inversion of a matrix as encountered in variational basis-set methods involves accumulation of numerical errors. In the present iterative scheme, inversion of a matrix is avoided and this improves significantly both the precision and the rate of convergence.

Although there are multiple Green's functions on the right-hand sides of Eqs. (4.79) and (4.80), evaluation of the numerators of the fractions in these equations involves a matrix vector multiplication, or equivalently, the calculation of N integrals, where N is the number of off-shell points. Once these integrals are stored, calculation of the denominators of these fractions requires the evaluation of one more integral. Thus, the numerical task for obtaining one iteration of the variational principle is equivalent to calculating a single iteration of the half-shell Lippmann–Schwinger equation. The procedure is much more complicated if a fully variational iterative calculation using Eq. (4.78) is performed.

4.5 VARIATIONAL PRINCIPLES OF HIGHER ORDER

The Schwinger and the Kohn–Newton variational principles refer to the matrix elements of t and $t - V$, or equivalently, of K and $K - V$, respectively. Evidently, one can continue this progression and construct variational principles for t or K minus the first n terms of the Born series, $n = 2,3,\ldots$. The next possibility in this progression is the variational principle for $t - V - VG_0^{(+)}V$. It is of advantage to pursue this procedure for weak potentials and at intermediate and higher energies. In both cases the Born series presumably converges and the first few terms of this series present a good approximation to the exact solution. Then the difference of the type $t - V - VG_0^{(+)}V$ is a small quantity. Under such a situation the exact treatment of the first few terms of the Born series and a variational treatment of the small difference should lead to a good approximation to the t or K matrix.

To derive a variational principle for the difference $t - V - VG_0^{(+)}V$, we introduce the functions

$$|\hat{\psi}_p^{(+)}\rangle = VG_0^{(+)}V|\psi_p^{(+)}\rangle, \tag{4.81}$$

$$\langle\hat{\psi}_{p'}^{(-)}| = \langle\psi_{p'}^{(-)}|VG_0^{(+)}V. \tag{4.82}$$

The iterated forms of Lippmann–Schwinger equations (1.93) and (1.94) can be written as

$$t = V + VG_0^{(+)}V + VG_0^{(+)}VG_0^{(+)}t = V + VG_0^{(+)}V + tG_0^{(+)}VG_0^{(+)}V. \tag{4.83}$$

In terms of functions (4.81) and (4.82), the following identities can be obtained from the momentum-space matrix elements of Eq. (4.83):

$$\langle p'|t|p\rangle = \langle p'|(V + VG_0^{(+)}V)|p\rangle + \langle p'|VG_0^{(+)}|\hat{\psi}_p^{(+)}\rangle \tag{4.84}$$

$$= \langle p'|(V + VG_0^{(+)}V)|p\rangle + \langle\hat{\psi}_{p'}^{(-)}|G_0^{(+)}V|p\rangle \tag{4.85}$$

$$= \langle p'|(V + VG_0^{(+)}V)|p\rangle + \langle\hat{\psi}_{p'}^{(-)}|(V^{-1} - G_0^{(+)})|\hat{\psi}_p^{(+)}\rangle. \tag{4.86}$$

Equation (4.86) follows from a definition of the quantities involved. Combining the three expressions above for the t matrix, the following trilinear expression is obtained in the usual fashion:

$$[\langle p'|t|p\rangle] = \langle p'|(V + VG_0^{(+)}V)|p\rangle + \langle p'|VG_0^{(+)}|\hat{\psi}_p^{(+)}\rangle$$
$$+ \langle \hat{\psi}_{p'}^{(-)}|G_0^{(+)}V|p\rangle$$
$$- \langle \hat{\psi}_{p'}^{(-)}|(V^{-1} - G_0^{(+)})|\hat{\psi}_p^{(+)}\rangle. \tag{4.87}$$

Expression (4.87) is stationary with respect to variations of the functions $|\hat{\psi}_p^{(+)}\rangle$ and $\langle\hat{\psi}_{p'}^{(-)}|$ and is a variational principle for $t - V - VG_0^{(+)}V$.

A variational basis-set method can be formulated from the variational principle (4.87) by expanding the trial functions $|\hat{\psi}_p^{(+)}\rangle$ and $\langle\hat{\psi}_{p'}^{(-)}|$ in basis sets. The functions $|\hat{\psi}_p^{(+)}\rangle$ and $\langle\hat{\psi}_{p'}^{(-)}|$ defined by (4.81) and (4.82) are $\mathcal{L}^2$; hence $\mathcal{L}^2$ basis functions should produce rapid convergence. If the trial functions in Eq. (4.87) are expanded in a basis set consisting of n $\mathcal{L}^2$ functions $|v_i\rangle, i = 1, 2, ..., n$, and their transpose, we obtain

$$[\langle p'|t|p\rangle] = \langle p'|(V + VG_0^{(+)}V)|p\rangle$$
$$+ \sum_{i,j=1}^{n} \langle p'|VG_0^{(+)}|v_i\rangle U_{ij}^{(n)} \langle v_j|G_0^{(+)}V|p\rangle, \tag{4.88}$$

where U^{-1} is defined by Eq. (4.38).

Another form of the variational principle for the difference $t - V - VG_0^{(+)}V$ can be obtained by using the scattered waves $|\eta_p^{(+)}\rangle$ and $\langle\eta_{p'}^{(-)}|$ of Eqs. (4.61) and (4.62), respectively. In terms of functions of Eqs. (4.81) and (4.82), these functions can be expressed as

$$|\hat{\psi}_p^{(+)}\rangle = V|\eta_p^{(+)}\rangle, \tag{4.89}$$

$$\langle\hat{\psi}_{p'}^{(-)}| = \langle\eta_{p'}^{(-)}|V. \tag{4.90}$$

Using Eqs. (4.89) and (4.90), the variational principle (4.87) can be rewritten as

$$[\langle p'|t|p\rangle] = \langle p'|(V + VG_0^{(+)}V)|p\rangle + \langle p'|VG_0^{(+)}V|\eta_p^{(+)}\rangle$$
$$+ \langle\eta_{p'}^{(-)}|VG_0^{(+)}V|p\rangle - \langle\eta_{p'}^{(-)}|(V - VG_0^{(+)}V)|\eta_p^{(+)}\rangle. \tag{4.91}$$

In this case the exact scattered waves $|\eta_p^{(+)}\rangle$ and $\langle\eta_{p'}^{(-)}|$ are not $\mathcal{L}^2$. Variational principle (4.91) involves the free Green's function, and hence $\mathcal{L}^2$ trial functions can be used in it. However, as the exact functions are not $\mathcal{L}^2$, for obtaining rapid convergence non-$\mathcal{L}^2$ sine and cosine functions should be included in addition to the $\mathcal{L}^2$ basis functions.

Variational principle (4.91) is particularly useful for numerical calculation. If we expand the trial functions $|\eta_p^{(+)}\rangle$ and $\langle\eta_{p'}^{(-)}|$ in basis sets, variational principle (4.91) yields the following convenient degenerate kernel form:

$$[\langle p'|t_n|p\rangle] = \langle p'|(V + VG_0^{(+)}V)|p\rangle + \sum_{i,j=1}^{n} \langle p'|VG_0^{(+)}V|f_i\rangle D_{ij}^{(n)} \langle f_j|VG_0^{(+)}V|p\rangle, \tag{4.92}$$

where the matrix D^{-1} is defined by Eq. (4.21).

The procedure can be continued and the next variational principle in this progression can be obtained in terms of the functions (4.81) and (4.82). Using iterations for the t-matrix Lippmann–Schwinger equations (1.93) and (1.94),

$$t = V + VG_0^{(+)}V + VG_0^{(+)}VG_0^{(+)}V + VG_0^{(+)}VG_0^{(+)}VG_0^{(+)}t \tag{4.93}$$

$$= V + VG_0^{(+)}V + VG_0^{(+)}VG_0^{(+)}V + tG_0^{(+)}VG_0^{(+)}VG_0^{(+)}V, \tag{4.94}$$

and the functions (4.81) and (4.82), we obtain the following variational principle:

$$\begin{aligned}[\langle p'|t|p\rangle] = {} & \langle p'|(V + VG_0^{(+)}V + VG_0^{(+)}VG_0^{(+)}V)|p\rangle \\ & + \langle\hat{\psi}_{p'}^{(-)}|G_0^{(+)}VG_0^{(+)}V|p\rangle \\ & + \langle p'|VG_0^{(+)}VG_0^{(+)}|\hat{\psi}_p^{(+)}\rangle \\ & - \langle\hat{\psi}_{p'}^{(-)}|(G_0^{(+)} - G_0^{(+)}VG_0^{(+)})|\hat{\psi}_p^{(+)}\rangle. \end{aligned} \tag{4.95}$$

The stationary property of expression (4.95) with respect to variations of trial functions (4.81) and (4.82) can be established by using the definition of the functions involved. Variational principle (4.95) has the form of the usual trilinear combination and is a variational principle for $\langle p'|t - (V + VG_0^{(+)}V + VG_0^{(+)}VG_0^{(+)}V)|p\rangle$.

If trial functions in variational principle (4.95) are expanded in terms of a basis set consisting of n $\mathcal{L}^2$ functions $|f_i\rangle, i = 1, 2, ..., n$, and their transpose, this variational principle can be rewritten in the following degenerate kernel form:

$$[\langle p'|t_n|p\rangle] = \langle p'|(V + VG_0^{(+)}V + VG_0^{(+)}VG_0^{(+)}V)|p\rangle$$
$$+ \sum_{i,j=1}^{n} \langle p'|VG_0^{(+)}VG_0^{(+)}|f_i\rangle P_{ij}^{(n)} \langle f_j|G_0^{(+)}VG_0^{(+)}V|p\rangle, \quad (4.96)$$

where P^{-1} is defined by Eq. (4.70). The trial functions (4.81) and (4.82) in this case are $\mathcal{L}^2$ and the basis functions should also be $\mathcal{L}^2$.

This procedure can be continued and one can write variational principles for higher-order corrections to the Born–Neumann series. In each order more than one variational principle can be written. For example, distinct variational principles (4.87) and (4.91) refer to the same quantity $t - V - VG_0^{(+)}V$.

The variational principles of this section can be reformulated for the K-matrix elements by replacing the outgoing-wave Green's function by the principal-value Green's function. The phase shifts obtained from the K-matrix version are real. The t-matrix version of the variational principle does not satisfy unitarity and leads to complex phase shifts even for real Hermitian potentials.

4.6 COMPLEX AND REAL KOHN VARIATIONAL PRINCIPLES

The usual Kohn variational basis-set method for calculating the on-shell K-matrix elements has the remarkable advantage of dealing with simple real integrals, but it is often plagued by the presence of spurious singularities. Here we present a slightly modified version of this method that can be used for calculating the off-shell amplitudes. Such a method, employing boundary conditions for the outgoing-wave (standing-wave) Green's function can be used for calculating the off-shell t (K)-matrix elements, which maintains most of the simple features of the usual Kohn variational principle [6]. In the modified version for the t matrix the spurious singularities become rare. The modified t (K)-matrix variational principle, called the *complex* (*real*) *Kohn variational principle*, deals with complex (real) algebra.

The Kohn variational principle is based on the stationary expression (4.63) for the t matrix or the following stationary expression for the K matrix [6]:

$$[\langle p'|K|p\rangle] = \langle p'|V|p\rangle + \langle p'|V|\eta_p^{\mathcal{P}}\rangle + \langle \eta_{p'}^{\mathcal{P}}|V|p\rangle - \langle \eta_{p'}^{\mathcal{P}}|(E - H_0 - V)|\eta_p^{\mathcal{P}}\rangle, \qquad (4.97)$$

where the function $\eta_p^{\mathcal{P}}$ of Eq. (4.97) is the scattered wave with standing-wave boundary condition. Like the Newton variational principle, Kohn variational principle (4.97) can be considered to be a variational principle for $K - V$.

In the usual fashion, from Eq. (4.63) or (4.97) the following rank-n form for the difference $t - V$ or $K - V$ could be arrived at by expanding the trial functions $|\eta_p\rangle$ and $\langle \eta_{p'}|$ in basis sets and exploiting the stationary property

$$[\langle p'|\mathcal{O}_n|p\rangle] = \langle p'|V|p\rangle + \sum_{i,j=1}^{n} \langle p'|V|f_i\rangle Z_{ij}^{(n)} \langle f_j|V|p\rangle, \qquad (4.98)$$

with

$$(Z^{-1})_{ji} = \langle f_j|(E - H)|f_i\rangle, \qquad i, j = 1, ..., n, \qquad (4.99)$$

where $\mathcal{O}$ stands for either the t or the K matrix. From Eq. (4.97) an identical rank-n expression results for the K matrix. But different basis functions are to be used for the t- or the K-matrix elements. Although the expansion functions of Eq. (4.73) can be $\mathcal{L}^2$, Eq. (4.98) cannot be used in its most simple interpretation, namely, with a set of $\mathcal{L}^2$ basis functions. The Green's function does not explicitly appear in Eq. (4.98). Hence, if a finite $\mathcal{L}^2$ basis set is used in this equation, the asymptotic boundary condition of scattering will not be included. In order that the approximate solution of Eq. (4.98) satisfy appropriate boundary condition, nonnormalizable basis functions consistent with proper asymptotic boundary condition are to be chosen. Consequently, to calculate the t (K) matrix, one of the f_i's of Eq. (4.98) should be taken to satisfy the complex outgoing-wave (real standing-wave) asymptotic boundary condition of the scattered wave, and the other functions could be $\mathcal{L}^2$.

The Kohn variational principle (4.98) can also be derived from the following operator identity for the t matrix:

$$t = V + V \times \frac{1}{E - H + i0} \times V. \tag{4.100}$$

If we introduce the presumably complete set of states $\sum_{i=1}^{n} |f_i\rangle\langle f_i|$, $n \to \infty$, which satisfies the appropriate boundary condition, in the two places indicated by the $\times$'s in expression (4.100), variational principle (4.98) can be derived.

Equation (4.98) results if the following stationary rank-n approximation for the Green's function $(E - H + i0)^{-1}$ is used in Eq. (4.100) [6]:

$$G_n^{(+)} = \sum_{i,j=1}^{n} |f_i\rangle Z_{ij}^{(n)} \langle f_j|. \tag{4.101}$$

The stationary property of Eq. (4.101) can be seen if in Eq. (3.2) the number c is taken to be the operator $(E - H)$. Then the variational estimate for an arbitrary matrix element of this operator between states $\langle\alpha|$ and $|\beta\rangle$ can be written as

$$\langle\alpha|G^{(+)}|\beta\rangle = \langle\alpha|G_1^{(+)}|\beta\rangle + \langle\alpha|G_2^{(+)}|\beta\rangle - \langle\alpha|G_1^{(+)}(E - H)G_2^{(+)}|\beta\rangle, \tag{4.102}$$

where $G_1^{(+)}$ and $G_2^{(+)}$ are two independent estimates of the Green's function. If we redefine the trial functions by $|\hat{\beta}\rangle \equiv G_2^{(+)}|\beta\rangle$ and $\langle\hat{\alpha}| \equiv \langle\alpha|G_1^{(+)}$, the variational expression becomes

$$\langle\alpha|G^{(+)}|\beta\rangle = \langle\alpha|\hat{\beta}\rangle + \langle\hat{\alpha}|\beta\rangle - \langle\hat{\alpha}|(E - H)|\hat{\beta}\rangle. \tag{4.103}$$

If we consider the following expansions for the trial functions:

$$|\hat{\beta}\rangle = \sum_{i=1}^{n} a_i |f_i\rangle, \tag{4.104}$$

$$\langle\hat{\alpha}| = \sum_{j=1}^{n} \langle f_j| b_j, \tag{4.105}$$

in Eq. (4.103) and make use of the stationary property, Eq. (4.101) is obtained.

In the case of the t matrix, the configuration space-matrix element of the outgoing-wave Green's function (4.101) is given by

$$\langle r|G_n^{(+)}|r'\rangle = \sum_{i,j=1}^{n} \langle r|f_i\rangle Z_{ij}^{(n)} \langle f_j|r'\rangle. \tag{4.106}$$

The asymptotic behavior (1.80) of the outgoing-wave Green's function (1.77) can be expressed as

$$\lim_{r\to\infty} \langle r|G^{(+)}|r'\rangle \to \frac{\exp(ikr)}{kr} \times \text{function}(r'). \tag{4.107}$$

This behavior can be reproduced by the rank-n expression (4.106) if one of the expansion functions $f_i(r) \equiv \langle r|f_i\rangle = \langle r|f_i\rangle$ is taken to carry the complex outgoing-wave asymptotic behavior of the Green's function. This is achieved if one of the following possible choices for $f_1(r)$ is employed [6]:

$$f_1(r) = \langle r|f_1\rangle = \langle f_1|r\rangle = \exp(ikr)[1 - \exp(-\beta r)] \tag{4.108}$$

$$= \exp(ikr) - \exp(-\beta r) \tag{4.109}$$

$$= \cos(kr)[1 - \exp(-\beta r)] + i\sin(kr). \tag{4.110}$$

It should be recalled that the functions f_i are used to expand the scattered wave corresponding to $\psi_L(r)$ of Eq. (1.37). This scattered wave is $\exp(ikr)$, and this asymptotic behavior is included in f_1. In order that expansion (4.106) satisfy the correct outgoing-wave boundary condition, both $\langle r|f_1\rangle$ and $\langle f_1|r\rangle$ should be equal and given by the expressions above and should not be complex conjugates of each other as is usually implied by the bra-ket notation of quantum mechanics. The remaining functions could be taken to be $\mathcal{L}^2$. Two possible choices for them, for partial wave L, are

$$f_i(r) = r^{(i-1+L)} \exp(-\alpha r) \tag{4.111}$$

$$= r^{(L+1)} \exp[-(i-1)\alpha r], \tag{4.112}$$

with $i = 2, ..., n$ [see Eqs. (3.75) and (3.76)]. Choices (4.111) and (4.112) are consistent with condition (1.35). Any of the three choices for f_1 above could be used with any of the two choices for f_i, $i > 1$.

If a standing-wave boundary condition is used, one gets a calculational scheme for the K matrix. The asymptotic behavior of the standing-wave Green's function (1.133) can be reproduced by the rank-n expression (4.106) if one of the expansion functions is chosen to carry this asymptotic condition. This is achieved if we take f_1, for example, to be given by Eq. (3.73). The remaining functions are taken to be $\mathcal{L}^2$, as in Eq. (4.111) or (4.112). Explicitly, in this approach the K matrix is calculated via Eqs (4.98) and (4.99), but with expansion functions (3.73) and (4.111) or (4.112).

It is possible to demonstrate the formal equivalence between the present on-shell standing-wave Kohn variational result (4.98) with Eq. (3.109) in the case of orthonormalized functions. We assume that the projection operator on the $\mathcal{L}^2$ functions is termed P and that on the non-$\mathcal{L}^2$ cosine function of Eq. (3.73) is termed Q, so that $PQ = QP = 0$. The plane-wave $j_L(kr) = \langle r|k\rangle_L$ is assumed to be orthogonal to the cosine function (3.73). Hence, $\langle k|VP = \langle k|HP = \langle k|(H-E)P$, $QVP = QHP = Q(H-E)P$. These conditions of orthonormality are necessary for showing the equivalence, but they are usually not satisfied in numerical calculations.

Then the Feshbach projection operator technique [23] leads to the well-known identity for the K-matrix version of Eq. (4.98):

$$\begin{aligned} K = {} & V + VG_{PP}V + (V + VG_{PP}V) \\ & \times Q\frac{1}{E - H_{QQ} - H_{QP}G_{PP}H_{PQ}}Q(V + VG_{PP}V), \end{aligned} \tag{4.113}$$

where

$$G_{PP} = P(E - H_{PP})^{-1}P, H_{PQ} = PVQ, H_{PQ} = PVQ = P(H - E)Q,$$

and so on. Using the orthonormality properties above, the on-shell plane-wave matrix element of this K matrix becomes

$$\begin{aligned} \langle k|K|k\rangle = {} & \langle k|(\hat{H} + \hat{H}G_{PP}\hat{H})|k\rangle + \langle k|(\hat{H} + \hat{H}G_{PP}\hat{H})|C\rangle \\ & \times \frac{1}{(E - H)_{QQ} - \hat{H}_{QP}G_{PP}\hat{H}_{PQ}}\langle C|(\hat{H} + \hat{H}G_{PP}\hat{H})|k\rangle, \end{aligned} \tag{4.114}$$

where $\hat{H} = (H - E)$ and C is the cosine function (3.73).

If we compare Eq. (4.114) with Eq. (3.109), we identify

$$\langle k|[\hat{H} + \hat{H}G_{PP}\hat{H}]|k\rangle = \frac{M_{00}}{k^2}, \tag{4.115}$$

$$\langle k|[\hat{H} + \hat{H}G_{PP}\hat{H}]|C\rangle = \frac{M_{10}}{k^2}, \tag{4.116}$$

$$\langle C|(E - H)_{QQ} - \hat{H}_{QP}G_{PP}\hat{H}_{PQ}|C\rangle = -\frac{M_{11}}{k^2}. \tag{4.117}$$

If we now use Eq. (1.152) and recall that $\lambda = \tan\delta$, the equivalence between Eqs. (4.114) and Eq. (3.109) follows. Hence, the present off-shell Kohn variational principle is equivalent to the on-shell version of Chapter 3 provided that orthogonal functions are used. As overlapping nonorthogonal functions are usually employed, these two forms may not be numerically equivalent. However, the present real Kohn variational scheme for the K matrix is numerically simpler in requiring the matrix inverse of a single real matrix involving several $\mathcal{L}^2$ and a single non-$\mathcal{L}^2$ function. In the implementation of Chapter 3 these two types of functions are treated separately by a two-step procedure. The present calculation scheme uses Eq. (4.98), whereas the on-shell method of Chapter 3 uses the equivalent relation (4.113).

In application of the complex Kohn variational principle for the t matrix, use of a decomposition of the type (4.114) is advantageous. This procedure treats the $\mathcal{L}^2$ and non-$\mathcal{L}^2$ functions separately and avoids the inversion of a complex matrix, as we see below. Miller and Janson op de Haan [6] advocated a closely related S-matrix form of the complex Kohn variational principle and showed how the result could be expressed in terms of the inverse of a real matrix of dimension $n - 1$, thus avoiding evaluation of the inverse of the complex matrix Z in Eq. (4.98). The rank-n degenerate S matrix of the complex Kohn variational principle is given by

$$[S_n(k) - 1] \equiv -ik[\langle k|t_n|k\rangle] = -ik[\langle k|V|k\rangle + \sum_{i,j=1}^{n} \langle k|V|f_i\rangle Z_{ij}^{(n)} \langle f_j|V|k\rangle]. \tag{4.118}$$

This S matrix is not explicitly unitary but can be used in practical numerical calculations. It is useful to partition the $n \times n$ matrix Y

$(= Z^{-1})$ defined by

$$Y_{ij} \equiv \langle f_i|(E - H)|f_j\rangle \tag{4.119}$$

into the 1×1 $i = j = 1$ block and the real $(n-1) \times (n-1)$ $i, j = 2, ..., n$, block and the coupling off-diagonal terms. Denoting the former 1×1 block by the projection operator Q and the latter $(n-1) \times (n-1)$ block by P, and assuming orthogonality $PQ = QP = 0$, one has for the t matrix the formal Feshbach identity (4.113) [23]. Then the expression for the S matrix can be written as

$$[S(k)] = 1 - ik[\langle k|V|k\rangle + \mathcal{V}^T \cdot \mathcal{Y}^{-1} \cdot \mathcal{V} + (\mathcal{V}_0 - \mathcal{V}^T \cdot \mathcal{Y}^{-1} \cdot \mathcal{Y}_0)$$

$$\times (\mathcal{Y}_{00} - \mathcal{Y}_0^T \cdot \mathcal{Y}^{-1} \cdot \mathcal{Y}_0)^{-1}(\mathcal{V}_0 - \mathcal{Y}_0^T \cdot \mathcal{Y}^{-1} \cdot \mathcal{V})], \tag{4.120}$$

where $\mathcal{Y}$ is the $(n-1) \times (n-1)$ real block of the matrix (4.119), $\mathcal{Y}_0$ is the complex $(n-1) \times 1$ rectangular matrix $\langle f_i|E - H|f_1\rangle$, $i = 2, ..., n$, $\mathcal{V}$ is the real $(n-1) \times 1$ rectangular matrix $\langle f_i|V|k\rangle$, $i = 2, ..., n$, $\mathcal{V}_0 = \langle f_1^*|V|k\rangle = \langle k|V|f_1\rangle$ and $\mathcal{Y}_{00} = \langle f_1^*|(E - H)|f_1\rangle$ are complex, and the superscript ^{T} denotes a transpose (without complex conjugation). If one plans to carry out the calculation at many energies, it may be useful to diagonalize the $(n-1) \times (n-1)$ real Hamiltonian matrix $\langle f_i|H|f_j\rangle$, $i, j = 2, ..., n$, so as to simplify the task of matrix inversion in Eq. (4.120). In this application of the complex Kohn variational method, the inversion of a large complex matrix is avoided.

The standing-wave Kohn variational principle of the present section and the usual Kohn-type variational principles of Chapter 3 yield unitary results. However, the t-matrix elements obtained from the complex Kohn variational principle do not automatically satisfy unitarity. In the converged calculation such phase shifts should be real, and the magnitude of the imaginary part of the phase shifts gives an idea about the degree of convergence of the solution. Hence the S matrix calculated from the complex Kohn variational principle (4.118) is distinct from the S matrix calculated from the real Kohn variational principle, that is,

$$S(k) = \frac{1 - ik\langle k|K_n|k\rangle}{1 + ik\langle k|K_n|k\rangle}, \tag{4.121}$$

where K_n is calculated from Eq. (4.98) but with real basis functions satisfying principal-value boundary conditions. The S matrix (4.118) is unitary, whereas the S matrix (4.121) is not.

Takatsuka and McKoy [3] pointed out that in the case of standing-wave boundary conditions it is possible to use Eqs. (4.98) and (4.99) with only $\mathcal{L}^2$ functions, provided that as in the R-matrix method, the configuration space is partitioned into two regions—interior and exterior—with a continuous log derivative of the wave function at the boundary. Then they show how to construct the basis functions that provide this continuity. This procedure leads to a practical numerical method for calculation of the K matrix using only $\mathcal{L}^2$ basis functions.

McCurdy et al. [24] pointed out the relation of the complex Kohn variational principle to the Kapur–Peierls form of R-matrix theory. The Kapur–Peierls form of R-matrix theory [24], which imposes outgoing-wave boundary conditions on the wave function at a finite r, can be put in a form very similar to Eq. (4.100). Its usual version involves surface integrals, which when converted into integrals over a finite volume leads to the following expression for the t matrix with the modified Hamiltonian $H + L$ [25]:

$$t = V + V \times \frac{1}{E - H - L + i0} \times V, \tag{4.122}$$

where the Bloch operator

$$L = \delta(r - a)\left(\frac{\partial}{\partial r} - ik\right) \tag{4.123}$$

is not Hermitian. Thus for a spectral representation of the operator $(E - H - L + i0)^{-1}$ one needs a biorthogonal set of eigenfunctions. This leads to a complex symmetric matrix representation of this operator as in the complex Kohn variational principle, and the dual eigenvector is the complex conjugate of the direct one. Equations (4.100) and (4.122) are identical in the limit $a \to 0$ or if the potential V vanishes outside $r = a$. Hence, it is expected that if identical expansion functions are used, both the complex Kohn variational principle and R-matrix theory should yield results of similar precision.

4.7 VARIATIONAL PRINCIPLES WITH DISTORTED WAVES

The use of distorted waves has a long history in treating reactions in nuclear, atomic, and molecular physics. The intercluster potential between two objects could often be very strong, so that the Born–Neumann series of scattering diverges strongly at medium to low energies. In such a situation, any solution algorithm based on the convergence properties of the Born–Neumann series is not very attractive. It is then advantageous to break up the entire intercluster interaction into two parts. In place of formulating the scattering theory in terms of the momentum eigenstates of the free Hamiltonian, it is of advantage to formulate it in terms of eigenstates of some other reference Hamiltonian which includes a major part of the intercluster potential and with which the reference problem can easily be solved. The eigenstates of the reference Hamiltonian are called the distorted waves. This reference potential is usually taken to be a simple model potential such as a square-well potential. All the subtleties of the original potential should remain in the residual potential. The purpose of the separation is to make the residual potential weak. The scattering theory is then formulated with the residual potential in terms of the distorted eigenstates of the reference Hamiltonian. The first terms of the Born–Neumann series with the residual potential in terms of a distorted wave may give a good description of scattering when the original Born–Neumann series diverges. One can employ distorted waves in the formulation of various variational principles. Such a treatment facilitates the numerical task.

Usually, the full Hamiltonian is divided into kinetic energy (H_0) and potential V parts: $H = H_0 + V$. The scattering problem is then naturally formulated in terms of the reference Hamiltonian H_0, which has plane-wave solutions at positive energies. For approximate numerical study, however, this approach has the disadvantage that for many problems of physical interest, higher-order terms of the Born–Neumann series play an essential (nonperturbative) role in the scattering process. This has a consequence on the convergence of the variational principles. For example, in complex-Kohn variational method (4.98) or in the higher-order variational method (4.96), the terms of the Born–Neumann series outside the sum is a poor representation of the t matrix, and the rank-n separable part is responsible for generating most of the t matrix. This possibly implies that n should be large for an accurate description of scattering. If the reference problem is formulated in terms of distorted waves, the distorted-wave Born term could be made a good approximation to the exact t matrix, and the remaining terms of the Born–Neumann series play an insignificant (perturbative) role in scattering process. Consequently, the responsibility of the rank-n separable term in the complex Kohn or higher-order variational principles will be somewhat reduced in producing the scattering observables, and a small number of basis functions can be used in the variational method.

The distorted-wave approach depends on decomposition of the Hamiltonian [2,26,27]:

$$H = H_a + V^a, \tag{4.124}$$

with the reference Hamiltonian defined by

$$H_a = H_0 + V_a, \tag{4.125}$$

so that $V = V_a + V^a$. It is presumed that the reference Hamiltonian permits an easy solution of the scattering problem, which is a good approximation to the full scattering solution. The full scattering solution is formulated using the known solution of the reference Hamiltonian, and the presumably small residual interaction V^a is treated variationally.

The usual Lippmann–Schwinger equation is formulated in terms of the free Green's function $G_0^{(+)}(E) = (E - H_0 + i0)^{-1}$. If we define the distorted-wave Green's function by

$$G_a^{(+)}(E) = (E - H_a + i0)^{-1}, \tag{4.126}$$

then we have the identities

$$I + G^{(+)}V = (I + G^{(+)}V^a)(I + G_a^{(+)}V_a), \tag{4.127}$$

$$I + VG^{(+)} = (I + V_a G_a^{(+)})(I + V^a G^{(+)}), \tag{4.128}$$

where $G^{(+)}$ is the full Green's function $G^{(+)} = (E - H + i0)^{-1}$. Using these identities the full t matrix defined by $t = V + VG^{(+)}V$ can be written as

$$t = t_a + t^a, \tag{4.129}$$

where

$$t_a = V_a + V_a G_a^{(+)} V_a = V_a + V_a G_0^{(+)} t_a = V_a + t_a G_0^{(+)} V_a, \tag{4.130}$$

$$t^a = (I + V_a G_a^{(+)}) t_A (I + G_a^{(+)} V_a), \tag{4.131}$$

with

$$t_A = V^a + V^a G_a^{(+)} t_A = V^a + t_A G_a^{(+)} V^a. \tag{4.132}$$

The equations above can be verified by using definitions of the various Green's functions and identities of type (1.90) and (1.92). We would like to formulate the scattering problem in terms of the outgoing and incoming distorted waves, defined by

$$|p^{(+)}\rangle = (I + G_a^{(+)} V_a)|p\rangle, \tag{4.133}$$

$$\langle p'^{(-)}| = \langle p'|(I + V_a G_a^{(+)}), \tag{4.134}$$

so that the on-shell t matrix becomes

$$\langle p'|t|p\rangle = \langle p'|V_a|p^{(+)}\rangle + \langle p'^{(-)}|t_A|p^{(+)}\rangle. \tag{4.135}$$

This is the Gell–Mann–Goldberger two-potential formula [27]. The first term on the right-hand side of Eq. (4.135) is the distorted-wave t matrix for the reference potential and is supposed to be known. The presumably small last term in this equation is to be calculated variationally.

Let us consider a couple of examples where this two-potential formulation is useful. (1) If the interaction potential has a long-range (repulsive) Coulomb tail, one can take the reference potential V_a to be the Coulomb potential, for which the solution is exactly known. The remaining short-range potential V^a is to be treated variationally. This procedure has been used to treat the proton–proton scattering problem variationally [26]. (2) In the case of a complicated short-range potential, one can take the reference potential V_a to be a simple (say, a square-well) potential that approximates the original potential well and for which the scattering solution is presumably known or can easily be found. The remaining small potential V^a is then treated variationally.

We can formulate all the variational principles of the present chapter in terms of distorted waves, and we illustrate a couple of them. To write the Schwinger variational principle for the last term of Eq. (4.135), we note that the operator t_A of this equation satisfies a formal Lippmann–Schwinger equation (4.132). We can define the scattering states in terms of distorted waves by

$$|\psi_p^{(+)}\rangle = |p^{(+)}\rangle + G_a^{(+)} V^a |\psi_p^{(+)}\rangle, \tag{4.136}$$

$$\langle \psi_{p'}^{(-)}| = \langle p'^{(-)}| + \langle \psi_{p'}^{(-)}| V^a G_a^{(+)}. \tag{4.137}$$

In terms of these scattering states, the t-matrix element $\langle p'^{(-)}|t_A|p^{(+)}\rangle$ can be written as

$$[\langle p'^{(-)}|t_A|p^{(+)}\rangle] = \langle p'^{(-)}|V^a|\psi_p^{(+)}\rangle = \langle \psi_{p'}^{(-)}|V^a|p^{(+)}\rangle$$
$$= \langle \psi_{p'}^{(-)}|V^a - V^a G_a^{(+)} V^a|\psi_p^{(+)}\rangle. \tag{4.138}$$

Then the following Schwinger variational principle for the residual term in Eq. (4.135) is obtained:

$$[\langle p'^{(-)}|t_A|p^{(+)}\rangle] = \langle p'^{(-)}|V^a|\psi_p^{(+)}\rangle + \langle \psi_{p'}^{(-)}|V^a|p^{(+)}\rangle$$
$$- \langle \psi_{p'}^{(-)}|V^a - V^a G_a^{(+)} V^a|\psi_p^{(+)}\rangle. \tag{4.139}$$

In this equation the distorted waves $\langle p'^{(-)}|$ and $|p^{(+)}\rangle$ are supposed to be known, and the functions $\langle \psi_{p'}^{(-)}|$ and $|\psi_p^{(+)}\rangle$ are the unknown trial functions. The similarity of the distorted-wave Schwinger variational principle (4.139), with the plain-wave version (4.5) is explicit. If we set $V_a = 0$, these two forms become identical.

To derive a finite-rank version of the distorted-wave Schwinger variational principle, the trial functions are expanded in basis set $f_i, i = 1, 2, ..., n$. Using the stationary property of Eq. (4.139) with respect to variations of the expansion coefficients in the usual fashion, one could obtain the rank-n form for the residual t matrix:

$$[t_A] = \sum_{i,j=1}^{n} V^a|f_i\rangle D_{ij}^{(n)} \langle f_j|V^a, \tag{4.140}$$

where

$$(D^{-1})_{ji} = \langle f_j|(V^a - V^a G_a^{(+)} V^a)|f_i\rangle. \tag{4.141}$$

Equations (4.140) and (4.141) are very similar to the Schwinger variational basis-set equations (4.24) and (4.21) for potential scattering.

In a similar fashion one can derive other variational principles in the distorted-wave representation. The distorted-wave Newton variational principle can be written as

$$[\langle p'^{(-)}|t_A|p^{(+)}\rangle] = \langle p'^{(-)}|V^a|p^{(+)}\rangle + \langle p'^{(-)}|V^a G_a^{(+)}|\chi_p^{(+)}\rangle$$

$$+ \langle \chi_{p'}^{(-)}|G_a^{(+)} V^a|p^{(+)}\rangle$$

$$- \langle \chi_{p'}^{(-)}|(G_a^{(+)} - G_a^{(+)} V^a G_a^{(+)})|\chi_p^{(+)}\rangle, \tag{4.142}$$

where the form factors $\langle \chi_{p'}^{(-)}|$ and $|\chi_p^{(+)}\rangle$ are now defined by

$$|\chi_p^{(+)}\rangle = V^a|\psi_p^{(+)}\rangle, \tag{4.143}$$

$$\langle \chi_{p'}^{(-)}| = \langle \psi_{p'}^{(-)}|V^a. \tag{4.144}$$

The stationary property of Eq. (4.142) can be established in exactly the same way as that of the Newton variational principle (4.60) of potential scattering.

The form factors (4.143) and (4.144) are $\mathcal{L}^2$. To derive a variational basis set method, one can expand these form factors into $\mathcal{L}^2$ basis sets. Then using the stationary property of the variational principle (4.142), with respect to the coefficients of the expansion, one can obtain the degenerate finite-rank version of the Newton variational principle, in close analogy with analogous Eqs. (4.73) and (4.70) of potential scattering:

$$[t_A] = V^a + \sum_{i,j=1}^{n} V^a G_a^{(+)}|f_i\rangle P_{ij}^{(n)} \langle f_j|G_a^{(+)} V^a, \tag{4.145}$$

where

$$(P^{-1})_{ji} = \langle f_j|(G_a^{(+)} - G_a^{(+)} V^a G_a^{(+)})|f_i\rangle, \tag{4.146}$$

respectively. These equations have been used successfully in scattering calculations and with $\mathcal{L}^2$ basis functions produce good convergence properties. Similar expressions can be written for the (complex) Kohn and higher-order variational principles.

4.8 INSERTION TECHNIQUES

Numerical application of the variational methods is performed by expanding the trial functions in appropriate basis sets. Then the numerical task is reduced to the solution of a linear algebraic set of equations of small dimension. Often, such algebraic treatment involves the evaluation of a single or multiple (infinite) integral over the free Green's function(s) and/or oscillating plane-wave states. Such integrations are easily performed in the case of potential scattering in both momentum and configuration spaces. In more complex situations, evaluation of these infinite integrals over oscillating functions could be quite complicated. The technique we describe below is of advantage in evaluating such integrals in configuration space. For example, in the symmetrical form of the Schwinger variational principle (4.24), the barrier to use of a straightforward calcuiational scheme in realistic atomic and molecular problems is evaluation of the integral containing the free Green's function in matrix D^{-1} of Eq. (4.21). Usually, additional approximations may help its evaluation. It has been pointed out that the calculation of such integrals becomes easier if a complete set of $\mathcal{L}^2$ basis functions is inserted in either side of the Green's function [28]. The denominator function of Eq. (4.21) is readily evaluated if the same $\mathcal{L}^2$ basis functions as used in the Schwinger variational principle are also used in both sides of the Green's function. Other $\mathcal{L}^2$ functions can also be used on both sides of the Green's function. Once these $\mathcal{L}^2$ functions are introduced, the infinite integrals over the Green's function can be cut off at some large distance, and this makes their evaluation easier. Usually, the potential is of longer range than the $\mathcal{L}^2$ basis functions and a direct evaluation of D^{-1} of Eq. (4.21) involves tedious integration over the Green's function to larger distances. We consider here several examples of such insertions.

Watson and McKoy [29] proposed an approximate form of the Schwinger variational principle (4.24), where the matrix elements (4.21) of D^{-1} are replaced by

$$(D^{-1})_{ji} = \langle f_j|V|f_i\rangle - \sum_{m,l=1}^{n} \langle f_j|V|h_m\rangle\langle h_m|G_0^{(+)}|h_l\rangle\langle h_l|V|f_i\rangle, \qquad i,j = 1, \ldots, n. \tag{4.147}$$

Here $h_m, m = 1, \ldots, n$, are a presumably real, complete, and orthonormal set of n $\mathcal{L}^2$ functions with the closure relation

$$\sum_{i=1}^{n} |h_i\rangle\langle h_i| = 1. \tag{4.148}$$

In actual implementation the set h_i (as well as the original $\mathcal{L}^2$ basis functions) should to be complete, linearly independent, and orthogonal or nearly so. In practice the functions h_i can be taken to be the same as the $\mathcal{L}^2$ basis functions f_i. However, in actual numerical application, these functions are usually taken to be linearly dependent and nonorthogonal. Equation (4.147) involves two insertions, and its use in the Schwinger variational method will be called a Schwinger variational method with double insertions.

Sometimes, to simplify numerical calculation, similar insertions are made in momentum-space matrix elements of the potential when the potential is known in configuration space. Transformation of such a potential to momentum space involves an integration over nonnormalizable momentum eigenstates. For a realistic problem such an integration is a difficult numerical task. Insertion of a complete set of $\mathcal{L}^2$ functions facilitates the task. Rescigno et al. [30] suggested, in addition to insertions of Eq. (4.147), two insertions in the potential terms of the Schwinger variational method. They rewrote the Schwinger variational principle (4.42) as

$$[\langle p'|t_n|p\rangle] = \sum_{i,j=1}^{n} \sum_{m,l=1}^{n} \langle p'|h_m\rangle\langle h_m|V|f_i\rangle D_{ij}^{(n)}\langle f_j|V|h_l\rangle\langle h_l|p\rangle, \tag{4.149}$$

where the D matrix is defined via (4.147). This approach involves four insertions of the basis set and is referred to as the Schwinger variational method with quadrupole insertions. In the original study, the K-matrix version of (4.149) was considered. In Eq. (4.149), matrix elements of both the potential and Green's function operators are required between $\mathcal{L}^2$ functions. No such matrix elements between non-$\mathcal{L}^2$ sine and cosine functions are required.

Similar insertions can also be made in other variational principles. For example, Staszewska and Truhlar [28] suggested replacing the function P^{-1} of Eq. (4.70) in Newton variational principle (4.73) by

$$(P^{-1})_{ji} = \langle f_j|G_0^{(+)}|f_i\rangle - \sum_{m,l=1}^{n} \langle f_j|G_0^{(+)}|h_m\rangle\langle h_m|V|h_l\rangle\langle h_l|G_0^{(+)})|f_i\rangle. \tag{4.150}$$

This procedure involves two insertions. The $\mathcal{L}^2$ functions h_i are chosen as in Eq. (4.147). After making these insertions we need to evaluate the matrix element of the Green's function only between $\mathcal{L}^2$ functions. At the same time, similar insertions can be made in the numerator functions of (4.73).

Staszewska and Truhlar [28] also suggested insertions in the simplest form of the Schwinger variational principle (4.34), which is rewritten as

$$[\langle p'|t_n|p\rangle] = \sum_{i,j=1}^{n} \sum_{m=1}^{n} \langle p'|h_m\rangle\langle h_m|V|f_i\rangle C_{ij}^{(n)} \langle v_j|p\rangle, \tag{4.151}$$

where

$$(C^{-1})_{ji} = \langle v_j|f_i\rangle - \sum_{m=1}^{n} \langle v_j|G_0^{(+)}|h_m\rangle\langle h_m|V|f_i\rangle, \qquad i,j = 1, ..., n. \tag{4.152}$$

After these insertions one needs to calculate matrix elements of the potential and Green's function only between $\mathcal{L}^2$ functions. Equations (4.151) and (4.152) correspond to insertions in the simplest version of the method of moments applied to the Lippmann–Schwinger equation. In fact, such insertions can be introduced in all variational principles. However, it is of great practical interest to find out how such insertions affect the rate of convergence.

4.9 VARIATIONAL PRINCIPLES FOR GENERAL MATRIX ELEMENTS

Variational principles for the t- or K-matrix elements have been very useful for finding the approximate solution of Schrödinger dynamics. Physical observables are then calculated from these t or K matrix elements. In some cases physical observables may be directly related to some other matrix element of the wave function. Then it might be convenient to write variational principles for these general matrix elements rather than for the t- or K-matrix elements. Next we consider variational principles for a general matrix element and show how to derive different variational principles for a t or K matrix as special cases. We also write a variational principle for the difference between the exact matrix element and the variational result. Such a subsidiary variational principle for the difference will help in finding the error in the variational evaluation of the original matrix element and will thus produce an improved result over the original variational calculation. This approach has interesting applications in studies of molecular photoionization cross sections where electronic transition matrix elements are required. In general, variational principles can be written down for many matrix elements of physical interest [31].

Let us consider the matrix elements for the scattering wave function $|\psi^{(+)}\rangle$, with outgoing-wave boundary condition, which satisfies the Lippmann–Schwinger equation

$$|\psi^{(+)}\rangle = |\phi\rangle + \mathcal{K}|\psi^{(+)}\rangle, \tag{4.153}$$

where $\mathcal{K} = G_0^{(+)}V$ and $|\phi\rangle$ is the incident plane wave. Let us also consider the conjugate equation

$$\langle\chi^{(-)}| = \langle R| + \langle\chi^{(-)}|\mathcal{K}, \tag{4.154}$$

where $\langle R|$ and $\langle\chi^{(-)}|$ are yet unknown quantities. In potential scattering we have seen that the conjugate equation is satisfied by $\langle R| = \langle\phi|$, the incident plane wave, and $\langle\chi^{(-)}| = \langle\psi^{(-)}|$, the scattering wave function with incoming-wave boundary condition. In a general quantum-mechanical problem, the unknown and inhomogeneous terms of the conjugate equation may or may not be related to those of the original equation. Using this flexibility in defining the conjugate equation, one can formulate a class of useful variational principles for calculating general matrix elements. We would like to establish variational principles for the following matrix element of the wave function $|\psi^{(+)}\rangle$ [13]:

$$M(R,\phi) \equiv \langle R|\psi^{(+)}\rangle = \langle\chi^{(-)}|\phi\rangle = \langle\chi^{(-)}|(I-\mathcal{K})|\psi^{(+)}\rangle. \tag{4.155}$$

The identities (4.155) can be derived using the defining Lippmann–Schwinger equations (4.153) and (4.154). With a different choice of R one can obtain a different matrix element.

The following stationary expression for $M(R,\phi)$ has the usual trilinear relation

$$[M(R,\phi)] = \langle R|\psi_t^{(+)}\rangle + \langle\chi_t^{(-)}|\phi\rangle - \langle\chi_t^{(-)}|(I-\mathcal{K})|\psi_t^{(+)}\rangle, \tag{4.156}$$

where the variational functions are $\langle \chi_t^{(-)}|$ and $|\psi_t^{(+)}\rangle$. The suffix t denotes trial and has been introduced for future convenience. The stationary property of Eq. (4.156) can be established by using Eqs. (4.153) and (4.154). Expression (4.156) can also be written in the fractional and finite-rank forms.

To obtain the finite-rank form, we have to expand the trial functions in a basis set, and after using the stationary property of Eq. (4.156), we obtain in usual fashion

$$M_0 = \sum_{i,j=1}^{n} \langle R|u_i\rangle D_{ij}^{(n)} \langle v_j|\phi\rangle, \tag{4.157}$$

where

$$(D^{-1})_{ji} = \langle v_j|(I - \mathcal{K})|u_i\rangle, \tag{4.158}$$

and u_i and v_j, $i, j = 1, 2, ..., n$, are two sets of basis functions. Expression (4.157) is the required finite-rank expansion. Here we are using a subscript zero on M to denote that this finite-rank estimate is a zeroth-order estimate and we shall find a correction to this zeroth-order estimate in the following.

In addition to calculating the finite-rank approximation (4.157), it is convenient to calculate the correction to this approximation. For this purpose we define the finite-rank operator T by [13]

$$T = \sum_{i,j=1}^{n} |u_i\rangle D_{ij}^{(n)} \langle v_j|. \tag{4.159}$$

Then the finite-rank expansion (4.157) can be written as

$$M_0 = \langle R|T|\phi\rangle. \tag{4.160}$$

From Eqs. (4.156), (4.157), and (4.159), we find that the trial functions $|\psi_t^{(+)}\rangle$ and $\langle \chi_t^{(-)}|$ are given by

$$|\psi_t^{(+)}\rangle = T|\phi\rangle, \qquad \langle \chi_t^{(-)}| = \langle R|T. \tag{4.161}$$

Next we would like to calculate an estimate of the error in the finite-rank approximation M_0 to $M(R, \phi)$. This error is given by

$$M_c \equiv M(R, \phi) - M_0 = \langle R|\psi^{(+)} - \psi_t^{(+)}\rangle = \langle \chi^{(-)} - \chi_t^{(-)}|\phi\rangle. \tag{4.162}$$

To estimate this error, we consider the zeroth-order solutions of Eqs. (4.153) and (4.154):

$$|\psi_0^{(+)}\rangle = |\phi\rangle + \mathcal{K}|\psi_t^{(+)}\rangle, \tag{4.163}$$

$$\langle\chi_0^{(-)}| = \langle R| + \langle\chi_t^{(-)}|\mathcal{K}, \tag{4.164}$$

and express the exact wave functions in terms of these solutions. From Eqs. (4.153) and (4.163) we obtain

$$(I - \mathcal{K})|\psi^{(+)}\rangle + \mathcal{K}|\psi_t^{(+)}\rangle = |\psi_0^{(+)}\rangle. \tag{4.165}$$

Using Eqs. (4.153) and (4.161), Eq. (4.165) can be rewritten as

$$(I + \mathcal{K}T)(I - \mathcal{K})|\psi^{(+)}\rangle = |\psi_0^{(+)}\rangle. \tag{4.166}$$

Equation (4.166) connects the zeroth-order and exact solutions $|\psi_0^{(+)}\rangle$ and $|\psi^{(+)}\rangle$, respectively. If we introduce the finite-rank operator X by

$$X \equiv \sum_{i,j=1}^{n} |u_i\rangle O_{ij}^{(n)} \langle v_j|, \tag{4.167}$$

with

$$(O^{-1})_{ji} = \langle v_j|u_i\rangle, \tag{4.168}$$

then this operator is related to the operator T of Eq. (4.159) by

$$T = X + T\mathcal{K}X = X + X\mathcal{K}T. \tag{4.169}$$

Equation (4.169) is formally similar to a Lippmann–Schwinger equation, which will aid in formal manipulations. In terms of X, Eq. (4.166) can be written as

$$[I - (\mathcal{K} + \mathcal{K}T\mathcal{K})(I - X)]|\psi^{(+)}\rangle = (I + \mathcal{K}T)(I - \mathcal{K})|\psi^{(+)}\rangle = |\psi_0^{(+)}\rangle. \tag{4.170}$$

In deriving Eq. (4.170), we have used Eq. (4.169).

In a similar fashion, one can obtain the following relation connecting the zeroth order and exact functions $\langle\chi^{(-)}|$ and $\langle\chi_0^{(-)}|$:

$$\langle\chi^{(-)}|[I-(I-X)(\mathcal{K}+\mathcal{K}T\mathcal{K})] = \langle\chi^{(-)}|(I-\mathcal{K})(I+T\mathcal{K}) = \langle\chi_0^{(-)}|. \tag{4.171}$$

Now the difference M_c of Eq. (4.162) can be written as

$$M_c = \langle\chi^{(-)}|(I-\mathcal{K})|\psi^{(+)}\rangle - \langle\chi^{(-)}|(I-\mathcal{K})T|\phi\rangle \tag{4.172}$$

$$= \langle\chi^{(-)}|[(I-\mathcal{K})-(I-\mathcal{K})T(I-\mathcal{K})]|\psi^{(+)}\rangle. \tag{4.173}$$

In deriving Eq. (4.172), we have used Eqs. (4.153), (4.154), and (4.161), and in deriving Eq. (4.173), we have used Eq. (4.153). By using Eqs. (4.166), (4.169), and (4.170), we obtain from Eq. (4.173)

$$M_c = \langle\chi^{(-)}|(I-X)|\psi_0^{(+)}\rangle. \tag{4.174}$$

In a similar fashion we can derive

$$M_c = \langle\chi_0^{(-)}|(I-X)|\psi^{(+)}\rangle. \tag{4.175}$$

Using Eqs. (4.170) and (4.171), we can rewrite Eqs. (4.174) and (4.175) as

$$M_c = \langle\chi^{(-)}|(I-X)[I-(\mathcal{K}+\mathcal{K}T\mathcal{K})(I-X)]|\psi^{(+)}\rangle. \tag{4.176}$$

Using the formal solutions of Eqs. (4.170) and (4.171) for $|\psi^{(+)}\rangle$ and $\langle\chi^{(-)}|$, in Eqs. (4.174) and (4.175), respectively, we obtain

$$M_c = \sum_{n=0}^{\infty}\langle\chi_0^{(-)}|(I-X)[(\mathcal{K}+\mathcal{K}T\mathcal{K})(I-X)]^n|\psi_0^{(+)}\rangle \tag{4.177}$$

$$= \sum_{n=0}^{\infty}\langle\chi_0^{(-)}|[(I-X)(\mathcal{K}+\mathcal{K}T\mathcal{K})]^n(I-X)]|\psi_0^{(+)}\rangle \tag{4.178}$$

We discuss two practical ways of computing M_c, both based on Schwinger variational principles. The first is a variationally stable method and the second is a iterative method. Because of the similarity with the method of Padé approximants, the first method is also called a variational Padé approximant method.

In Eqs. (4.174), (4.175), and (4.176), the zeroth-order estimates $\langle\chi_0^{(-)}|$ and $|\psi_0^{(+)}\rangle$ are considered known, whereas the exact functions $\langle\chi^{(-)}|$ and $|\psi^{(+)}\rangle$ are considered unknown. We consider the following variational principle for the quantity M_c of (4.176):

$$[M_c] = \langle\chi_0^{(-)}|(I-X)|\psi^{(+)}\rangle + \langle\chi^{(-)}|(I-X)|\psi_0^{(+)}\rangle$$

$$- \langle\chi^{(-)}|[(I-X) - (I-X)(\mathcal{K}+\mathcal{K}T\mathcal{K})(I-X)]|\psi^{(+)}\rangle, \quad (4.179)$$

where $\langle\chi^{(-)}|$ and $|\psi^{(+)}\rangle$ are new trial functions. This is a Schwinger functional of the difference M_c. The stationary property of the trilinear expression (4.179) follows from Eqs. (4.170) and (4.171).

We have proposed variational principle (4.156) for the general matrix element $M(R,\phi)$ of Eq. (4.155). An approximate calculational scheme of this quantity can be performed via finite-rank expression (4.157). We have also proposed variational expression (4.179) for the difference M_c of Eq. (4.162). By using Eq. (4.179), we can obtain an estimate of the error in the original variational calculation of $M(R,\phi)$ based on the finite-rank variational expression (4.157). With this estimate of error, we can obtain an improved result for $M(R,\phi)$. Next we describe two ways of computing the correction term.

An estimate for the error in $M(R,\phi)$ of Eq. (4.155) can be obtained when the trial functions $|\psi^{(+)}\rangle$ and $\langle\chi^{(-)}|$ of Eq. (4.179) are constructed as linear combinations of following basis functions, respectively:

$$|f_i\rangle = [(\mathcal{K}+\mathcal{K}T\mathcal{K})(I-X)]^i|\psi_0^{(+)}\rangle, \qquad i = 0, 1, \ldots, \quad (4.180)$$

$$\langle g_j| = \langle\chi_0^{(-)}|[(I-X)(\mathcal{K}+\mathcal{K}T\mathcal{K})]^i, \qquad j = 0, 1, \ldots. \quad (4.181)$$

The set of functions above are suggested by Eqs. (4.177) and (4.178). When the complete set of functions (4.180) and (4.181) are used in the stationary expression (4.179), the exact result for the difference M_c is obtained. If the trial functions $\langle\chi^{(-)}|$ and $|\psi^{(+)}\rangle$ are expanded in basis functions $\langle g_j|$ and $|f_i\rangle$, respectively, and the stationary property of Eq. (4.179) is used, we obtain a finite-rank form for the difference M_c. Then the variational approximation to M, obtained by using this finite-rank approximation to M_c, is given by

$$M_P = M_0 + \sum_{i,j=0}^{N} \langle\chi_0^{(-)}|(I-X)|f_i\rangle C_{ji}^{(N)} \langle g_j|(I-X)|\psi_0^{(+)}\rangle, \quad (4.182)$$

$$(C^{-1})_{ji} = \langle g_j|(I-X)[I - (\mathcal{K}+\mathcal{K}T\mathcal{K})(I-X)]|f_i\rangle. \quad (4.183)$$

Instead of calculating M_0 and the remainder term separately via Eqs. (4.157) and (4.182), the correction scheme can be incorporated in the Schwinger expression (4.157). First, the initial Schwinger basis functions are to be used in Eq. (4.157). Once these functions are exhausted, the functions of basis sets $|f_i\rangle$ and $\langle g_j|$ of Eqs. (4.180) and (4.181) are to be included in place of the basis sets $|u_i\rangle$ and $\langle v_j|$ of Eq. (4.157), respectively. The entire scheme then appears as a unified Schwinger variational method. This method has also been called the variational Padé method.

Another approach for obtaining approximation to M_c is the iterative Schwinger method, where the basis functions in Eq. (4.157) are augmented from the following sets constructed iteratively:

$$|\psi_N^{(+)}\rangle = |\phi\rangle + \sum_{i,j} \mathcal{K}|u_i\rangle D_{ij}\langle v_j|\phi\rangle, \tag{4.184}$$

$$\langle \chi_N^{(-)}| = \langle R| + \sum_{i,j} \langle R|u_i\rangle D_{ij}\langle v_j|\mathcal{K}, \tag{4.185}$$

where $|\psi_0^{(+)}\rangle$ and $\langle\chi_0^{(-)}|$ are $|\psi_0^{(+)}\rangle$ and $\langle\chi_0^{(-)}|$ of Eqs. (4.163) and (4.164). The functions $|\psi_0^{(+)}\rangle$ and $\langle\chi_0^{(-)}|$ are obtained if only the initial basis functions $|u_i\rangle$ and $\langle v_j|$ are used in Eqs. (4.184) and (4.185), respectively. If more functions are included in the expansions of Eqs. (4.184) and (4.185), improved $|\psi^{(+)}\rangle$ and $\langle\chi^{(-)}|$ can be obtained. From Eqs. (4.163) and (4.164) we find that the inclusion of functions $|\psi_0^{(+)}\rangle$ and $\langle\chi_0^{(-)}|$ in the variational result (4.157) should lead to an iterative improvement. The next functions to be included in Eq. (4.157) are $|\psi_1^{(+)}\rangle$ and $\langle\chi_1^{(-)}|$ constructed from Eqs. (4.184) and (4.185), where the functions $|\psi_0^{(+)}\rangle$ and $\langle\chi_0^{(-)}|$ are included in addition to the original functions on the right-hand side of these equations. This procedure can be continued indefinitely by including higher-order correction functions $|\psi_N^{(+)}\rangle$ and $\langle\chi_N^{(-)}|$ in the Schwinger expression (4.157) when the sum in Eqs. (4.184) and (4.185) runs over all the initial functions as well as the functions $|\psi_i^{(+)}\rangle$ and $\langle\chi_i^{(-)}|$, $i = 1, 2, \ldots, (N-1)$, respectively.

Both the iterative methods described above proceed indefinitely for calculation of the correction term M_c. But in practice few terms are expected to lead to good improvement. The two methods are quite similar in spirit. For $N = 1$ they yield identical results, but they are different for a general N.

In the case of potential scattering, the foregoing methods have interesting consequences. In that case, in Eq. (4.154) one can identify

$$\langle R| = \langle \phi|V. \tag{4.186}$$

Then consistency with the t-matrix Lippmann–Schwinger equations (1.93) and (1.94) leads to

$$\langle \chi^{(-)}| = \langle \phi|t = \langle \psi^{(-)}|V. \tag{4.187}$$

If we recall that $\mathcal{K} = G_0^{(+)}V$, Eq. (4.154) is identified as the Lippmann–Schwinger equation (1.94) and Eq. (4.156) as the Schwinger variational principle (4.12) or (4.5). The procedure described above then provides definite prescriptions for improvement on the Schwinger variational result.

The scheme above is appropriate to derive the Schwinger variational principle from a general variational principle and also to find improvement over it. One can also derive Newton variational principle from a general scheme. Such a scheme can be obtained if we partially expand the unknown function $|\psi^{(+)}\rangle$ of Eq. (4.153) in an iterative Neumann series and use it in expression (4.155) for $M(R, \phi)$ to obtain

$$M(R, \phi) = \sum_{k=0}^{m-1} \langle R|\mathcal{K}^k|\phi\rangle + I_m(R, \phi), \tag{4.188}$$

where

$$I_m(R, \phi) = \langle R|\mathcal{K}^m|\psi^{(+)}\rangle. \tag{4.189}$$

Equation (4.188) is an identity. The first term on the right-hand side of this equation should be treated exactly, and the last term $I_m(R, \phi)$ of this equation given by Eq. (4.189) is to be treated variationally. For this purpose we introduce the function $\langle \chi_{(m)}^{(-)}|$ by

$$\langle \chi_{(m)}^{(-)}|(I - \mathcal{K}) = \langle R|\mathcal{K}^m. \tag{4.190}$$

In terms of this function, the quantity $I_m(R, \phi)$ can be rewritten as

$$I_m(R, \phi) = \langle \chi_{(m)}^{(-)}|\phi\rangle = \langle \chi_{(m)}^{(-)}|(I - \mathcal{K})|\psi^{(+)}\rangle. \tag{4.191}$$

By combining Eqs. (4.189) and (4.191), a trilinear variational relation for $I_m(R,\phi)$ can be written as

$$[I_m(R,\phi)] = \langle R|\mathcal{K}^m|\psi^{(+)}\rangle + \langle\chi_{(m)}^{(-)}|\phi\rangle - \langle\chi_{(m)}^{(-)}|(I-\mathcal{K})|\psi^{(+)}\rangle. \quad (4.192)$$

In Eq. (4.192) the functions $\langle R|$ and $|\phi\rangle$ are considered known. The functions $|\psi^{(+)}\rangle$ and $\langle\chi_{(m)}^{(-)}|$ are unknowns that are to be expanded in basis sets. If we expand these unknown functions in basis sets and use the stationary property of Eq. (4.192), the following finite-rank expression is obtained:

$$[I_m(R,\phi)] = \sum_{i,j=1}^{n} \langle R|\mathcal{K}^m|u_i\rangle D_{ij}^{(n)} \langle v_j|\phi\rangle, \quad (4.193)$$

where D_{ji}^{-1} is given by Eq. (4.158) and u_i and v_j, $i,j = 1,2,\ldots,n$, are two sets of basis functions.

By redefining the expansion functions in Eq. (4.193), one can obtain a class of variational expressions for $I_m(R,\phi)$. Let us illustrate this for potential scattering with $\langle R|$ defined by Eq. (4.186). The first term in the sequence corresponds to taking $m=0$. For $m=0$, the first term on the right-hand side of Eq. (4.188) is zero. Then Eq. (4.193) becomes identical to Eq. (4.34) and the usual Schwinger variational principle is obtained.

The next term in the progression corresponds to taking $m=1$, with $\langle R|$ defined by Eq. (4.186). In this case if one defines $V|u_i\rangle = |f_i\rangle$ and $\langle v_j| = \langle f_j|G_0^{(+)}V$, Eqs. (4.188) and (4.193) lead to

$$M(R,\phi) = \langle\phi|V|\phi\rangle + \sum_{i,j=1}^{n} \langle\phi|VG_0^{(+)}|f_i\rangle P_{ij}^{(n)} \langle f_j|G_0^{(+)}V|\phi\rangle, \quad (4.194)$$

with P^{-1} defined by Eq. (4.70). Equation (4.194) is now identified with the Newton variational method of Eq. (4.73). For $m=2$ it can be shown that Eqs. (4.188) and (4.193) lead to the higher-order variational principle (4.92). In a similar fashion other higher-order variational principles for the t matrix can be recovered for $m>2$.

The matrix element $M(R,\phi)$ considered above can also be used to calculate quantities other than the t matrix. If $\langle R|$ is taken to be the configuration space eigenfunction (e.g., $\langle R| = \langle\mathbf{r}|$), $M(R,\phi)$ is a variational functional for the configuration space representation for the wave function. By considering different values of m and different basis functions, we obtain a class of variational principles for the scattering wave function.

If we take $\langle R| = \langle\phi_0|\mathbf{p}$, the quantity $M(R, \phi) = \langle\phi_0|\mathbf{p}|\psi\rangle$ becomes the dipole matrix element required to compute photoionization cross sections. Here ϕ_0 is a bound orbital of a molecular electronic wave function. The function $\mathbf{p}$ is either $\mathbf{r}$, which gives a dipole length matrix element, or $\nabla_{\mathbf{r}}/e$, with e the photon energy. The second choice gives a dipole velocity matrix element, where ψ is the photoelectron continuum orbital. For $m = 0$ one can define $|f_i\rangle = |u_i\rangle$ and $\langle f_j|V = \langle v_j|$, then Eqs. (4.188) and (4.193) become [13]

$$[M(R, \phi)] = \sum_{i,j=1}^{n} \langle\phi_0|\mathbf{p}G_0^{(+)}V|f_i\rangle D_{ij}^{(n)}\langle f_j|V|\phi\rangle, \tag{4.195}$$

where D^{-1} is now defined by Eq. (4.21). This expression for $M(R, \phi)$ yields the required dipole matrix element. In a similar fashion, higher-order variational expressions can be written down for $m > 1$.

4.10 VARIATIONAL PRINCIPLES FOR THE Γ MATRIX

The variational principles for the t or K matrix with explicit dependence on the Green's function are usually transformed into algebraic basis-set calculational schemes. The resulting numerical work in momentum space involves dealing with complex algebra and/or principal-value prescription in treatment of the Green's function. Troubles associated with complex algebra and/or principal-value treatment can be avoided by dealing with the real but nonsymmetric Γ matrix and considering variational principles for the Γ matrix.

One can write Schwinger- or Newton-type variational principles for the Γ matrix. The nonsymmetric aspect of this matrix does not cause any practical problem. First, we derive a Schwinger-type variational principle for the Γ matrix introduced in Section 2.4 with $\bar{k} = k$. Let us consider the quantities $|\eta_{k,p}\rangle$ and $\langle\hat{\eta}_{k,p'}|$, defined by

$$V|p\rangle = (\mathcal{V} - \mathcal{V}G_0\mathcal{V})|\eta_{k,p}\rangle, \tag{4.196}$$

$$\langle p'|\mathcal{V} = \langle\hat{\eta}_{k,p'}|(\mathcal{V} - \mathcal{V}G_0^{(+)}\mathcal{V}), \tag{4.197}$$

respectively, where $\mathcal{V}$ is defined by Eq. (2.27). The explicit energy dependence of various operators is suppressed for notational convenience. The potential $\mathcal{V}$ of Eq. (2.27) is energy (k) dependent, which is exhibited explicitly on the functions $|\eta_{k,p}\rangle$ and $\langle\hat{\eta}_{k,p'}|$. The operator $\mathcal{V}G_0$ is real and does not require outgoing-wave or principal-value prescription for its evaluation. Using the formal solution $\Gamma = (I - \mathcal{V}G_0)^{-1}V$ of the Γ-matrix equation (2.28), one can derive the following identities:

$$\begin{aligned}\langle p'|\Gamma|p\rangle &= \langle p'|(I - \mathcal{V}G_0)^{-1}V|p\rangle \\ &= \langle p'|(I - \mathcal{V}G_0)^{-1}(\mathcal{V} - \mathcal{V}G_0\mathcal{V})|\eta_{k,p}\rangle \qquad (4.198) \\ &= \langle p'|\mathcal{V}|\eta_{k,p}\rangle = \langle\hat{\eta}_{k,p'}|(\mathcal{V} - \mathcal{V}G_0\mathcal{V})|\eta_{k,p}\rangle \\ &= \langle\hat{\eta}_{k,p'}|V|p\rangle. \qquad (4.199)\end{aligned}$$

In deriving these identities, definitions (4.196) and (4.197) and the Γ-matrix equation (2.28) have been used.

Using identity (4.199), the Schwinger variational principle for the Γ matrix can be written as

$$[\langle p'|\Gamma|p\rangle] = \langle p'|\mathcal{V}|\eta_{k,p}\rangle + \langle\hat{\eta}_{k,p'}|V|p\rangle - \langle\hat{\eta}_{k,p'}|(\mathcal{V} - \mathcal{V}G_0\mathcal{V})|\eta_{k,p}\rangle, \qquad (4.200)$$

where $|\eta_{k,p}\rangle$ and $\langle\hat{\eta}_{k,p'}|$ are the unknown trial functions. Using definition (2.28) of the Γ matrix and Eqs. (4.196) and (4.197), the stationary property of Eq. (4.200) can be established.

The trilinear form (4.200) for the variational principle for the matrix Γ can also be written in the following fractional form:

$$[\langle p'|\Gamma|p\rangle] = \frac{\langle p'|\mathcal{V}|\eta_{k,p}\rangle\langle\hat{\eta}_{k,p'}|V|p\rangle}{\langle\hat{\eta}_{k,p'}|(\mathcal{V} - \mathcal{V}G_0\mathcal{V})|\eta_{k,p}\rangle}. \qquad (4.201)$$

A rank-n version of the variational principle (4.200) can be derived if one expands the unknown functions $|\eta_{k,p}\rangle$ and $\langle\hat{\eta}_{k,p'}|$ in basis sets and makes use of the stationary property. Then the following rank-n form for the Γ matrix is derived in the usual fashion:

$$[\langle p'|\Gamma_n|p\rangle] = \sum_{i,j=1}^{n}\langle p'|\mathcal{V}|u_i\rangle D_{ij}^{(n)}\langle w_j|V|p\rangle, \qquad (4.202)$$

where

$$(D^{-1})_{ji} = \langle w_j|(\mathcal{V} - \mathcal{V}G_0\mathcal{V})|u_i\rangle, \qquad i, j = 1, ..., n, \tag{4.203}$$

and $\langle w_j|$ and $|u_i\rangle$ are basis functions. For $u_i = w_i$, the matrix $D_{ij}^{(n)}$ is symmetric and the asymmetry of the Γ matrix is manifested by the asymmetric form factors $\langle p'|\mathcal{V}|u_i\rangle$ and $\langle w_m|V|p\rangle$.

In a similar way, the Newton variational principle for the Γ matrix can also be obtained. The Newton variational principle is a variational principle for $\Gamma - V$. To obtain this variational principle, we define

$$V|p\rangle = (I - A)|\xi_{k,p}\rangle, \tag{4.204}$$

$$\langle p'|A = \langle\hat{\xi}_{k,p'}|(I - A), \tag{4.205}$$

where $A \equiv \mathcal{V}G_0$ is defined by Eq. (2.26).

Using the following formal solution of the Γ-matrix equation (2.30):

$$\langle p'|(\Gamma - V)|p\rangle = \langle p'|(I - \mathcal{V}G_0)^{-1}\mathcal{V}G_0V|p\rangle \tag{4.206}$$

and definitions (4.204) and (4.205), one can derive the following identities:

$$\langle p'|(\Gamma - V)|p\rangle = \langle p'|(I - \mathcal{V}G_0)^{-1}\mathcal{V}G_0(I - A)|\xi_{k,p}\rangle \tag{4.207}$$

$$= \langle p'|A|\xi_{k,p}\rangle \tag{4.208}$$

$$= \langle\hat{\xi}_{k,p'}|(I - A)|\xi_{k,p}\rangle = \langle\hat{\xi}_{k,p'}|V|p\rangle. \tag{4.209}$$

In deriving these identities, definitions (4.204) and (4.205) and the Γ-matrix equation (2.30) have been used. Using these identities, the Newton variational principle for the Γ matrix can be written as

$$[\langle p'|\Gamma|p\rangle] = \langle p'|V|p\rangle + \langle p'|A|\xi_{k,p}\rangle + \langle\hat{\xi}_{k,p'}|V|p\rangle - \langle\hat{\xi}_{k,p'}|(I - A)|\xi_{k,p}\rangle, \tag{4.210}$$

where $|\xi_{k,p}\rangle$ and $\langle\hat{\xi}_{k,p'}|$ are unknown trial functions. Allowing small variations of the functions $|\xi_{k,p}\rangle$ and $\langle\hat{\xi}_{k,p'}|$, the stationary property of Eq. (4.210) can be verified.

The variational principle (4.210) can also be written in the following equivalent fractional form:

$$[\langle p'|\Gamma|p\rangle] = \langle p'|V|p\rangle + \frac{\langle p'|A|\xi_{k,p}\rangle\langle\hat{\xi}_{k,p'}|V|p\rangle}{\langle\hat{\xi}_{k,p'}|(I-A)|\xi_{k,p}\rangle}. \tag{4.211}$$

A finite-rank version of variational principle (4.210) can be derived by expanding the unknown functions $|\xi_{k,p}\rangle$ and $\langle\hat{\xi}_{k,p'}|$ in a finite basis set and using the stationary property. The resultant degenerate kernel form of the Γ matrix can be written as

$$[\langle p'|\Gamma_n|p\rangle] = \langle p'|V|p\rangle + \sum_{i,j=1}^{n}\langle p'|A|u_i\rangle D_{ij}^{(n)}\langle w_j|V|p\rangle, \tag{4.212}$$

where

$$(D^{-1})_{ji} = \langle w_j|(I-A)|u_i\rangle, \qquad i,j = 1,\ldots,n, \tag{4.213}$$

and $\langle w_j|$ and $|u_i\rangle$ are basis functions.

One of the ideas behind constructing the Γ matrix was to make the kernel A of the real nonsingular integral equation (2.28) weak with a proper choice of the function γ, so that this equation has a convergent iterative Neumann series solution. When this happens, $\mathcal{V}$ is supposed to be small and so should be the last term in Eq. (4.212). As the potential term is explicitly included in the stationary form (4.212) for the Γ matrix and the last term of this equation is constructed in terms of the potential $\mathcal{V}$, this form is supposed to converge rapidly.

The procedure above can be continued further by including more and more terms of the Neumann series exactly in the variational expression. For example, one can consider the following formal solution to the Γ-matrix equation (2.30):

$$\Gamma = V + AV + (I-A)^{-1}A^2V \tag{4.214}$$

for constructing a variational principle for $\Gamma - V - AV$. For this purpose we introduce the functions $|\chi_{k,p}\rangle$ and $\langle\hat{\chi}_{k,p'}|$, defined by

$$AV|p\rangle = (I-A)|\chi_{k,p}\rangle, \tag{4.125}$$

$$\langle p'|A = \langle\hat{\chi}_{k,p'}|(I-A). \tag{4.216}$$

Using these functions we have the following identities:

$$\langle p'|\Gamma - V - AV|p\rangle = \langle\hat{\chi}_{k,p'}|AV|p\rangle = \langle p'|A|\chi_{k,p}\rangle$$
$$= \langle\hat{\chi}_{k,p'}|(I - A)|\chi_{k,p}\rangle. \qquad (4.217)$$

With the use of this identity, the variational principle for the matrix element of $\Gamma - V - AV$ becomes

$$[\langle p'|\Gamma|p\rangle] = \langle p'|(V + AV)|p\rangle + \langle p'|A|\chi_{k,p}\rangle + \langle\hat{\chi}_{k,p'}|AV|p\rangle$$
$$- \langle\hat{\chi}_{k,p'}|(I - A)|\chi_{k,p}\rangle, \qquad (4.218)$$

where $|\chi_{k,p}\rangle$ and $\langle\hat{\chi}_{k,p'}|$ are the unknown trial functions. Allowing small variations of the functions $\langle\hat{\chi}_{k,p'}|$ and $|\chi_{k,p}\rangle$ around their exact values, the stationary property of Eq. (4.218) can be verified. Variational principle (4.218) can also be written in the following fractional form:

$$[\langle p'|\Gamma|p\rangle] = \langle p'|(V + AV)|p\rangle + \frac{\langle p'|A|\chi_{k,p}\rangle\langle\bar{\chi}_{k,p'}|AV|p\rangle}{\langle\bar{\chi}_{k,p'}|(I - A)|\chi_{k,p}\rangle}. \qquad (4.219)$$

From Eq. (4.218) we could obtain a practical basis-set method. Expanding the unknown functions in this equation in basis sets and using the stationary property, variational principle (4.218) can be written in the following rank-n form:

$$[\langle p'|\Gamma_n|p\rangle] = \langle p'|(V + AV)|p\rangle + \sum_{i,m=1}^{n} \langle p'|A|u_i\rangle D_{im}^{(n)}\langle w_m|AV|p\rangle, \qquad (4.220)$$

where the matrix D is defined by (4.213) and $\langle w_j|$ and $|u_i\rangle$ are basis functions. The expression (4.220) seems to be particularly suitable for numerical calculation. If the Γ matrix is properly constructed so that the kernel A is weak, the first term on the right-hand side of Eq. (4.220) is a good approximation to the Γ matrix, and the last term is the variational expression for the remainder. As the variational principle is built on a zeroth-order term which is already a good approximation to the Γ matrix, the inclusion of the variational expression for the remainder should lead to good convergence.

The degenerate kernel results (4.212) and (4.220) can also be derived from the following formal solutions for the Γ matrix:

$$\Gamma = V + A \times \frac{1}{I - A} \times V, \tag{4.221}$$

$$\Gamma = V + AV + A \times \frac{1}{I - A} \times AV, \tag{4.222}$$

respectively, by introducing a complete set of states $\sum_i |u_i\rangle\langle u_i|$ and $\sum_j |w_j\rangle\langle w_j|$ at places marked by $\times$'s. However, such a derivation does not exhibit the variational nature of the solution. This was noted in connection with Eqs. (3.24) and (3.39) for solution of the general Fredholm equation (3.6).

All the variational principles considered in this section involves the free Green's function explicitly, and hence the basis functions can be taken to be of the $\mathcal{L}^2$ type and the trouble associated with non-$\mathcal{L}^2$ functions can be avoided.

4.11 ANOMALOUS BEHAVIOR

Until very recently it was believed that all variational principles presented in this chapter do not possess anomalous behavior. The numerical basis-set implementation of these methods involves inversion of a matrix. The anomalous behavior refers to the appearance of spurious poles in the matrix inversion at specific energies that should produce a discontinuous jump in the phase shift. Such behavior is absent in the exact solution. In the usual Kohn variational principles of Chapter 3, similar anomalous behaviors were observed.

The situation under which an anomaly appears becomes clear when the variational principles are written in fractional forms or in forms that require the inversion of a single matrix element. The spurious zero of the matrix element to be inverted corresponds to an anomaly. In the basis-set method this corresponds to a spurious zero of a determinant. For example, let us consider the fractional form of the Schwinger variational principle [e.g., Eq. (4.6)]. As this form is completely equivalent to the trilinear form (4.5) and the finite-rank form (4.24), consideration of this does not imply any specialization. Anomalies may appear when the function ψ is approximated by a trial function. There is no reason why for arbitrary trial functions $\psi = \psi_t$, the denominator $\langle \psi_t | (V - VG_0^{(+)} V) | \psi_t \rangle$ should not vanish at specific positive energies that should lead to anomalies.

Apagyi et al. [7] were the first to find anomalies in any of the variational principles of this chapter. In their study of electron–hydrogen–atom scattering in singlet and triplet states, they noted that certain spurious behavior appears in the phase shifts computed by the Schwinger variational method in the momentum space. The anomaly implies a sudden extra drop of the phase shift through π as energy is increased, so that the phase-shift difference between 0 and ∞ energies $[\delta(0) - \delta(\infty)]$ increases by π in the presence of a single anomaly [8]. In a different context, in their study of nucleon–nucleon scattering using the method of separable expansion (or the Schwinger variational method), Haidenbauer and Plessas [32] found that the phase shift suddenly increases by π with energy, representing a resonancelike behavior. These anomalies correspond to vanishing or nearly vanishing of the denominator of Eq. (4.6) or of the determinant of the D^{-1} matrix of (4.21) in the rank-n form (4.24) of the Schwinger variational method.

The t matrices (4.6) and (4.24) are solutions of the Lippmann–Schwinger equation with the finite-rank nonlocal Hermitian potentials (4.7) and (4.25), respectively. For local Hermitian potentials the domain of energies for which bound and scattering states may appear is completely distinct; there cannot be any square-integrable bound state in the continuum above the threshold for scattering. However, this is not true, in general, for nonlocal finite-rank potentials of type (4.7) or (4.25). Such a finite-rank potential can sustain square-integrable continuum bound states at positive energies, where the homogeneous version of the Lippmann–Schwinger equation permits a nontrivial solution, so that there is no unique scattering solution [8]. When the denominator function of Eq. (4.6) or the determinant of the D^{-1} matrix of (4.21) vanishes, the anomalies of the Schwinger variational principle correspond to these continuum bound states of the finite-rank potentials (4.7) and (4.25) in approximate (unconverged) calculation [8]. They are spurious resonances with zero widths. The converged calculation in the Schwinger variational method corresponds to taking the trial function equal to the exact scattering function with the outgoing-wave boundary condition. Once this choice of trial function is made, the Schwinger variational principle becomes an identity and there are no anomalies.

At the anomalies of the Schwinger method, the phase shift increases abruptly (or discontinuously) by π [8]. However, if a continuous curve for the phase shift versus energy is made, the difference $\delta(0) - \delta(\infty)$ shows an additional jump of π, reflecting the presence of the continuum bound state consistent with the modified Levinson's theorem [8]:

$$\delta(0) - \delta(\infty) = N\pi, \tag{4.223}$$

where N is the total number of bound and continuum bound states. For a slightly different trial function, there are no zeros in the denominator, which, however, remains very small. Then the continuum bound state or the zero-width resonance move into the complex energy plane as a resonance of finite width and the phase shift increases rapidly (but not discontinuously) by π at the resonance energy [8]. In such situations the resulting t or K matrices are not really infinite but very large, and the resultant phase shift exhibits anomalous behavior. Such behavior was observed by Heidenbauer and Plessas [32] in coupled-channel nucleon–nucleon scattering.

Let us next see why the appearance of anomalies in the Schwinger method is plausible. To illustrate this, we consider the simplest fractional form (4.14) of the Schwinger variational principle involving the Green's function. The condition for a zero of the denominator of (4.14) for an arbitrary trial function $|\psi_t\rangle$ was originally thought to be

$$(I - G_0^{(+)} V)|\psi_t\rangle = 0. \tag{4.224}$$

For a real Hermitian local potential this equation does not permit solution in the scattering region and is satisfied only at the bound-state poles at negative energies. However, the anomalies do not correspond to Eq. (4.224) but rather to

$$\langle\psi_t|(I - G_0^{(+)} V)|\psi_t\rangle = 0. \tag{4.225}$$

For arbitrary trial functions, Eq. (4.225) can be satisfied without satisfying (4.224). A similar argument holds in the case of all the variational principles involving the Green's function in the form factors.

If both the potential V and the trial function ψ_t are of definite sign, the matrix element, such as in Eq. (4.225), has not been found to vanish [7,8]. Actually, in numerical application, simple (exponential) trial functions of definite (positive) sign are used. If, in addition, the potential is also taken to be of definite sign (purely attractive or purely repulsive), no zero of the matrix element (4.225) has been found and hence no anomalies. Anomalies have been found for trial functions of definite sign if the potential is not of definite sign (e.g., if it is partly attractive and partly repulsive). In both nuclear and atomic physics there are potentials with a soft core or with short-range repulsive and long-range attractive parts. As the use of such potentials generates a positive and a negative part in the matrix element of Eq. (4.225), cancellation between them leads to anomalies. In all the variational principles presented in this chapter, anomalies have been observed in basis-set calculation with basis functions of definite sign only for potentials with partly attractive and partly repulsive parts.

All variational principles in this chapter can be written either in t-matrix form or in K-matrix form. In the case of the Schwinger variational principle, these two forms yield identical results (phase shifts). Hence, the K-matrix form of the Schwinger variational principle shows the same anomalies as those in the t-matrix form of the Schwinger variational principle.

Other variational principles of this chapter can also exhibit anomalies [9]. There are two types of variational principles: one type involving the free Green's function and the Kohn type involving the Hamiltonian. Let us consider first the type involving the Green's function. This type includes the Newton variational principle and variational principles of higher orders. In both cases there are a t-matrix form and a K-matrix form. Unlike in the case of the Schwinger variational principle, these two forms yield distinct phase shifts. Anomalies appear in these cases due to the vanishing of a denominator function. In the basis-set method it corresponds to the vanishing of a determinant. As in the Schwinger variational principle, such a denominator or determinant can vanish for a particular trial function at a specific energy. In the case of the t-matrix form of the Newton variational principle, such a denominator is a complex number and one has to have simultaneous vanishing of its real and imaginary parts [9,10]. In the case of the K-matrix form, there is no imaginary part and vanishing of the real part alone of such denominators leads to anomalies [9]. As the anomalies are accidental in nature, it will be rarer to have such denominators vanish in t-matrix form than in K-matrix form, as the former requires simultaneous vanishing of the uncorrelated real and imaginary parts of a matrix element at the same energy. In practice, they have been seen to be rare in both cases and can be distinguished from real resonances. Once a more complete set of expansion functions is employed, such anomalies should disappear, whereas a real resonance should establish itself at a specific energy.

As the denominator of the complex Kohn variational principle does not involve the Green's function, the appearance of anomalies in this case deserves some comment [9,10]. The anomalies of the complex Kohn variational principle correspond to the vanishing of the denominator $\langle \eta_t | (E - H_0 - V) | \eta_t \rangle$ of Eq. (4.65) for an arbitrary complex trial function $|\eta_t\rangle$. For a complex function, the eigenfunction-eigenvalue problem

$$(E - H_0 - V)|\eta_t\rangle = 0 \tag{4.226}$$

cannot be satisfied for real Hermitian potentials at positive energies. However, there is no reason why the matrix element appearing in the denominator of Eq. (4.63) could not vanish at positive energies, thus leading to anomalies of the Kohn variational principle. Actually, one could have

$$\langle \eta_t | (E - H_0 - V) | \eta_t \rangle = 0 \tag{4.227}$$

even if Eq. (4.226) is not satisfied.

Unlike the case of the Schwinger variational method, the anomalies of other methods do not correspond to a specific behavior of phase shifts near the anomaly. In these other variational methods, the phase shift does not show a jump of π as in the Schwinger method. In each case they show a peculiar singular behavior [9]. It has been demonstrated numerically that such anomalies, though rare, could appear in all these variational methods for potentials with partly attractive and partly repulsive parts. No anomalies have been found for potentials with a definite sign.

Anomalies may appear in all the variational principles presented in this chapter. All claims to date that they do not exist in these variational principles are false. However, they may not be of concern in actual numerical calculations. They should become narrower and finally are supposed to disappear as the number of basis functions is increased to span the whole space.

4.11.1 Anomaly-Free Kohn Method

Of the different variational principles the real Kohn variational principle of Section 4.6 does not require integration over Green's functions and is simple and advantageous. However, anomalies appear more frequently in this method.

A closely related basis-set method for the K matrix, which avoids spurious singularities but maintains all the advantages and simplicities of the real Kohn method, has been proposed by Takatsuka and McKoy. We consider an on-shell version of the method below, although it can be modified for off-shell elements. The method uses the fractional form of the stationary expression (4.65) for the standing-wave boundary condition [13]:

$$[\langle k|K|k\rangle] = \langle k|V|k\rangle + \frac{\langle k|V|\eta_k^{\mathcal{P}}\rangle\langle\eta_k^{\mathcal{P}}|V|k\rangle}{\langle\eta_k^{\mathcal{P}}|(E - H_0 - V)|\eta_k^{\mathcal{P}}\rangle}, \tag{4.228}$$

where the superscript $\mathcal{P}$ denotes that a principal-value prescription has been used in the calculation of the scattered waves. The denominator of the last term of Eq. (4.228) leads to spurious singularities. Let us consider the following on-shell functional F:

$$\langle k|F|k\rangle \equiv \langle k|V|\tilde{\eta}_x^{\mathcal{P}}\rangle = \langle\tilde{\eta}_x^{\mathcal{P}}|V|k\rangle = \langle\tilde{\eta}_x^{\mathcal{P}}|(E - H_0 - V - dX)|\tilde{\eta}_x^{\mathcal{P}}\rangle, \tag{4.229}$$

where d is an arbitrary parameter, $|\tilde{\eta}_x^{\mathcal{P}}\rangle = x|\eta_k^{\mathcal{P}}\rangle$, x is a nonzero constant, and

$$X \equiv V|k\rangle\langle k|V. \tag{4.230}$$

Then one can formulate the following stationary expression for $\langle k|F|k\rangle$:

$$[\langle k|F|k\rangle] = \frac{\langle k|V|\tilde{\eta}_x^{\mathcal{P}}\rangle\langle\tilde{\eta}_x^{\mathcal{P}}|V|k\rangle}{\langle\tilde{\eta}_x^{\mathcal{P}}|(E - H_0 - V - dX)|\tilde{\eta}_x^{\mathcal{P}}\rangle}. \tag{4.231}$$

The stationary property of this expression follows from the definition of the quantities involved. The functional F of (4.231) reduces to the last term of Eq. (4.228) when $d = 0$. As the matrix elements of E, H_0, and V in the denominator of Eq. (4.231) are all finite, and the term involving d is of definite sign, for a large enough d it is possible to avoid all poles—true and spurious—in $\langle k|F|k\rangle$. The two parameters d and x of Eq. (4.231) are to be adjusted in such a fashion that the functional F remains stationary with variations of $\langle\tilde{\eta}_x^{\mathcal{P}}|$ and $|\tilde{\eta}_x^{\mathcal{P}}\rangle$.

The stationary property of F of Eq. (4.231) leads to

$$(E - H - dX)|\tilde{\eta}_x^{\mathcal{P}}\rangle = V|k\rangle \tag{4.232}$$

or

$$(E - H)|\tilde{\eta}_x^{\mathcal{P}}\rangle = [1 + d\langle k|V|\tilde{\eta}_x^{\mathcal{P}}\rangle]V|k\rangle. \tag{4.233}$$

Recalling that

$$|\eta_k^{\mathcal{P}}\rangle \equiv G_0^{\mathcal{P}} V|\psi_k^{\mathcal{P}}\rangle = |\psi_k^{\mathcal{P}}\rangle - |k\rangle, \tag{4.234}$$

we find that the Schrödinger equation is written as

$$(E - H)|\eta_k^{\mathcal{P}}\rangle = V|k\rangle. \tag{4.235}$$

Demanding the equivalence of Eq. (4.233) with (4.235), we see that $|\tilde{\eta}_x^{\mathcal{P}}\rangle = x|\eta_k^{\mathcal{P}}\rangle$ with

$$x = [1 - d\langle k|V|\eta_k^{\mathcal{P}}\rangle]^{-1}. \tag{4.236}$$

Therefore, the new function $|\tilde{\eta}_x^{\mathcal{P}}\rangle$ is proportional to $|\eta_k^{\mathcal{P}}\rangle$. However, the constant factor x cancels out from the functional (4.231). Hence, functional (4.231) is equivalent to the last term of the Kohn variational principle (4.228) provided that Eq. (4.236) holds.

The resultant F can be written as

$$\langle k|F|k\rangle = x\langle k|V|\eta_k^{\mathcal{P}}\rangle = \frac{\langle k|V|\eta_k^{\mathcal{P}}\rangle}{1 - d\langle k|V|\eta_k^{\mathcal{P}}\rangle} \tag{4.237}$$

or

$$\langle k|V|\eta_k^{\mathcal{P}}\rangle = \frac{\langle k|F|k\rangle}{1 + d\langle k|F|k\rangle}, \tag{4.238}$$

where we have used Eq. (4.236). Then the on-shell K matrix is calculated through

$$\langle k|K|k\rangle = \langle k|V|k\rangle + \langle k|V|\eta_k^{\mathcal{P}}\rangle = \langle k|V|k\rangle + \frac{\langle k|F|k\rangle}{1 + d\langle k|F|k\rangle}. \tag{4.239}$$

All resonances are avoided by this procedure because the denominator of Eq. (4.231) does not have zero for large d. However, the true resonance poles must appear in the result. The true resonances are given by a zero of the denominator in Eq. (4.239):

$$1 + d\langle k|F|k\rangle = 0 \tag{4.240}$$

and not by the zeros in the denominator of Eq. (4.231).

A basis-set calculational scheme can be developed in this approach. If we expand the unknown functions $|\tilde{\eta}_x^{\mathcal{P}}\rangle$ in the stationary expression (4.231) in terms of the functions u_i, $i = 1, ..., n$, we get the following rank-n expression:

$$[\langle k|F_n|k\rangle] = \sum_{i,j}^{n} \langle k|V|u_i\rangle A_{ij}^{(n)} \langle u_j|V|k\rangle, \tag{4.241}$$

with

$$(A^{-1})_{ji} = \langle u_j|E - H - dX|u_i\rangle. \tag{4.242}$$

For $d = 0$ the present formulation reduces to the usual Kohn method. In the present approach X is a positive semidefinite operator, hence for large d all the poles—true and spurious—of the operator A above can be removed. The true resonances correspond to Eq. (4.240).

4.12 NOTES AND REFERENCES

[1] Schwinger (1947a,b).

[2] See, for example, Newton (1982).

[3] Takatsuka and McKoy (1981a).

[4] Adhikari and Sloan (1975a–d); contains extensive numerical studies of Schwinger variational methods to single- and coupled-channel problems.

[5] Sloan and Brady (1972), Brady and Sloan (1972,1974); these seem to be the first numerical studies of the Newton method to single- and coupled-channel problems.

[6] Mito and Kamimura (1976), Miller and Jansen op de Haar (1987); these are two pioneering studies of the complex Kohn variational method.

[7] Apagyi et al. (1988); this is the first study to detect anomalies in any of the variational methods discussed in this chapter.

[8] Adhikari (1990).

[9] Adhikari (1991, 1992a,b).

[10] Adhikari (1992a), Lucchese (1989).

[11] Schwartz (1961a).

[12] Fonseca (1986a,b).

[13] Lucchese et al. (1986); this is a useful review of applications of the Schwinger variational method to problems in atomic and molecular physics.

[14] Amado (1963, 1969), Levinger (1974); these are useful reviews of the numerical solution of few-nucleon scattering problems.

[15] For a review, see Adhikari and Kowalski (1991).

[16] Harms (1970), Levinger (1974).

[17] Weinberg (1963, 1964), Levinger (1974).

[18] Ernst et al. (1973a,b).

[19] Plessas (1986), Plessas and Haidenbauer (1987).

[20] Fonseca and Lim (1985), Sofianos et al. (1979).

[21] Pearce (1987).

[22] In this section we follow the approach of Adhikari (1996).

[23] Feshbach (1958,1962).

[24] McCurdy et al. (1987), Kapur and Peierls (1938), Lane and Robson (1966).

[25] Burke (1977).

[26] Adhikari (1976), this seems to be the first study on the distorted-wave Schwinger variational principle.

[27] Gell-Mann and Goldberger (1953).

[28] Staszewska and Truhlar (1987); this is an useful review of different basis-set methods.

[29] Watson and McKoy (1979).

[30] Rescigno et al. (1976).

[31] Gerjuoy et al. (1983); this review considers the use of variational principles in a variety of problems.

[32] Haidenbauer and Plessas (1983).

CHAPTER 5

VARIATIONAL METHODS FOR REALISTIC PROBLEMS

5.1 INTRODUCTION

So far we have seen developments and applications of variational methods in potential scattering. Most of the realistic scattering processes in nuclear, atomic, and molecular physics can hardly be modeled as potential scattering. In these problems there could be a large number of two-cluster channels even after neglecting the multicluster channels. Tractable two-cluster multichannel models are required before applying variational methods to these problems. In the present chapter we consider a few multichannel reductions of these scattering processes and the use of variational principles for their solution.

The close-coupling (CC) approximation scheme of Section 1.6 provides a theoretical framework for deriving multichannel equations. This scheme has proved to be very useful for electron–light–atom and positron–light–atom scattering at low energies where the target wave function has a simple form [1]. One can use the CC equations directly in the application of different variational methods. In Section 5.2.1 we derive multichannel Schwinger variational principle based on the CC equations. The CC equations also have limitations. Although one can get a reasonable description of scattering using the CC approach at low and medium energies, the final convergence could be slow. Also, in most complex problems, such as electron–molecule scattering, atom–diatom scattering, and photoionization of molecules, the relevant wave functions are unknown. Under these situations one needs, in addition, useful models for relevant bound-state wave functions, and the CC formulation turns out to be complicated. Consequently, the CC scheme may be less attractive under many situations.

Several new calculational schemes have been formulated by different workers. These schemes reduce the actual physical problem to a set of coupled momentum-space scattering equations which can be solved by the use of variational principles or other methods. Hence, these schemes lead to multichannel scattering equations for different processes and variational methods for their solution.

Three major variational principles were considered in Chapter 4: the Schwinger, Newton, and complex kohn variational methods. Each of these approaches can be used in the study of realistic scattering problems, reducing them to practical basis-set methods. McKoy and collaborators [2] used the Feshbach projection operator technique to derive an expression for the multichannel Schwinger variational functional. Then they made a basis-set expansion of the scattering function and reduced their Schwinger variational expression to a set of algebraic equations. They also showed how to formulate their calculational scheme in terms of distorted waves and applied their approach to different realistic problems in atomic and molecular physics. We describe this effort in Section 5.2.2.

Using the Fock coupling scheme and distorted-wave representation, Kouri, Truhlar and collaborators [3] derived the multi-channel Newton variational principle for both inelastic and rearrangement collisions. They also made a basis-set expansion of the wave functions and reduced the Newton variational principle to a set of algebraic equations. They applied this method to realistic problems in molecular physics. Their approach is described in Section 5.3. This approach is closely related to the CC equations of Chapter 1.6. Miller and collaborators [4] extended the complex Kohn variational method to multichannel problems and applied their approach successfully to realistic problems in atomic and molecular physics. This formulation is described in Section 5.4. In formulating the multichannel models one often requires model wave functions for the target and projectiles. Hartree–Fock orbitals have often been used for this purpose. We present a brief account of these orbitals in Section 5.5.

Numerical studies have been made of different scattering processes by using these multichannel formulations. The solution scheme of these formulations takes advantage of methods and techniques based on, or related to, different variational principles developed in Chapter 4. In the following we describe these formulations and their eventual use in the solution of scattering problems.

5.2 SCHWINGER VARIATIONAL METHOD

The Schwinger variational principle can be generalized to the case of multichannel scattering. There are many ways of deriving two-cluster multichannel scattering equations. For example, such a model could be derived employing the CC approach, Feshbach projection operator approach, or from a multiparticle description of scattering, just to name a few. In the following we present two useful ways of formulating the multichannel Schwinger variational principle.

5.2.1 Close-Coupling Equations

We develop the Schwinger variational principle using the multichannel CC equations of Section 1.6.3. The CC equations have the same formal structure as the potential scattering equations. In the multichannel CC equations, all operators have channel indices in addition to the momentum variables. Hence, the variational principles for the CC equations can be written in close analogy with the variational principles of potential scattering. For example, the Schwinger variational principle in this case can be written as

$$[\langle p'_\beta|t|p_\alpha\rangle] = \langle p'_\beta|V_{\beta\alpha}|\psi^{(+)}_{p_\alpha}\rangle + \langle\psi^{(-)}_{p'_\beta}|V_{\beta\alpha}|p_\alpha\rangle - \langle\Psi^{(-)}_{p'_\beta}|(V_{\beta\alpha} - \sum_{\gamma=1}^{N} V_{\beta\gamma}G^{(+)}_{0\gamma}V_{\gamma\alpha})|\Psi^{(+)}_{p_\alpha}\rangle, \tag{5.1}$$

where the subscripts β and α on the plane waves and the scattering states refer to channels. The exact scattering states obey the formulas

$$|\Psi^{(+)}_{p_\alpha}\rangle = |p_\alpha\rangle + \sum_{\gamma=1}^{N} G^{(+)}_{0\alpha}V_{\alpha\gamma}|\Psi^{(+)}_{p_\gamma}\rangle, \tag{5.2}$$

$$\langle\Psi^{(-)}_{p'_\beta}| = \langle p'_\beta| + \sum_{\gamma=1}^{N}\langle\Psi^{(-)}_{p'_\gamma}|V_{\gamma\beta}G^{(+)}_{0\beta}. \tag{5.3}$$

The stationary property of Eq. (5.1) follows when approximate wave functions are employed. In fractional form, variational principle (5.1) becomes

$$[\langle p'_\beta|t|p_\alpha\rangle] = \frac{\langle p'_\beta|V_{\beta\alpha}|\psi^{(+)}_{p_\alpha}\rangle\langle\psi^{(-)}_{p'_\beta}|V_{\beta\alpha}|p_\alpha\rangle}{\langle\psi^{(-)}_{p'_\beta}|(V_{\beta\alpha} - \sum_{\gamma=1}^{N} V_{\beta\gamma}G^{(+)}_{0\gamma}V_{\gamma\alpha})|\Psi^{(+)}_{p_\alpha}\rangle}. \tag{5.4}$$

To perform a numerical calculation one needs to expand each of the scattering states in a basis set:

$$|\psi^{(+)}_{p_\alpha}\rangle = \sum_{i=1}^{n} a_{\alpha i}(p_\alpha)|f_{\alpha i}\rangle, \tag{5.5}$$

$$\langle\psi^{(-)}_{p'_\beta}| = \sum_{j=1}^{n}\langle f_{\beta j}|b_{\beta j}(p'_\beta), \tag{5.6}$$

where $a_{\alpha i}$ and $b_{\beta j}$ are coefficients to be determined by the variational requirement and $|f_{\alpha i}\rangle$ is a set of expansion states for channel α.

To determine the coefficients $a_{\alpha i}$ and $b_{\beta j}$ we substitute Eqs. (5.5) and (5.6) into the Schwinger variational principle (5.1) and require that the resulting expression be stationary with respect to independent variations of $a_{\alpha i}$ and $b_{\beta j}$. Then in the usual fashion one obtains the following finite-rank version of the multichannel Schwinger variational principle:

$$[\langle p'_\beta|t_n|p_\alpha\rangle] = \sum_{i,j=1}^{n}\sum_{\gamma,\sigma=1}^{N}\langle p'_\beta|V_{\beta\gamma}|f_{\gamma i}\rangle D^{(n)}_{\gamma i\sigma j}\langle f_{\sigma j}|V_{\sigma\alpha}|p_\alpha\rangle, \tag{5.7}$$

where

$$(D^{-1})_{\sigma j\gamma i} = \langle f_{\sigma j}|(V_{\sigma\gamma} - \sum_{\kappa=1}^{N} V_{\sigma\kappa}G^{(+)}_{0\kappa}V_{\kappa\gamma})|f_{\gamma i}\rangle, \qquad i,j = 1,\ldots,n. \tag{5.8}$$

These results are very similar to those of the single-channel case [e.g., Eqs. (4.24) and (4.21)]. However, the dimension of the space is increased in the multichannel case. The scattering wave function for each channel is expanded in terms of several functions. The total number of expansion functions in all channels define the dimension of the matrix D above. If an equal number n of functions is used in each N channel, the dimension of the D matrix is $\mathcal{N} = nN$. In Eq. (5.8), explicitly, we have

$$\langle f_{\sigma j}|V_{\sigma\kappa}G_{0\kappa}^{(+)}V_{\kappa\gamma}|f_{\gamma i}\rangle = \frac{2}{\pi}\int q^2\,dq\,\frac{\langle f_{\sigma j}|V_{\sigma\kappa}|q\rangle\langle q|V_{\kappa\gamma}|f_{\gamma i}\rangle}{k_\kappa^2 - q^2 + i0}. \tag{5.9}$$

Equations (5.7) and (5.8) form a useful basis-set calculational method for multichannel scattering based on the Schwinger variational principle.

5.2.2 Feshbach Projection Operator Technique

Next, we describe an approach utilizing Feshbach projection operators [5] for deriving multichannel scattering equations from a many-body problem [2]. Once a multichannel scattering equation is obtained, it can be solved with the use of the Schwinger variational principle.

In the CC equations, the same multichannel form exists even when some of the channels are energetically closed. Consequently, in applications of the variational methods, similar mathematical and numerical difficulties appear independent of whether or not some of the channels are closed. The Feshbach projection operator technique provides one way of simplifying the calculation when some of the channels are closed. In this approach the number of the channels of the model include only some (or all) of the open channels. Closed channels are not included. Let us consider scattering of a light elementary particle (electron) by a target composed of N electrons. The Hamiltonian for this problem is

$$H = H_0 + V \equiv (H_N + T_{N+1}) + V, \tag{5.10}$$

where H_N is the target Hamiltonian, T_{N+1} the kinetic energy of the projectile, V the interaction potential, and N the number of target electrons. For the electron–molecule system, the potential V is given by

$$V = \sum_{i=1}^{N} \frac{e^2}{r_{i,N+1}} - \sum_{\alpha} \frac{Z_\alpha e^2}{r_{\alpha,N+1}}, \tag{5.11}$$

where the first term represents the repulsion between N target electrons at $\mathbf{r} = \mathbf{r}_i, i = 1, 2, ..., N$, and the incident electron at $\mathbf{r} = \mathbf{r}_{N+1}$, and the second term represents the attraction between the incident electron and the nuclei at $\mathbf{r} = \mathbf{r}_\alpha$, with e the electronic charge and $Z_\alpha e$ the nuclear charge. In Eq. (5.11), $r_{i,N+1} = |\mathbf{r}_i - \mathbf{r}_{N+1}|$, $r_{\alpha,N+1} = |\mathbf{r}_\alpha - \mathbf{r}_{N+1}|$.

We introduce Feshbach projection operators P and Q for open and closed channels defined in the Hilbert space of $N+1$ particles, respectively. The operators P and Q satisfy $P + Q = 1$, $PQ = QP = 0$. The Schrödinger equation for the open-channel space is then given by [5]

$$[P\hat{H}P + P\hat{H}Q(Q\hat{H}Q)^{-1}Q\hat{H}P]|\Psi^{(+)}\rangle = 0, \tag{5.12}$$

with $\hat{H} = (E - H)$. These P space equations have been used in applications of the multichannel Kohn and Schwinger variational methods [2,6]. However, there is one nontrivial problem with this approach. The elastic optical potential $P\hat{H}Q(Q\hat{H}Q)^{-1}\ Q\hat{H}P$ may have singularities at real positive energies. These singularities may be related to physical resonances of the system; they could also be spurious. The spurious resonances are an added difficulty to potential scattering.

Lucchese et al. [2,7] suggested a useful and practical method that avoids the above-mentioned singularities. They take P as the projector onto the target eigenstates ϕ_n for open channels only:

$$P = \sum_n |\phi_n\rangle\langle\phi_n|. \tag{5.13}$$

With this operator, a projected Lippmann–Schwinger equation for scattering initiated in the target state n_0 can be written as

$$P|\Psi_{n_0}^{(+)}\rangle = |\Phi_{n_0}\rangle + G_{PP}^{(+)} V|\Psi_{n_0}^{(+)}\rangle, \tag{5.14}$$

where

$$|\Phi_{n_0}\rangle = |\phi_{n_0}\rangle \exp(i\mathbf{k}_{n_0} \cdot \mathbf{r}_{N+1}), \tag{5.15}$$

is the usual incident plane wave and

$$G_{PP}^{(+)} = P(E - PH_0P + i0)^{-1}P \tag{5.16}$$

is the projected Green's function on the target states. The Hamiltonians, potentials, and energies are all expressed in units of $\hbar^2/2\mu$, where μ is the reduced mass of the system.

Now operating by the potential V on Eq. (5.14), we obtain

$$(VP - VG_{PP}^{(+)}V)|\Psi_{n_0}^{(+)}\rangle = V|\Phi_{n_0}\rangle. \tag{5.17}$$

Equation (5.17) is an open-channel projected equation and does not contain closed-channel information. Because of the presence of the projection operator P, the quantity $VP - VG_{PP}V$ of Eq. (5.17) is not time-reversal symmetric, as the closed-channel contributions are projected out of this equation in a nonsymmetrical fashion. To restore symmetry, we consider the following projected Schrödinger equation:

$$[\hat{H} - a(P\hat{H} + \hat{H}P)]|\Psi_{n_0}^{(+)}\rangle = a(VP - PV)|\Psi_{n_0}^{(+)}\rangle, \tag{5.18}$$

with a an arbitrary parameter. Equation (5.18) can be verified using the Schrödinger equation $\hat{H}|\Psi_{n_0}^{(+)}\rangle = 0$ and the fact that the projection operator P commutes with the target Hamiltonian: $PH_0 = H_0P$. A linear combination of Eqs. (5.17) and (5.18) leads to the following integrodifferential equation with symmetric operators:

$$\left[\frac{PV + VP}{2} - VG_{PP}^{(+)}V + \frac{\hat{H} - a(P\hat{H} + \hat{H}P)}{2a}\right]|\Psi_{n_0}^{(+)}\rangle = V|\Phi_{n_0}\rangle. \tag{5.19}$$

By construction Eq. (5.19) determines the full multichannel scattering wave function. For Fermion collisions, the constant a can be determined from the condition that the problem leads to a unique solution. Let us generalize the outgoing-wave equation (5.19) for both outgoing and incoming waves:

$$A^{(\pm)}|\Psi_{n_0}^{(\pm)}\rangle \equiv \left[\frac{PV + VP}{2} - VG_{PP}^{(\pm)}V + \frac{\hat{H} - a(P\hat{H} + \hat{H}P)}{2a}\right]|\Psi_{n_0}^{(\pm)}\rangle \tag{5.20}$$

$$= V|\Phi_{n_0}\rangle. \tag{5.21}$$

The amplitude for transition from the target state n_0 to state n is defined by

$$T_{nn_0} = \langle\Psi_n^{(-)}|V|\Phi_{n_0}\rangle = \langle\Phi_n|V|\Psi_{n_0}^{(+)}\rangle. \tag{5.22}$$

Because of Eq. (5.21), this amplitude can also be written as

$$T_{nn_0} = \langle \Psi_n^{(-)} \left| \left[\frac{PV + VP}{2} - VG_{PP}V + \frac{\hat{H} - a(P\hat{H} + \hat{H}P)}{2a} \right] \right| \Psi_{n_0}^{(+)} \rangle. \tag{5.23}$$

From Eqs. (5.22) and (5.23), a symmetrized multichannel Schwinger variational functional for the transition operator can be written as

$$[T_{nn_0}] = \frac{\langle \Psi_n^{(-)} | V | \Phi_{n_0} \rangle \langle \Phi_n | V | \Psi_{n_0}^{(+)} \rangle}{\langle \Psi_n^{(-)} | [(PV + VP)/2 - VG_{PP}V + \{\hat{H} - a(P\hat{H} + \hat{H}P)\}/(2a)] | \Psi_{n_0}^{(+)} \rangle}, \tag{5.24}$$

where $\langle \Psi_n^{(-)}|$ and $|\Psi_{n_0}^{(+)}\rangle$ are the unknown trial functions. The stationary property of Eq. (5.24) follows with the use of Eq. (5.21). For obtaining a unique solution from this functional, one should have $\langle \Psi_n^{(-)} | A^{(+)} \Psi_n^{(+)} \rangle = \langle A^{(-)} \Psi_n^{(-)} | \Psi_n^{(+)} \rangle$. Lucchese et al. showed that for electrons, the choice $a = (N+1)/2$ is required to satisfy this condition [2,7].

For numerical applications, one makes an expansion of the unknown scattering function in terms of some known $\mathcal{L}^2$ basis set $|g_i\rangle$. Consequently, as in the case of potential scattering, the multichannel Schwinger variational principle can be reduced to solution of the following set of linear equations:

$$[T_{nn_0}] = \sum_{i,j} \langle \Phi_n | V | g_i \rangle C_{ij} \langle g_j | V | \Phi_{n_0} \rangle, \tag{5.25}$$

where $(C^{-1})_{ji} = \langle g_j | A^{(+)} | g_i \rangle$.

Simplifications arise in the case of elastic scattering when one makes a static exchange approximation to this formulation. The total wave function is taken as

$$|\Psi_0^{(+)}\rangle = \mathcal{A} | \phi_0 \psi^{(+)} \rangle, \tag{5.26}$$

where the zero subscript denotes the target ground state, ϕ_0 its (Hartree–Fock) wave function (see Section 5.5), $\psi^{(+)}$ the scattering orbital, and $\mathcal{A}$ the antisymmetrizer. Then integration over the target coordinates yields

$$[T_{nn_0}] = \frac{\langle \psi^{(-)} | V_s | k_{n_0} \rangle \langle k_n | V_s | \psi^{(+)} \rangle}{\langle \psi^{(-)} | (V_s - V_s g_0^{(+)} V_s) | \psi^{(+)} \rangle}, \tag{5.27}$$

where V_s is the static exchange potential and $g_0^{(+)}$ is the Green's function in this space. As expected, the result is analogous to that of potential scattering.

Technical complications arise in the case of electron–molecule collisions and in molecular photoionization. In these cases the physical scattering potential has parts corresponding to completely different ranges. The direct molecular Hartree–Fock potential is of long range, whereas the exchange and polarization potentials are of much shorter range. These two types of potentials can be treated by a two-potential formalism of Section 4.7. Such a two-potential approach has been used in studies of electron–molecule collisions employing the Kohn variational principle [8]. This approach has also been extended to the case of multichannel Schwinger variational principle and is useful in studies of electronic excitation of dipole-allowed states [2].

The present approach can be generalized to yield a distorted wave formulation (see Section 4.7). First the scattering problem is solved with the direct potential. The solutions are called the distorted waves for the problem. The residual potential is then treated in the distorted-wave basis and the Schwinger variational principle is applied to the residual problem. The direct problem for the construction of distorted waves is defined by the following Hamiltonian:

$$\tilde{H}_0 = H_N + T_{N+1} + PVP, \tag{5.28}$$

where the potential, Hamiltonian, and projection operators are defined as in Eqs. (5.10) and (5.13). The potential V of Eq. (5.28) may include, in addition to a direct local part, a model local exchange and a local polarization part. As the Hamiltonian $\tilde{H}_0$ contains only local potentials, solution of the scattering problem with this Hamiltonian can be obtained routinely.

The projected wave function satisfies an equation similar to (5.14):

$$P|\Psi_{n_0}^{(+)}\rangle = |\chi_{n_0}^{(+)}\rangle + G_R^{(+)} V_R |\Psi_{n_0}^{(+)}\rangle, \tag{5.29}$$

where $V_R = V - PVP$ and

$$G_R^{(+)} = P(E - P\tilde{H}_0 P + i0)^{-1} P \tag{5.30}$$

is the projected Green's function for the Hamiltonian $\tilde{H}_0$ of Eq. (5.28). The states $\chi_{n_0}^{(+)}$ are the distorted waves. The multichannel Schwinger variational principle can now be constructed in terms of the distorted waves as in Section 4.7. In the distorted-wave basis, the amplitude for transition from state n_0 to n is now given by

$$T_{nn_0}^{d} = \langle \Psi_n^{(-)} | V_R | \chi_{n_0}^{(+)} \rangle = \langle \chi_n^{(-)} | V_R | \Psi_{n_0}^{(+)} \rangle \tag{5.31}$$

$$= \langle \Psi_n^{(-)} | \left[\frac{PV_R + V_R P}{2} - V_R G_R^{(+)} V_R + \frac{\hat{H} - a(P\hat{H} + \hat{H}P)}{2a} \right] | \Psi_{n_0}^{(+)} \rangle. \tag{5.32}$$

Then one can construct the following Schwinger variational functional for the transition amplitude due to the nonlocal short-range potential V_R:

$$[T_{nn_0}]_s = \frac{\langle \Psi_n^{(-)} | V_R | \chi_{n_0}^{(+)} \rangle \langle \chi_n^{(-)} | V_R | \Psi_{n_0}^{(+)} \rangle}{\langle \Psi_n^{(-)} | [(PV_R + V_R P)/2 - V_R G_R^{(+)} V_R + \{\hat{H} - a(P\hat{H} + \hat{H}P)\}/(2a)] | \Psi_{n_0}^{(+)} \rangle}, \tag{5.33}$$

with $a = (N+1)/2$. Here $\langle \Psi_n^{(-)} |$ and $| \Psi_{n_0}^{(+)} \rangle$ are the unknown trial functions. The derivation of Eq. (5.33) follows exactly the same steps as the derivation of Eq. (5.24). For a numerical application of Eq. (5.33), the trial functions are then expanded in a set of $\mathcal{L}^2$ functions and this problem is reduced to a set of algebraic equations as in Eq. (5.25).

5.3 NEWTON VARIATIONAL METHOD

Here we present a formalism [3], due to Kouri, Truhlar, and collaborators, for treating (chemical) reactions with general basis functions, including electronically excited states. The approach is especially well suited for atom–diatom scattering. The scattering observables are calculated by the use of a generalized Newton variational principle on a distorted-wave basis.

5.3.1 Inelastic Scattering

First, we consider the simple case of inelastic and nonreactive scattering of atom A with diatomic molecule BC, where only excitations of target and projectile are allowed. In the present notation this corresponds to partition $a = 1$. In this case the Hamiltonian can be partitioned as

$$H = H^A + V^{\text{int}} \equiv (H_0 + V^{\text{vib}}) + V^{\text{int}}, \tag{5.34}$$

where H_0 is the kinetic energy, $V \equiv (V^{\text{vib}} + V^{\text{int}})$ the full potential of the system, V^{vib} the long-range potential responsible for binding of the clusters, and V^{int} the remaining interaction potential. The asymptotic Hamiltonian H^A responsible for binding satisfies $H^A\Phi_n = E\Phi_n$, where Φ_n is the product of the bound states of the target and the projectile and a plane wave of relative motion between them and n labels the different states of the projectile and the target.

In the next stage the interaction potential is broken up into a distortion (V^D) and a coupling (V^C) potential: $V^{\text{int}} = V^D + V^C$. The distortion potential V^D is usually a simple potential that takes most of the scattering into account. Then the distorted-wave Hamiltonian H^D is defined by

$$H = H^D + V^C \equiv (H_0 + V^{\text{vib}} + V^D) + V^C. \tag{5.35}$$

The distorted-wave states $|\Phi_{n_0}^{(+)}\rangle$ satisfy $(E - H^D)|\Phi_{n_0}^{(+)}\rangle = 0$ and have the structure $|\Phi_{n_0}^{(+)}\rangle = |\phi_{n_0}\rangle|\phi^{(+)}\rangle$, where $|\phi_{n_0}\rangle$ represents the bound states of the clusters in channel n_0 and $|\phi^{(+)}\rangle$ is the distorted wave in this channel. Here n_0 is the incident channel, and the energies and Hamiltonians have been expressed in units of $\hbar^2/2\mu$, where μ is the appropriate reduced mass.

The scattering states $|\Psi_{n_0}^{(+)}\rangle$ satisfy the Lippmann–Schwinger equations

$$|\Psi_{n_0}^{(+)}\rangle = |\Phi_{n_0}^{(+)}\rangle + G^{D(+)}V^C|\Psi_{n_0}^{(+)}\rangle, \tag{5.36}$$

where $G^{D(+)} = (E - H^D + i0)^{-1}$. Equation (5.36) is the usual multiparticle Lippmann–Schwinger equation with a noncompact kernel and has a nonunique solution. By projecting this equation on channel states $|\Phi_{n_0}^{(+)}\rangle$, Fredholm equations with a unique solution can be obtained.

In applying a variational principle to solve (5.36), one could proceed differently. First, formal variational principles are written. Then by expanding the unknown function in $\mathcal{L}^2$ basis functions, a well-defined set of linear equations are derived. The appropriate matrix element of the transition operator, given by

$$T_{nn_0} \equiv \langle \Phi_n^{(-)} | T | \Phi_{n_0}^{(+)} \rangle = \langle \Phi_n^{(-)} | V^C | \Psi_{n_0}^{(+)} \rangle, \tag{5.37}$$

satisfies the following generalized Newton variational principle:

$$[T_{nn_0}] = \langle \Phi_n^{(-)} | V^C | \Phi_{n_0}^{(+)} \rangle + \langle \Phi_n^{(-)} | V^C G^{D(+)} | \chi_{n_0}^{(+)} \rangle + \langle \chi_n^{(-)} | G^{D(+)} V^C | \Phi_{n_0}^{(+)} \rangle - \langle \chi_n^{(-)} | (G^{D(+)} - G^{D(+)} V^C G^{D(+)}) | \chi_{n_0}^{(+)} \rangle, \tag{5.38}$$

where the first term on the right-hand side of Eq. (5.38) is the distorted-wave Born amplitude and

$$|\chi_n^{(\pm)}\rangle \equiv V^C |\Psi_n^{(\pm)}\rangle \tag{5.39}$$

is the amplitude density. In Eq. (5.38) $\langle \chi_n^{(-)}|$ and $|\chi_{n_0}^{(+)}\rangle$ are the unknown trial functions. The stationary property of Eq. (5.38) follows as in the case of potential scattering.

Closely related to the generalized Newton variational principle above, one can write an outgoing-wave variational principle for the transition-matrix element by introducing the outgoing/incoming waves as

$$|\eta_n^{(\pm)}\rangle = G^{D(\pm)} |\chi_n^{(\pm)}\rangle. \tag{5.40}$$

Using these waves, the generalized Newton variational principle can be rewritten in the form of the following Kohn-type outgoing-wave variational principle:

$$[T_{nn_0}] = \langle \Phi_n^{(-)} | V^C | \Phi_{n_0}^{(+)} \rangle + \langle \Phi_n^{(-)} | V^C | \eta_{n_0}^{(+)} \rangle + \langle \eta_n^{(-)} | V^C | \Phi_{n_0}^{(+)} \rangle - \langle \eta_n^{(-)} | (E - H) | \eta_{n_0}^{(+)} \rangle. \tag{5.41}$$

The stationary property of the generalized outgoing-wave variational principle (5.41) follows as in the case of potential scattering. Outgoing-wave variational principle (5.41) should be compared to single-channel equation (4.63). There are no Green's functions in variational principle (5.41). To use it one must employ basis functions with proper boundary conditions. If outgoing-wave boundary conditions are employed, this method is a generalization of the complex Kohn variational principle of Section 4.6 and becomes identical to the multichannel complex Kohn variational principle of Section 5.4, which can be used to calculate the multichannel t-matrix elements. We shall come to a detailed discussion of this method in Section 5.4. However, if standing-wave boundary conditions are employed, this method yields multichannel K-matrix elements.

5.3.2 Reactive Scattering

We now turn to a description of reactive scattering of atom A with diatomic molecule BC, where rearrangement transitions are explicitly taken into account. Now in place of Eq. (5.34), the following three partitions of the Hamiltonian are allowed for three values of partition a:

$$H = H_a^A + V_a^{\text{int}} \equiv (H_0 + V_a^{\text{vib}}) + V_a^{\text{int}}. \tag{5.42}$$

The three allowed values of a correspond to three arrangements of the atoms: $a = 1$ for arrangement $A + BC$, $a = 2$ for $B + CA$, and $a = 3$ for $C + AB$. Each of these partitions is appropriate for more than one channel, corresponding to excited states of fragments.

The asymptotic Hamiltonian responsible for the binding of the clusters in partition a, H_a^A is defined in Eq. (5.42) and satisfies the eigenfunction–eigenvalue problem $H_a^A \Phi_{an} = E\Phi_{an}$, where n is now the channel index in partition a. Each partition may correspond to more than one channel. For example, n varies from 1 to n_1 for $a = 1$, from $(n_1 + 1)$ to n_2 for $a = 2$, and so on. In the following material we shall, however, carry the dummy partition index a with the channel index to denote that channel n belongs to partition a. The cluster states $|\Phi_{an}\rangle$ are the product of the bound states of the two fragments of channel $\alpha \equiv an$ and the plane wave of relative motion between them.

Then, as in the inelastic case, the interaction potential in partition a is broken up into a distortion (V_a^D) and a coupling (V^a) potential by $V_a^{\text{int}} = V_a^D + V^a$. The distorted-wave Hamiltonian H_a is defined by the partition

$$H = H_a + V^a \equiv (H_0 + V_a^{\text{vib}} + V_a^D) + V^a. \tag{5.43}$$

The partitions of the Hamiltonian by Eq. (5.43) is the generalization of Eq. (5.35) in the absence of rearrangement.

The distorted-wave states $|\Phi_{a_0 n_0}^{(+)}\rangle$ for an initial channel n_0 in partition a_0 satisfy $(E - H_{a_0})|\Phi_{a_0 n_0}^{(+)}\rangle = 0$, where n_0 is the initial channel belonging to partition a_0. The full outgoing-wave scattering states $|\Psi^{a_0 n_0(+)}\rangle$ satisfy the following set of decoupled Lippmann–Schwinger equations:

$$|\Psi^{a_0 n_0(+)}\rangle = \delta_{aa_0}|\Phi^{(+)}_{a_0 n_0}\rangle + G_a^{(+)} V^a |\Psi^{a_0 n_0(+)}\rangle, \tag{5.44}$$

where $G_a^{(+)} = (E - H_a + i0)^{-1}$. Equations (5.44) are the usual multichannel Lippmann–Schwinger equations of Section 1.6.1 and have noncompact kernel and nonunique solution. Compactness can be obtained by projecting them on appropriate channel states.

A variational calculational scheme can be developed by expanding the full wave function in components:

$$|\Psi^{a_0 n_0(+)}\rangle = \sum_a |\Psi_a^{a_0 n_0(+)}\rangle. \tag{5.45}$$

The wave-function components satisfy the following coupled set of generalized Lippmann–Schwinger equations:

$$|\Psi_a^{a_0 n_0(+)}\rangle = \delta_{aa_0}|\Phi^{(+)}_{a_0 n_0}\rangle + G_a^{(+)} \sum_{a'} U_{aa'} |\Psi_{a'}^{a_0 n_0(+)}\rangle, \tag{5.46}$$

where

$$U_{aa'} = V^a - (E - H_a)\bar{\delta}_{aa'} \tag{5.47}$$

are the usual Fock coupling potentials. Equation (5.46) is a generalization of Eqs. (1.180) and (1.181) to the case where there could be more than one channel in a given partition. However, Eqs. (1.180) and (1.181) are written in terms of the incident plane wave, whereas in Eq. (5.46) a distorted-wave incident state is employed.

For developing the generalized Newton variational principle, we introduce a reactive amplitude density in partition a_0 for initial channel n_0 by

$$|\chi_a^{a_0 n_0(+)}\rangle = \sum_{a'} U_{aa'} |\Psi_{a'}^{a_0 n_0(+)}\rangle. \tag{5.48}$$

The transition matrix connecting an initial state n_0 in partition a_0 to a final state n in partition a is defined as

$$\langle \Phi^{(-)}_{an} | T^{aa_0}_{nn_0} | \Phi^{(+)}_{a_0 n_0}\rangle = \langle \Phi^{(-)}_{an} | \chi_a^{a_0 n_0(+)}\rangle = \sum_{a'} \langle \Phi^{(-)}_{an} | U_{aa'} | \Psi_{a'}^{a_0 n_0(+)}\rangle. \tag{5.49}$$

Then by multiplying Eq. (5.46) from the left by appropriate elements of potential U and using Eq. (5.49), one can derive the following Lippmann–Schwinger equations for the transition operators:

$$T_{nn_0}^{aa_0} = U_{aa_0} + \sum_{a'} U_{aa'} G_{a'}^{(+)} T_{n'n_0}^{a'a_0}. \tag{5.50}$$

In close analogy with potential scattering, in this case a generalized Newton variational principle can be written as

$$\begin{aligned}
[\langle \Phi_{an}^{(-)} | T_{nn_0}^{aa_0} | \Phi_{a_0 n_0}^{(+)} \rangle] = {} & \langle \Phi_{an}^{(-)} | U_{aa_0} | \Phi_{a_0 n_0}^{(+)} \rangle + \sum_{a'} \langle \Phi_{an}^{(-)} | U_{aa'} G_{a'}^{(+)} | \chi_{a'}^{a_0 n_0 (+)} \rangle \\
& + \sum_{a'} \langle \chi_{a'}^{an(-)} | G_{a'}^{(+)} U_{a'a_0} | \Phi_{a_0 n_0}^{(+)} \rangle \\
& - \sum_{a'a''} \langle \chi_{a'}^{an(-)} | (G_{a'}^{(+)} \delta_{a'a''} - G_{a'}^{(+)} U_{a'a''} G_{a''}^{(+)}) | \chi_{a''}^{a_0 n_0 (+)} \rangle,
\end{aligned} \tag{5.51}$$

where $\langle \chi_{a'}^{an(-)} |$ and $| \chi_{a''}^{a_0 n_0 (+)} \rangle$ are unknown trial functions.

It is possible to write the final result in terms of the partition potentials, V^a, of Eq. (5.43). With $U_{aa'}$ defined by Eq. (5.47), we have for both $a = a'$ and $\neq a'$,

$$U_{aa'} | \Phi_{a'n'}^{(+)} \rangle = (H - H_{a'}) | \Phi_{a'n'}^{(+)} \rangle = V^{a'} | \Phi_{a'n'}^{(+)} \rangle, \tag{5.52}$$

$$U_{aa'} G_{a'}^{(+)} = (H - H_{a'}) G_{a'}^{(+)} - \bar{\delta}_{aa'} = V^{a'} G_{a'}^{(+)} - \bar{\delta}_{aa'}. \tag{5.53}$$

Then a symmetrized version of the generalized on-shell Newton variational principle (5.51) for reactive scattering can be written as

$$\begin{aligned}
[\langle \Phi_{an}^{(-)} | T_{nn_0}^{aa_0} | \Phi_{a_0 n_0}^{(+)} \rangle] = {} & \langle \Phi_{an}^{(-)} | V^{a_0} | \Phi_{a_0 n_0}^{(+)} \rangle \\
& - \sum_{a'a''} \langle \chi_{a'}^{an(-)} | (G_{a'}^{(+)} - G_{a'}^{(+)} V^{a''} G_{a''}^{(+)}) | \chi_{a''}^{a_0 n_0 (+)} \rangle \\
& + \sum_{a'} [\langle \Phi_{an}^{(-)} | V^{a} G_{a'}^{(+)} | \chi_{a'}^{a_0 n_0 (+)} \rangle \\
& + \langle \chi_{a'}^{an(-)} | G_{a'}^{(+)} V^{a_0} | \Phi_{a_0 n_0}^{(+)} \rangle].
\end{aligned} \tag{5.54}$$

Equation (5.54) is a formal expression of the multichannel Newton variational principle with distorted wave in terms of (physical) channel potentials (5.11). As in the case of reactive scattering, one can also derive a multichannel Kohn variational principle in this case. If an outgoing-wave boundary condition is used, one obtains a different derivation of the multichannel complex-Kohn variational principle with rearrangement, which we discuss in the next section.

To develop an algebraic calculational scheme based on Eq. (5.38) or (5.54), we expand the amplitude densities $|\chi^{(+)}\rangle$ and $\langle\chi^{(-)}|$ in terms of $\mathcal{L}^2$ basis functions $|g_i\rangle$. Let us illustrate this in the case of Eq. (5.38). The index i includes complete specification of the indices for translational, vibrational, and rotational basis functions. Using these expansions in (5.38) and demanding that the result remain stationary with respect to the coefficients, we obtain the degenerate expression

$$[T_{nn_0}] = \langle\Phi_n^{(-)}|V^C|\Phi_{n_0}^{(+)}\rangle + \sum_{i,j}\langle\Phi_n^{(-)}|V^C G^{D(+)}|g_j\rangle C_{ji}\langle g_i|G^{D(+)}V^C|\Phi_{n_0}^{(+)}\rangle, \tag{5.55}$$

where

$$(C^{-1})_{ij} = \langle g_i|(G^{D(+)} - G^{D(+)}V^C G^{D(+)})|g_j\rangle. \tag{5.56}$$

A similar degenerate kernel expression can be obtained from Eq. (5.54). In that case one will have a summation over channel index and variational functions as in Eq. (5.7). The final result Eq. (5.55), is a generalization of the potential scattering result discussed in Section 4.3. A similar expansion can be derived for the Kohn-type multichannel variational principle (5.41). The success of a calculational scheme using an algebraic basis-set equation such as Eq. (5.55) depends on an appropriate choice of of the basis functions, for a discussion of which we refer the interested reader to review articles [3].

5.3.3 Photodissociation of Triatomic Molecule

Photodissociation with initial state selection for studying atom–diatom molecular dynamics may probe a single total angular momentum state of the target triatomic molecule. However, the computational task for this problem continues to be challenging, especially when there are multiple dissociation channels, whether or not they are energetically open.

We consider an application of the Newton variational principle for studying photodissociation [9] in the weak-photon field limit in the ground or a single excited electronic state for a single breakup channel. The Schrödinger equation for the state vector on the electronically excited surface is nonhomogeneous and is given by

$$(E - H_0)|\Psi^{(+)n_0}\rangle = V|\Psi^{(+)n_0}\rangle + X|\xi^{n_0}\rangle, \tag{5.57}$$

where $|\xi^{n_0}\rangle$ is the initial state for the system on the lower electronic surface, X the dipole coupling responsible for the excitation to the dissociative excited electronic state, H_0 the Hamiltonian for the asymptotic photodisintegration fragments, and V the interaction between the fragments. Here $|\Psi^{(+)n_0}\rangle$ is the scattering state produced by photodissociation. The quantum index n_0 is a collective index for all the quantum numbers of the initial bound state $|\xi^{n_0}\rangle$.

In the usual Lippmann–Schwinger equation there is an inhomogeneous term corresponding to the initial plane-wave state. In the case of photodissociation there is no such term, and one can convert Eq. (5.57) to the integral form

$$|\Psi^{(+)n_0}\rangle = G_0^{(+)}X|\xi^{n_0}\rangle + G_0^{(+)}V|\Psi^{(+)n_0}\rangle, \tag{5.58}$$

where the free outgoing-wave Green's function $G_0^{(+)}$ is defined by $G_0^{(+)} = (E - H_0 + i0)^{-1}$.

The transition amplitude for dissociating into a final two-cluster state $\langle\Phi_n|$ is

$$T_{nn_0} = \langle\Phi_n|[X|\xi^{n_0}\rangle + V|\Psi^{(+)n_0}\rangle] \equiv \langle\Phi_n|\chi_{n_0}^{(+)}\rangle. \tag{5.59}$$

Here $\chi_{n_0}^{(+)}$ is an amplitude density. The state $|\Phi_n\rangle$ satisfies $(E - H_0)|\Phi_n\rangle = 0$ and is the product of bound states of the final clusters and the relative plane wave between them. If we define the amplitude density $|\chi_{n_0}^{(+)}\rangle$ via $|\Psi^{(+)n_0}\rangle = G_0^{(+)}|\chi_{n_0}^{(+)}\rangle$, then from Eqs. (5.58) and (5.59) we get the following inhomogeneous integral equation for the amplitude density:

$$|\chi_{n_0}^{(+)}\rangle = X|\xi^{n_0}\rangle + VG_0^{(+)}|\chi_{n_0}^{(+)}\rangle. \tag{5.60}$$

Using this amplitude density, the transition amplitude given by Eq. (5.59) can be rewritten as

$$T_{nn_0} = \langle\Phi_n|X|\xi^{n_0}\rangle + \langle\Phi_n|VG_0^{(+)}|\chi_{n_0}^{(+)}\rangle. \tag{5.61}$$

The scattering wave function $\langle\Psi^{(-)n}|$ with incoming-wave boundary conditions satisfies the following Lippmann–Schwinger equation:

$$\langle\Psi^{(-)n}| = \langle\Phi_n| + \langle\Psi^{(-)n}|VG_0^{(+)}. \tag{5.62}$$

Thus all states involving the time-reversed boundary condition originate from eigenstates of H_0. Then from Eq. (5.62) we have

$$\langle\Psi^{(-)n}|X = \langle\Phi_n|X + \langle\Psi^{(-)n}|VG_0^{(+)}X. \tag{5.63}$$

If we introduce the amplitude density $\langle\chi_n^{(-)}|$ via $\langle\chi_n^{(-)}| = \langle\Psi^{(-)n}|V$, the transition amplitude can be written as

$$T_{nn_0} \equiv \langle\Psi^{(-)n}|X|\xi^{n_0}\rangle = \langle\Phi_n|X|\xi^{n_0}\rangle + \langle\chi_n^{(-)}|G_0^{(+)}X|\xi^{n_0}\rangle. \tag{5.64}$$

From Eqs. (5.60) and (5.64) we obtain

$$T_{nn_0} = \langle\Phi_n|X|\xi^{n_0}\rangle + \langle\chi_n^{(-)}|(G_0^{(+)} - G_0^{(+)}VG_0^{(+)})|\chi_{n_0}^{(+)}\rangle. \tag{5.65}$$

Equations (5.61), (5.64), and (5.65) are three identities for the t matrix. From these equations we obtain the following generalized Newton variational principle for the present problem:

$$\begin{aligned}[T_{nn_0}] = {} & \langle\Phi_n|X|\xi^{n_0}\rangle + \langle\chi_n^{(-)}|G_0^{(+)}X|\xi^{n_0}\rangle + \langle\Phi_n|VG_0^{(+)}|\chi_{n_0}^{(+)}\rangle \\ & - \langle\chi_n^{(-)}|(G_0^{(+)} - G_0^{(+)}VG_0^{(+)})|\chi_{n_0}^{(+)}\rangle,\end{aligned} \tag{5.66}$$

where $\langle\chi_n^{(-)}|$ and $|\chi_{n_0}^{(+)}\rangle$ are the unknown trial functions.

A Kohn variational principle can be written in the present case in terms of the scattered waves. The scattered wave $|\eta_{n_0}^{(+)}\rangle$ associated with photodissociation and $|\eta_n^{(-)}\rangle$ associated with normal incoming-wave boundary condition are given, respectively, by

$$|\eta_{n_0}^{(+)}\rangle \equiv |\Psi^{(+)n_0}\rangle = G_0^{(+)}|\chi_{n_0}^{(+)}\rangle, \tag{5.67}$$

$$\langle\eta_n^{(-)}| \equiv \langle\Psi^{(-)n}| - \langle\Phi_n| = \langle\chi_n^{(-)}|G_0^{(+)}. \tag{5.68}$$

In terms of these scattered waves, the generalized Kohn variational functional for the t matrix (5.66) is given by

$$[T_{nn_0}] = \langle \Phi_n | X | \xi^{n_0} \rangle + \langle \eta_n^{(-)} | X | \xi^{n_0} \rangle + \langle \Phi_n | V | \eta_{n_0}^{(+)} \rangle \\ - \langle \eta_n^{(-)} | (E - H) | \eta_{n_0}^{(+)} \rangle, \tag{5.69}$$

where $\langle \eta_n^{(-)} |$ and $| \eta_{n_0}^{(+)} \rangle$ are the unknown trial functions.

To perform a variational calculation using Newton and Kohn variational principles (5.66) and (5.69), respectively, one can expand the unknown trial functions in terms of a known set of functions. In the case of the Newton variational principle (5.66), all basis functions can be taken to be $\mathcal{L}^2$ in nature. In the case of the Kohn variational principle (5.69), a few of these functions are to be taken as non-$\mathcal{L}^2$ type so as to provide the correct boundary conditions.

The formulation above can be generalized to the case where more than one asymptotic clustering or fragmentation channel is allowed in the photodissociation process. Then the full Hamiltonian can be partitioned as in Eq. (5.43), where H_a is the Hamiltonian for the clusters, possibly including some distortion potential, and V^a is the interaction among clusters in partition a. In this case we label the channel and partition indices as in Section 5.3.2. Then we decompose the full wave function into components according to an as in Eq. (5.45), so that Lippmann–Schwinger equation (5.58) can now be written as the following coupled set of equations:

$$|\Psi_a^{(+)a_0 n_0}\rangle = G_a^{(+)} X | \xi^{n_0} \rangle + G_a^{(+)} \sum_{a'} U_{aa'} |\Psi_{a'}^{(+)a_0 n_0}\rangle, \tag{5.70}$$

where $G_a^{(+)} = (E - H_a + i0)^{-1}$ and the Fock coupling potentials, $U_{aa'}$, are defined by (5.47). The wave-function components, $|\Psi_a^{(+)a_0 n_0}\rangle$, are as defined in Eq. (5.45).

Now the generalized amplitude density for photodissociation is defined by

$$|\chi_a^{(+)n_0}\rangle = X | \xi^{n_0} \rangle + \sum_{a'} U_{aa'} |\Psi_{a'}^{(+)a_0 n_0}\rangle, \tag{5.71}$$

so that $|\Psi_a^{(+)a_0 n_0}\rangle = G_a^{(+)} |\chi_a^{(+)n_0}\rangle$. Then one has the following integral equation for the amplitude density:

$$|\chi_a^{(+)n_0}\rangle = X | \xi^{n_0} \rangle + \sum_{a'} U_{aa'} G_{a'}^{(+)} |\chi_{a'}^{(+)n_0}\rangle. \tag{5.72}$$

Equation (5.72) is a generalization of Eq. (5.61) to this case. The transition amplitude for going from an initial bound state $|\xi^{n_0}\rangle$ to the nth scattering state in partition a is given by

$$T_{nn_0} \equiv \langle \Phi_{an}^{(-)} | \chi_a^{(+)n_0} \rangle = \langle \Phi_{an}^{(-)} | X | \xi^{n_0} \rangle + \sum_{a'} \langle \Phi_{an}^{(-)} | U_{aa'} G_{a'}^{(+)} | \chi_{a'}^{(+)n_0} \rangle. \tag{5.73}$$

Here $\langle \Phi_{an}^{(-)}|$ is the adjoint of the scattering solution of the distorted Hamiltonian with the incoming-wave boundary condition $(E - H_a)|\Phi_{an}^{(-)}\rangle = 0$. In the single partition case, an undistorted Hamiltonian was used; consequently, one had the plane wave $|\Phi_n\rangle$ in place of the distorted wave $|\Phi_{an}^{(-)}\rangle$. Equation (5.73) should be compared with Eq. (5.61) in the absence of rearrangement channels.

The usual scattering solution $|\Psi^{(-)an}\rangle$ of the full Hamiltonian with the incoming-wave boundary condition can also be decomposed into components:

$$|\Psi^{(-)an}\rangle = \sum_{a'} |\Psi_{a'}^{(-)an}\rangle, \tag{5.74}$$

which satisfy the following Lippmann–Schwinger equations:

$$|\Psi_{a''}^{(-)an}\rangle = \delta_{a''a} |\Phi_{an}^{(-)}\rangle + G_{a''}^{(-)} \sum_{a'} U_{a''a'} |\Psi_{a'}^{(-)an}\rangle, \tag{5.75}$$

in close analogy with Eq. (5.46). The amplitude density $|\chi_{a''}^{(-)an}\rangle$ is defined by

$$|\chi_{a''}^{(-)an}\rangle \equiv \sum_{a'} U_{a''a'} |\Psi_{a'}^{(-)an}\rangle = U_{a''a} |\Phi_{an}^{(-)}\rangle + \sum_{a'} U_{a''a'} G_{a'}^{(-)} |\chi_{a'}^{(-)an}\rangle. \tag{5.76}$$

In terms of the incoming-wave scattering solutions the transition amplitude can also be written as

$$T_{nn_0} = \langle \Psi^{(-)an} | X | \xi^{n_0} \rangle = \sum_{a'} \langle \Psi_{a'}^{(-)an} | X | \xi^{n_0} \rangle. \tag{5.77}$$

Using Eq. (5.75), this equation can be rewritten as

$$T_{nn_0} = \langle \Phi_{an}^{(-)} | X | \xi^{n_0} \rangle + \sum_{a'',a'} \langle \Psi_{a'}^{(-)an} | U_{a'a''} G_{a''}^{(+)} X | \xi^{n_0} \rangle. \tag{5.78}$$

Next, by Eq. (5.76), this becomes

$$T_{nn_0} = \langle \Phi_{an}^{(-)} | X | \xi^{n_0} \rangle + \sum_{a'} \langle \chi_{a'}^{(-)an} | G_{a'}^{(+)} X | \xi^{n_0} \rangle. \tag{5.79}$$

Equations (5.73) and (5.79) are two identities for the t matrix.

Using Eq. (5.72) in the last term in Eq. (5.79), we get

$$T_{nn_0} = \langle \Phi_{an}^{(-)} | X | \xi^{n_0} \rangle + \sum_{a'',a'} \langle \chi_{a''}^{(-)an} | (G_{a''}^{(+)} - G_{a''}^{(+)} U_{a''a'} G_{a'}^{(+)}) | \chi_{a'}^{(+)n_0} \rangle. \tag{5.80}$$

Then the following stationary expression for the t matrix can be written in the usual fashion by adding Eqs. (5.73) and (5.79) to it and subtracting Eq. (5.80):

$$\begin{aligned}[T_{nn_0}] = {} & \langle \Phi_{an}^{(-)} | X | \xi^{n_0} \rangle - \sum_{a'',a'} \langle \chi_{a''}^{(-)an} | (G_{a''}^{(+)} - G_{a''}^{(+)} U_{a''a'} G_{a'}^{(+)}) | \chi_{a'}^{(+)n_0} \rangle \\ & + \sum_{a} \langle \chi_{a}^{(-)an} | G_{a}^{(+)} X | \xi^{n_0} \rangle + \sum_{a'} \langle \Phi_{an}^{(-)} | U_{aa'} G_{a'}^{(+)} | \chi_{a'}^{(+)n_0} \rangle. \end{aligned} \tag{5.81}$$

This is the generalized Newton variational principle for the t matrix for photodissociation with multiple fragmentation pathways. In Eq. (5.81), $\langle \chi_{a''}^{(-)an} |$ and $| \chi_{a'}^{(+)n_0} \rangle$ are the unknown trial functions. The stationary property of Eq. (5.81) follows from the definition of these functions. A basis-set calculational scheme can be developed in this case by expanding the amplitude densities in basis sets and exploiting the stationary property as in the single-channel case.

5.4 COMPLEX KOHN VARIATIONAL METHOD

The complex Kohn variational principle of Section 4.6 can be generalized for solving multichannel scattering problems. Following Miller [4], we show how this generalization can be made for nonreactive and reactive scattering problems.

First, we consider nonreactive scattering of the type (1.2). We present a t-matrix description of the method, and an S-matrix description can be made straightforwardly. The method is applicable for any nonreactive process, such as $e + H \rightarrow e + H^*$, where e stands for the electron and H the hydrogen atom. We also consider the following atom-diatom scattering $A + BC \to A + (BC)^*$ and allow for vibrational excitation of the cluster $(BC)^*$ in the final state. Then the Hamiltonian can be written as

$$H = -\frac{\hbar^2}{2\mu}\frac{d^2}{dR^2} + h + V(r, R), \tag{5.82}$$

where h is the vibrational Hamiltonian for diatom BC, r the vibrational coordinate (relative separation between atoms B and C), and R is the translational coordinate (of atom A). The incoming plane-wave states are taken as

$$\Phi_n(r, R) = \phi_n(r)\sin(k_n R), \tag{5.83}$$

where $\phi_n(r)$ is the nth vibrational state of BC and k_n is the channel wave vector, as in Eq. (1.183). Then the t matrix for the $n_1 \to n_2$ transition is

$$t_{n_2,n_1} = \langle\Phi_{n_2}|V|\Phi_{n_1}\rangle + \langle\Phi_{n_2}|VG^{(+)}(E)V|\Phi_{n_1}\rangle. \tag{5.84}$$

Next we use the following finite-rank expansion for the full Green's function $G^{(+)}$ in the product space (r, R) [see Eq. (4.101)]:

$$G_A^{(+)} = \sum_{i,j}^{n} |\eta_i\rangle Z_{ij}^{(n)}\langle\eta_j|, \tag{5.85}$$

where $(Z^{-1})_{ji} = \langle\eta_j|(E - H)|\eta_i\rangle$. Now η_i is a function of r and R.

In the case of single-channel potential scattering the pure plane-wave states are to be replaced by products of plane waves and the vibrational states $\phi_{n_1}(r)$. Consequently, the variational expansion functions $\eta(r, R)$ should be products of the vibrational states $\phi_n(r)$ and single-channel basis functions $f_{in}(R)$ given by Eqs. (4.108) to (4.112):

$$\eta(r, R) = \sum_{i,n} C_{in}\phi_n(r)f_{in}(R), \tag{5.86}$$

where C_{in} are the expansion coefficients. The index i refers to expansion functions. However, now one needs a subscript n to denote the different vibrational states and the corresponding wave number k_n.

Specifically, the states $f_{in}(R)$, $i = 1, 2, \ldots$, are taken as in Eqs. (4.108) to (4.112), with k replaced by k_n. The $i = 1$ function f_{in} is chosen to impose boundary conditions as in the single-channel case. The n dependence of the functions f_{in} for $i > 1$ is dummy. Using these expansion functions, an expression for the complex Kohn variational t matrix can be written as

$$[t_{n_2 n_1}] = \langle \Phi_{n_2} | V | \Phi_{n_1} \rangle + \sum_{i,n} \sum_{j,n'} \langle \Phi_{n_2} | V | f_{in} \phi_n \rangle D_{injn'} \langle f_{jn'} \phi_{n'} | V | \Phi_{n_1} \rangle, \tag{5.87}$$

with

$$(D^{-1})_{jn'in} = \langle f_{jn'} \phi_{n'} | (E - H) | f_{in} \phi_n \rangle, \tag{5.88}$$

where the inverse denotes a matrix inverse in channel and function space. For $i = 1$, the sum over n needs to include only the open vibrational states, but for $i > 1$ both open and closed channels must be included. Also, for $i > 1$ the functions $\phi_n(r)$ need not be the vibrational eigenfunctions; any complete set of functions in variable r can be used. Once the multichannel t matrix has been calculated, one can calculate the multichannel S matrix using, for example, Eq. (1.210).

A distorted-wave version of this method can easily be formulated, and this makes the method more efficient. Then the first term on the right-hand side of Eq. (5.87) is a better approximation to the multichannel t matrix. To achieve this, one writes

$$V(r, R) = V_0(R) + V_1(r, R), \tag{5.89}$$

where the distortion potential V_0 is responsible for elastic scattering. The functions $\Phi_n(r, R)$ are now defined as

$$\Phi_n(r, R) = \phi_n(r) \psi_n(R), \tag{5.90}$$

where $\psi_n(R)$ is the elastic scattering function for the distortion potential $V_0(R)$ for wave number k_n with asymptotic behavior

$$\lim_{R \to \infty} \psi_n(R) \sim \sin(k_n R + \delta_n), \tag{5.91}$$

where δ_n is the elastic scattering phase shift in channel n with distortion potential V_0. The distorted-wave expression for the t

matrix is again given by an expression similar to that of Eq. (5.87), with the distorted-wave functions now defined by Eq. (5.90) and V replaced by V_1. Hence, the formal modification needed for developing the distorted wave function is quite straightforward. The idea is to take a distortion potential V_0 for which the scattering problem could be solved essentially exactly, and the remaining potential $V_1(r, R)$ is to be treated variationally. This procedure is of advantage when the lowest-order term given by the distorted-wave Born approximation is in reasonable agreement with the exact result.

Also, for either the plane- or distorted-wave version, the inverse of the matrix of $E - H$ in combined channel and function space can be carried out as in the single-channel case by partitioning the $i = 1$ block from the $i > 1$ block as described in Section 4.6. This facilitates the computational work for the inversion of a complex matrix, where the $i > 1$ block in each channel is real. This procedure replaces the inverse of a complex matrix by that of a smaller real matrix and simple algebraic manipulations.

One can also formulate reactive scattering in conjunction with the complex Kohn variational principle. One can consider a rearrangement reaction of the type $A + BC \rightarrow C + AB$. The plane-wave representation for the reactive t matrix is generalized from Eq. (5.84) by employing the partition index $\gamma = a$ for $A + BC$, b for $B + CA$, and c for $C + AB$. Then in addition to a channel index one has a partition index for the functions involved and the t matrix can be written as

$$t_{\gamma_2 n_2, \gamma_1 n_1} = \langle \Phi_{\gamma_2 n_2} | V^{\gamma_1} | \Phi_{\gamma_1 n_1} \rangle + \langle \Phi_{\gamma_2 n_2} | V^{\gamma_2} G^{(+)}(E) V^{\gamma_1} | \Phi_{\gamma_1 n_1} \rangle, \quad (5.92)$$

where

$$\Phi_{\gamma n}(r_\gamma, R_\gamma) = \phi_{\gamma n}(r_\gamma) \sin(k_{\gamma n} R_\gamma). \quad (5.93)$$

Here (r_γ, R_γ) are the vibrational and translational coordinates for partition γ, V^γ is the interaction potential for partition γ, and $\phi_{\gamma n}$ is the nth vibrational state in partition γ.

The formal development is similar to the nonreactive case. In this case we have a partition index γ on all variables. The function of Eq. (5.86) is now expanded as

$$\eta(r, R) = \sum_{i,\gamma,n} C_{i\gamma n} \phi_{\gamma n}(r) f_{i\gamma n}(R), \quad (5.94)$$

where $C_{i\gamma n}$ are the expansion coefficients. The expansion functions $f_{i\gamma n}(R)$ are taken as in Eqs. (4.108) to (4.112). Then the degenerate complex Kohn variational form of the t matrix is given by

$$[t_{\gamma_2 n_2 \gamma_1 n_1}] = \langle \Phi_{\gamma_2 n_2} | V | \Phi_{\gamma_1 n_1} \rangle + \sum_{l,\gamma,n} \sum_{l',\gamma',n'} \langle \Phi_{\gamma_2 n_2} | V | f_{\gamma l} \phi_{\gamma n} \rangle$$

$$\times D_{l\gamma n, l'\gamma' n'} \langle f_{\gamma' l'} \phi_{\gamma' n'} | V | \Phi_{\gamma_1 n_1} \rangle, \tag{5.95}$$

where

$$(D^{-1})_{l\gamma n, l'\gamma' n'} = \langle f_{\gamma l} \phi_{\gamma n} | (E - H) | f_{\gamma' l'} \phi_{\gamma' n'} \rangle. \tag{5.96}$$

An equivalent expression for the S matrix can also be written down. This prescription can be used for the multichannel description of reactive scattering involving molecules.

5.5 HARTREE–FOCK CONTINUUM ORBITALS

Most of the variational calculations require approximate wave functions for atoms and molecules. The Hartree–Fock equations have often been used to describe the bound-state atomic and molecular orbitals. Such equations can also be formulated for scattering states. The orbitals so calculated can be used in the application of variational principles. Also, a version of the Schwinger variational method has been formulated by McKoy and collaborators [2] in the application of Hartree–Fock equations for photoionization and electron–molecule scattering studies. The formulation becomes somewhat simple when the molecular system is initially in a closed-shell state. We shall only consider this simple case here, although generalization to open-shell systems is possible [10].

In the simplest form of the Hartree–Fock theory, called the Hartree theory, a central-field approximation is made for each atom of the molecule. The spin of the electrons are neglected except insofar as they influence the filling of the orbitals. In the case of a neutral atom of nuclear charge Ze, the Hartree wave function is postulated to be

$$\psi = \phi_1(\mathbf{r}_1)\phi_2(\mathbf{r}_2) \cdots \phi_n(\mathbf{r}_n), \tag{5.97}$$

where each orbital ϕ_i is an eigenfunction of the single-particle

Hamiltonian

$$\mathcal{H}_i = -\frac{\hbar^2}{2m_e}\nabla_i^2 - \frac{Ze^2}{r_i} + V_i(\mathbf{r}_i), \tag{5.98}$$

where n is the number of atomic orbitals, m_e the electronic mass, e the electronic charge, and $\mathbf{r}_i$ the position of the ith electron. The term $V_i(\mathbf{r}_i)$ is the effective central potential felt by electron i due to the presence of other electrons. The full wave function ψ satisfies

$$\sum_{i=1}^{n} \mathcal{H}_i\psi = 0. \tag{5.99}$$

One needs to find a central potential $V_i(r_i)$ to be used in $\mathcal{H}_i$ and hence calculate the orbitals ϕ_i. In the Hartree method the central potential is taken to be

$$V_i(\mathbf{r}_i) = \sum_{j=1(j\neq i)}^{n} \int \phi_j^*(\mathbf{r}_j)\frac{e^2}{|\mathbf{r}_i - \mathbf{r}_j|}\phi_i(\mathbf{r}_i)d^3r_j. \tag{5.100}$$

The orbitals and hence the energies are then found by solving the following Hartree equations self-consistently:

$$\mathcal{H}_i\phi_i(\mathbf{r}_i) = E_i\phi_i(\mathbf{r}_i), \qquad i = 1, 2, ..., n. \tag{5.101}$$

Once the Hartree equations are solved, the atomic wave function is the product of the individual orbitals. The expectation value of the full Hamiltonian $\sum_{i=1}^{n}\mathcal{H}_i$ with respect to the atomic wave function is the approximate energy.

A more complete description of the atomic states is given by the Hartree–Fock theory, which treats the spin of the electron explicitly. The atomic wave function must explicitly be antisymmetric in electron coordinates. This is achieved by replacing the product form of the wave function by a Slater determinant,

$$\psi = |\phi_1(\mathbf{r}_1)\alpha\cdot\phi_1(\mathbf{r}_1)\beta\cdot\phi_2(\mathbf{r}_2)\alpha\cdot\phi_2(\mathbf{r}_2)\beta\cdots\phi_n(\mathbf{r}_n)\alpha\cdot\phi_n(\mathbf{r}_n)\beta|, \tag{5.102}$$

where α (β) is the spin-up (spin-down) spin function. In Eq. (5.102), in each orbital there are two electrons: one spin-up, another spin-down.

The Hartree–Fock wave functions and energies are determined by solving a coupled set of differential equations. The coupling between various atomic orbitals is introduced through an explicit account of the Pauli principle.

The foregoing treatment can be extended to the case of closed-shell molecules, where in the Hamiltonian $\mathcal{H}_i$ of Eq. (5.98), the second term on the right-hand side is to be replaced by a sum with contributions from different nuclear charge centers $Z_\gamma e$ of different atoms. The Hartree–Fock wave function for a closed-shell molecule is taken in the form (5.102) and that for scattering of an electron off this molecule is taken as

$$\psi_k = |\phi_1(\mathbf{r}_1)\alpha \cdot \phi_1(\mathbf{r}_1)\beta \cdot \phi_2(\mathbf{r}_2)\alpha \cdot \phi_2(\mathbf{r}_2)\beta \cdots \phi_n(\mathbf{r}_n)\alpha \cdot \phi_n(\mathbf{r}_n)\beta \cdot \phi_k(\mathbf{r}_k)\alpha|, \tag{5.103}$$

where the incident electron has been labeled k in Eq. (5.103). The new continuum orbital $\phi_k(\mathbf{r}_k)$ is now determined variationally by [2]

$$\langle \delta\psi_k|(H - E)|\psi_k\rangle = 0, \tag{5.104}$$

where the variations are defined by

$$\delta\psi_k = |\phi_1(\mathbf{r}_1)\alpha \cdot \phi_1(\mathbf{r}_1)\beta \cdot \phi_2(\mathbf{r}_2)\alpha \cdot \phi_2(\mathbf{r}_2)\beta \cdots \phi_n(\mathbf{r}_n)\alpha \cdot \phi_n(\mathbf{r}_n)\beta \cdot \delta\phi_k(\mathbf{r}_k)\alpha|. \tag{5.105}$$

The Hamiltonian for the electron–molecule system can be written as

$$H = \sum_{i=1}^{N}\left(-\frac{\hbar^2}{2m_e}\nabla_i^2 - \sum_{\alpha}\frac{Z_\alpha e^2}{r_{i\alpha}}\right) + \sum_{i<j}\frac{e^2}{|\mathbf{r}_i - \mathbf{r}_j|}, \tag{5.106}$$

where $Z_\alpha e$ are the nuclear charges, $N(= 2n + 1)$ the total number of electrons, and $r_{i\alpha}$ the distance of the ith electron from nucleus α. In this case a Hartree–Fock Hamiltonian H^{HF} can be introduced which defines the HF orbitals ϕ_i and the energies ϵ_i via the following equation in a self-consistent manner:

$$H^{\mathrm{HF}}\phi_i = \epsilon_i\phi_i, \qquad i = 1, 2, \ldots, n. \tag{5.107}$$

The procedure for calculation of the exact H^{HF} is beyond the scope of this book. In general, it is difficult to calculate the exact H^{HF}, and approximations are made in its evaluation. For closed-shell molecules, if exact H^{HF} is employed, the solutions of Eq. (5.107) with $\epsilon_i > 0$ are the scattering solutions, and those with $\epsilon_i < 0$ define the bound orbitals. These two types of solutions are orthogonal to each other.

If the orbitals ϕ_k and $\delta\phi_k$ are taken to be orthogonal to each other or to the orthonormal set of occupied target orbitals, Eq. (5.104) leads to

$$\langle (P\delta\phi_k)|(H^{\mathrm{HF}} - \epsilon)|P\phi_k\rangle = 0, \tag{5.108}$$

where

$$P = 1 - \sum_{i=1}^{n} |\phi_i\rangle\langle\phi_i|, \tag{5.109}$$

and $\epsilon = E - E^{\mathrm{HF}}$, where E^{HF} is the energy of the Hartree–Fock wave function (5.102). In case of scattering by a closed-shell system, if we take $P = 1$, Eq. (5.108) becomes the Hartree–Fock equation (5.107). Thus for scattering from a closed-shell system, it is not necessary to introduce the orthogonality constraints via projection operator P as above. However, when an approximate Hartree–Fock potential is employed, it is of advantage to enforce orthogonality.

Let us consider the problem of photoionization from a closed-shell molecule in the frozen-core Hartree–Fock (FCHF) approximation, which leads to similar equations. In this approximation the core does not take active part in the reaction and the outside electron is removed by photoionization. If an electron is removed from the orbital ϕ_n, the final-state wave function is taken to be [2]

$$\psi_k = \frac{1}{\sqrt{2}}\Big[|\phi_1\alpha\cdot\phi_1\beta\cdot\phi_2\alpha\cdot\phi_2\beta\cdots\phi_n\alpha\cdot\phi_k\beta| + |\phi_1\alpha\cdot\phi_1\beta\cdot\phi_2\alpha\cdot\phi_2\beta\ldots\phi_k\alpha\cdot\phi_n\beta|\Big]. \tag{5.110}$$

The continuum orbital ϕ_k is then obtained by solving Eq. (5.104) with Eq. (5.110) and

$$\delta\psi_k = \frac{1}{\sqrt{2}}\Big[|\phi_1\alpha\cdot\phi_1\beta\cdot\phi_2\alpha\cdot\phi_2\beta\ldots\phi_n\alpha\cdot\delta\phi_k\beta| + |\phi_1\alpha\cdot\phi_1\beta\cdot\phi_2\alpha\cdot\phi_2\beta\cdots\delta\phi_k\alpha\cdot\phi_n\beta|\Big]. \tag{5.111}$$

With the imposition of the orthogonality constraint, using the projection operator P of Eq. (5.109) and taking $\delta\phi_k = \phi_n$ the FCHF equation can be written as

$$\langle (P\delta\phi_k) | (H^{\mathrm{FCHF}} - \epsilon) | P\phi_k \rangle = 0, \tag{5.112}$$

where H^{FCHF} is the FCHF Hamiltonian. Again we do not give an exact expression for this Hamiltonian [2]. Unlike the case of electron scattering off a closed-shell molecule, the orthogonality constraint is needed to obtain the correct FCHF solution.

The straightforward procedure to transform the HF and FCHF equations into the form of a Lippmann–Schwinger equation is through the use of a pseudopotential. Then a one-electron Schrödinger equation of the form (5.108) or (5.112) is rewritten as [2]

$$(I - Q)L(I - Q)\phi_k = 0, \tag{5.113}$$

with $Q = I - P$ and

$$L \equiv H_0 + V - \epsilon = -\frac{\hbar^2}{2m_e}\nabla_k^2 + V - \epsilon, \tag{5.114}$$

where V is the self-consistent HF or FCHF one-particle potential. The effective Schrödinger equation (5.113) can be rewritten as

$$(\epsilon - H_0)\phi_k = (V - LQ - QL + QLQ)\phi_k, \tag{5.115}$$

which can be transformed to yield the following Lippmann–Schwinger equation:

$$\phi_k = \phi_k^0 + G_0 V_Q \phi_k, \tag{5.116}$$

with $V_Q = V - LQ - QL + QLQ$, ϕ_k^0 the free-particle solution, and $G_0 \equiv (\epsilon - H_0 + i0)^{-1}$. Once this Lippmann–Schwinger equation is obtained, variational principles can be constructed and used in the usual fashion [2].

5.6 NOTES AND REFERENCES

[1] Geltman (1969), Burke (1977), Bransden (1983).
[2] For a review, see Lucchese et al. (1986).

[3] For a review, see Tawa et al. (1994) and Sun et al. (1990)

[4] For a review, see Miller (1992,1994)

[5] Feshbach (1958,1962).

[6] Nesbet (1980).

[7] Lima and McKoy (1988).

[8] Rountree and Parnell (1977).

[9] Kouri and Truhlar (1989). This reference presents a derivation of the Newton variational principle for molecular photodissociation and also brings a complete reference of earlier works on this subject.

[10] Lucchese and McKoy (1980).

CHAPTER 6

NUMERICAL STUDIES

6.1 INTRODUCTION

We now present an account of the numerical applications of the methods presented so far. These applications are large in number, include a wide class of interesting problems in different areas, and extend in time over the last three decades. Hence, it is not possible to present an account of all of them. We describe only representative examples of each type. We classify this presentation in several categories. First, we present numerical applications in simple models, such as single-channel Yukawa and exponential potentials for electron–atom [1–3] and nucleon–nucleon [4] scattering. These models do not present a good description of physical observables but are used to test the numerical feasibility and accuracy obtained in various methods. Then we present applications to the numerical solution of some simple coupled-channel theoretical models. Finally, we present an application of the methods to realistic problems in nuclear [4,5], atomic [1–3] and molecular [6–9] physics. For the realistic problems, we shall be limited primarily to a discussion of applications of the numerical methods presented in this book and related methods, rather than presenting a comprehensive review of the subject under consideration. However, we present adequate references so that one can have easy access to reviews and original works on these topics [1–9].

All these methods involve the solution of a linear set of algebraic (matrix) equations. Basically, there are two types of methods. The first type [10] is designed to provide an accurate solution of a mathematically simple problem, for example, a two-particle scattering problem involving a small number of channels and angular momentum states. In this approach an accurate solution of the scattering equation is attempted via matrix inversion, iteration, or otherwise. These methods deal directly with the t-, K-, or Γ-matrix equations presented in Chapter 1. Independent of how the solution is obtained, after discretization of the integrals, these methods require the use of a relatively large number of integration mesh points per channel. If the number of coupled channels in the original problem is large, the resultant matrix equation could be of such a large dimension that these methods become unattractive. Under such a situation, methods of the second type (e.g., the algebraic variational basis-set methods [1–4]) are of interest. However, the purpose behind the use of the variational basis-set methods is not always to compete with the methods of the first type in numerical accuracy, but rather, to obtain a reasonably accurate result with relatively little numerical effort and/or to obtain convenient finite-rank expansions. Although the variational basis-set methods yield approximate solutions easily, convergence to the exact result could be slow. The reason for this could be manyfold. In some cases the use of inappropriate basis functions (e.g., $\mathcal{L}^2$ basis functions for expanding a non-$\mathcal{L}^2$ trial function) could be responsible for slow convergence. In other cases the complicated medium-range nature of the (coupling) potential could be responsible for slow convergence.

The first type of method mentioned above is efficient for obtaining high-precision numerical results for the solution of scattering problems. For obtaining accurate solutions, the iterative Neumann series method is often preferred to a direct solution, by matrix inversion or otherwise, of a large set of algebraic equations. If a convergent solution could be obtained with a small number of terms in the Neumann series, the accumulated error is expected to be small. If the Neumann series solution diverges or does not converge satisfactorily, the Padé technique [11,12] can be used to construct a solution of the problem from the divergent Neumann series solution. However, if a very large number of terms of a slowly convergent or divergent Neumann series (as in the Padé technique) are used, one expects low accuracy. Direct solution of a large set of matrix equation also leads to a large round-off error. In such cases a convergent Neumann series for the Γ matrix [10] of Section 2.4 can be used. In most problems the Γ-matrix equation leads to a convergent Neumann series when the Neumann series of the original

Lippmann–Schwinger equation is divergent. If convergence of the Γ-matrix equation is unsatisfactory, one can use the Padé technique to improve the convergence of the Neumann series of the Γ-matrix equation. All these iterative methods will have slower convergence if the original Born–Neumann series diverges very strongly, and this may set a limitation on the use of these methods. However, in practice this problem has not been of practical concern.

In nuclear physics, expansion methods [4,5], as obtained from the Schwinger variational principle, have been applied widely to nucleon–nucleon scattering involving short-range potential. This is because a separable expansion to the nucleon–nucleon t matrix greatly facilitates the task of solving the three- and four-nucleon scattering problem by reducing the dimensionality of the momentum-space scattering integral equations by one. The use of separable expansion on a two-body t matrix reduces the three-body problem to an effective two-body problem. Similarly, the use of separable expansion on both two- and three-body t matrices reduces the four-body problem to an effective two-body problem [5]. These simplifications have been exploited successfully in solving three- and four-nucleon problems.

The Schwinger basis-set method has also been used with great success to solve real-world multichannel atomic and molecular physics problems. Benchmark calculations of various physical processes in atomic and molecular physics that would otherwise be impossible to solve have been performed using Schwinger and Newton basis-set methods. The purpose behind use of a degenerate-kernel scheme (Kohn or Newton variational principles) or of a separable expansion scheme (a Schwinger-type variational principle) in two-, three-, and four-nucleon scattering is to find an approximate solution with the least effort. This is also the motivation behind use of these methods in atomic or molecular physics or chemistry. In the latter situations, an accurate numerical solution is neither called for, because of the drastic approximations made in formulating the scattering model, nor easily obtainable, because of the extreme dimensions of the problem.

In atomic/molecular physics, numerous applications of variational methods deserve special mention. Of these we shall focus on those scattering processes that have a simple target structure within each class. For example, in atomic physics we emphasize scattering involving hydrogen and helium atoms, and in molecular physics we emphasize scattering involving hydrogen molecule. In atomic/molecular physics the simplest projectiles are electrons, positrons, and photons. In addition, in molecular physics one could have an atomic projectile for studying atom-diatom scattering. All these problems have been studied by the present variational methods. If the target wave functions are reasonably well known, it is likely that simple realistic models can be formulated for these scattering processes, which can be solved by one (some) of the standard numerical methods, and the accuracy and usefulness of the (these) method(s) can be assessed. Problems in atomic physics are easier to handle than problems in molecular physics. Simple atomic physics problems were studied exhaustively by various workers in the 1960s and 1970s by using the on-shell variational methods [1–3]. The off-shell variational methods were popularized later and after adequate tests in simple model problems, they have been mostly applied to the scattering problems in molecular physics in the last 15 years [6–9]. In the following we present an account of these numerical applications.

In Section 6.2 we describe how to calculate explicitly the matrix elements of a basis-set method. In Section 6.3 we present numerical results for single-channel electron scattering with atomic Yukawa and exponential potentials. In Section 6.4 an account of multichannel model studies is presented. In Section 6.5 numerical applications of the variational methods to few-nucleon scattering problems are described. In Section 6.6 we describe calculations of electron and positron scattering with a hydrogen atom. In Section 6.7 calculations of scattering of electrons and positrons with a helium atom are described. In Section 6.8 electron scattering by heavier atoms is described. Results for molecular photoionization are described in Section 6.9. Finally, in Sections 6.10 and 6.11, we describe results of electron–molecule and atom–diatom collisions, respectively.

6.2 EXPLICIT MATRIX ELEMENTS

The formulations presented in earlier chapters can be used for practical calculation of scattering observables. All these methods finally reduce the original scattering problem to a set of linear algebraic equations. Several matrix elements need to be calculated to construct this set of equations. This calculation can be performed in either momentum space or configuration space. In the following we present explicit forms of these matrix elements.

In this section and the following we work explicitly in three dimensions. The basis functions are usually taken as linear combinations of the scattering wave function, as in Eqs. (4.39), (4.40), and

(4.41). The present wave function is $\psi^{(+)}(r)/kr$ of Section 1.3. Hence the basis functions of Eqs. (4.40) and (4.41) are to be divided by r before using the equations in this chapter. The relations between the momentum- and configuration-space representations of the basis function $|f_i\rangle$ are given by

$$\langle rL|f_i\rangle \equiv f_{iL}(r) = \frac{2}{\pi}\int_0^\infty j_L(pr)w_{iL}(p)p^2\,dp, \tag{6.1}$$

$$\langle pL|f_j\rangle \equiv w_{iL}(p) = \int_0^\infty j_L(pr)f_{iL}(r)r^2\,dr. \tag{6.2}$$

First we present the explicit matrix elements of the Schwinger variational method for the t matrix given by Eqs. (4.24) and (4.21). If we assume a local potential $V_L(r)$ in the Lth partial wave, the matrix elements in configuration space are given by

$$\langle p|V_L|f_i\rangle = \int_0^\infty j_L(pr)V_L(r)f_{iL}(r)r^2dr, \tag{6.3}$$

$$\begin{aligned}(D^{-1})_{ji} = \int_0^\infty f_{jL}(r)V_L(r)f_{iL}(r)r^2dr + ik\int_0^\infty\int_0^\infty r'^2\,dr'\,r^2\,dr\\ \times f_{jL}(r')V_L(r')j_L(kr_<)h_L^{(1)}(kr_>)V_L(r)f_{iL}(r),\end{aligned} \tag{6.4}$$

where the outgoing-wave Green's function (1.77) has been used. For calculating the K matrix the standing-wave Green's function (1.32) should be used. In configuration space, the double integral of Eq. (6.4) involving non-$\mathcal{L}^2$ functions is tedious to evaluate, and momentum-space treatment provides some advantage. In momentum space these matrix elements are given by

$$\langle p|V_L|f_i\rangle = \frac{2}{\pi}\int_0^\infty V_L(p,q)w_{iL}(q)q^2dq, \tag{6.5}$$

$$\begin{aligned}(D^{-1})_{ji} &= \left(\frac{2}{\pi}\right)^2\int_0^\infty\int_0^\infty p'^2dp'p^2dp\,w_{jL}(p)w_{iL}(p')\\ &\quad\times\left[V_L(p,p') - \frac{2}{\pi}\int_0^\infty \frac{V_L(p,q)V_L(q,p')}{k^2-q^2}q^2\,dq\right], \qquad (6.6)\\ &\equiv \frac{2}{\pi}\int_0^\infty p^2\,dp\left[w_{jL}(p)\langle p|V_L|f_i\rangle - \frac{\langle f_j|V_L|p\rangle\langle p|V_L|f_i\rangle}{k^2-p^2}\right], \qquad (6.7)\end{aligned}$$

where the outgoing-wave $i0$ (or standing-wave principal-value) prescription for the t (K) matrix should be used in the energy denominator. Equation (6.7) is particularly convenient for numerical calculation. Once the matrix elements $\langle p|V_L|f_i\rangle$ of Eq. (6.5) are calculated and stored, expression (6.7) can be calculated without ever requiring direct evaluation of a double integral.

Next we present the matrix elements encountered in numerical application of the Newton variational method via Eqs. (4.73) and (4.70). This method requires evaluation of a double integral involving two Green's functions which is efficiently handled in momentum space. This is why we present the matrix elements of this method only in momentum space. In this case

$$\langle p|V_L G_0|f_i\rangle = \frac{2}{\pi}\int_0^\infty \frac{V_L(p,q)w_{iL}(q)}{k^2-q^2}q^2 dq, \tag{6.8}$$

$$(P^{-1})_{ji} = \frac{2}{\pi}\int_0^\infty p^2 dp\left[\frac{w_{jL}(p)w_{iL}(p)}{k^2-p^2} - \frac{w_{jL}(p)\langle p|V_L G_0|f_i\rangle}{k^2-p^2}\right]. \tag{6.9}$$

In this case outgoing-wave $i0$ (or standing-wave principal value) prescription for the t (K) matrix should be used in the energy denominator. By storing the matrix elements of Eq. (6.8), direct evaluation of double integrals involving two Green's functions is avoided in Eq. (6.9).

Finally, we present the matrix elements of the Kohn variational method of Eqs. (4.98) and (4.99). In this case there are no Green's function and the configuration-space treatment is usually employed. The elements $\langle p|V_L|f_i\rangle$ of Eq. (4.98) are evaluated via Eq. (6.3). The matrix element of Eq. (4.99) is then given by

$$(Z^{-1})_{ji} = \int_0^\infty dr(rf_j(r))\left[k^2 + \frac{d^2}{dr^2} - V(r) - \frac{L(L+1)}{r^2}\right](rf_i(r)). \tag{6.10}$$

Usually, one uses analytical expansion functions, so that the second derivative in Eq. (6.10) can also be evaluated analytically. Here f_i of Eq. (6.10) is the function in three dimensions. Asymptotically, f_i have the three-dimensional scattered waves $\exp(ikr)/kr$ and $\cos(kr - L\pi/2)/kr$ for complex Kohn and Kohn variational principles, respectively. Functions f_i of Eqs. (4.108) to (4.110) have one-dimensional scattered waves $\exp(ikr)$ and $\cos(kr - L\pi/2)$ for complex Kohn and Kohn variational principles, respectively. Functions f_i of

Eqs. (4.108) to (4.110) are to be divided by r before being used in equations of this chapter.

We do not present explicit matrix elements of all the methods. The matrix elements encountered in other methods can be written down in a similar fashion.

6.3 SINGLE-CHANNEL MODEL STUDIES

In this section we present single-channel model studies of the various methods presented so far. Two of the potentials commonly used for this purpose are the following atomic exponential and Yukawa potentials for electron scattering:

$$V(r) = -\frac{\hbar^2}{ma_0^2}\exp\left(-\frac{r}{a_0}\right), \tag{6.11}$$

$$V(r) = -\frac{\hbar^2}{ma_0^2}\frac{\exp(-r/a_0)}{r/a_0}, \tag{6.12}$$

where a_0 is the Bohr radius of the hydrogen atom. These potentials have been used in numerous tests of various computational methods in atomic physics. Most of the tests considered S-wave scattering in momentum and configuration space. After factorization of $\hbar^2/2m$, the S-wave momentum-space matrix elements of the foregoing potentials are given in atomic units (a.u.) by

$$V_0(p,q) = -4[1+(p+q)^2]^{-1}[1+(p-q)^2]^{-1}, \tag{6.13}$$

$$V_0(p,q) = -\frac{1}{2pq}\ln\frac{1+(p+q)^2}{1+(p-q)^2}. \tag{6.14}$$

In atomic units the length is measured in units of a_0, as are the momentum-space matrix elements of the t, V, and K matrices. The wave number is given in units of a_0^{-1}.

The Bohr radius of the hydrogen atom is given by $a_0 = \hbar^2/me^2 = 0.5292$ Å, where m is the electronic mass and e the charge. The binding energy of the ground state of hydrogen atom is $\hbar^2/2ma_0^2 = e^2/2a_0 = 13.6$ eV. The atomic unit of energy is $e^2/a_0 = 27.2$ eV. Another frequently used energy unit is the rydberg (Ry): 1 Ry $= e^2/2a_0 = 13.6$ eV. Often, the electron/positron energy is expressed in terms of the wave number k, expressed in units of a_0 or a.u. Then the energy is k^2 Ry.

6.3.1 Variational Basis-Set Methods

Schwartz [13,14] calculated S-wave elastic scattering phase shifts for the atomic Yukawa potential (6.12) with the on-shell Kohn method given by Eqs. (3.110) and (3.111) and the (off-shell) Schwinger variational method given by Eqs. (4.37) and (4.38) for the K matrix. The results are compared with those obtained from the off-shell principal-value Kohn method of (4.98) and (4.99) for the K matrix [15]. Although for phase shifts the on- and off-shell Kohn methods yield identical results for orthonormal basis functions, the use of linearly dependent nonorthogonal basis functions may lead to different results in practice.

In applications of the Schwinger method given by Eqs. (4.37) and (4.38), Schwartz [14] used the following set of functions in momentum space for $L = 0$:

$$w_i(p) = (p^2 + \alpha^2)^{-i}, \qquad i = 1, 2, ..., n, \tag{6.15}$$

which in configuration space is equivalent to the following set of functions:

$$f_i(r) = r^{(i-2)} \exp(-\alpha r), \qquad i = 1, 2, ..., n. \tag{6.16}$$

If one transforms configuration-space functions (6.16) to momentum space, there is no one-to-one correspondence between each $w_i(p)$ and $f_i(r)$ for a fixed i, but after transformation to momentum space, $f_j(r)$ can be expressed as a linear combination of functions $w_i(p)$, $i = 1, ..., j$. For $L(\neq 0)$ wave scattering, one could use

$$f_{iL}(r) = r^{(i+L-2)} \exp(-\alpha r), \qquad i = 1, 2, ..., n. \tag{6.17}$$

These simple expansion functions facilitate calculation of the integrals and are often used in numerical application of variational methods.

In the Schwinger variational method (SVM) of Eqs. (4.37) and (4.38), the multiple integral involving the Green's function has a simple structure which is handled efficiently in momentum space. This approach requires the calculation of an integral involving the inverse potential operator, which is evaluated in configuration space for an everywhere attractive (or repulsive) potential. Here all integrals are single integrals without requiring evaluation of multiple integrals. Also, the unknown function is a $\mathcal{L}^2$ form factor, and the use of $\mathcal{L}^2$ basis functions has led to good convergence. Schwartz also used the on-shell version of the Kohn variational method (KVON) of Eqs. (3.110) and (3.111) with trial functions (3.73), (3.74), and (3.75) (see Section 3.31) for calculating phase shifts. For comparison we present the phase shifts obtained from the off-shell Kohn variational method (KVOF) for the K matrix of Eqs. (4.98) and (4.99) with basis functions (3.73) and (4.111) [15]. Although the on- and off-shell Kohn methods are formally equivalent for calculating phase shifts, numerically, the off-shell method is simpler, as noted in Section 4.6.

In these three methods the $\mathcal{L}^2$ basis functions have similar functional form. The Schwinger variational method (SVM) uses only $\mathcal{L}^2$ basis functions. The Kohn methods, KVON and KVOF, use, in addition, real non-$\mathcal{L}^2$ functions. The rate of convergence of a given method should be improved by varying the parameters α and β of the expansion functions. For the best convergence these parameters could be different and could depend on energy. However, in Table 1 the energy-independent choice $\alpha = \beta = 1.8$ a.u. has been used at all energies.

In Table 1 we present the numerical results for S-wave elastic scattering, phase shifts for electron scattering with the atomic Yukawa potential as calculated by various basis-set methods. A large

TABLE 1. S-Wave Scattering Length and Tangent of S-Wave Phase Shifts for the Yukawa Potential (6.12) for Different Values of n[a]

n	KVON $(k = 0)$[b]	SVM $(k = 0)$[b]	KVOF $(k = 0)$[c]	KVOF $(k = 0.15$ a.u.$)$[c]	KVOF $(k = 0.55$ a.u.$)$[c]
1	—	1.023067	1.0031559918	1.0022295713	0.9926326659
2	1.001285	1.012292	1.0012853635	1.0012824833	0.9986054340
3	1.000005	1.000114	1.0000050358	1.0000254684	0.9958296795
4	1.000000	1.000051	1.0000006054	1.0000004830	0.9999654332
6	1.000000	1.000011	1.0000000032	1.0000000067	0.9999999133
8	1.000000	1.000001	1.0000000001	1.0000000000	0.9999999792
10	1.000000	1.000000	1.0000000000	1.0000000000	1.0000000001

[a] n is the total number of $\mathcal{L}^2$ and non-$\mathcal{L}^2$ basis functions. The ratio of the calculated result to the converged result is shown in each case. The present calculation is performed with variational parameters $\beta = \alpha = 1.8$ a.u. Converged scattering length in this case is 7.911380206 a.u. The converged result at $k = 0.15$ a.u. is $\tan\delta = -1.386339121$, and at $k = 0.55$ a.u. is $\tan\delta = 5.797211392$.

[b] From Schwartz [14].

[c] From Adhikari [15].

number of studies in the literature have compared the rates of convergence of various basis-set methods [16]. In these studies, phase shifts obtained with a fixed number of $\mathcal{L}^2$ functions by different methods are compared. The non-$\mathcal{L}^2$ functions used in some of the methods, in addition to the $\mathcal{L}^2$ functions, are not counted, for the sake of comparison. For example, the results of the Kohn method with one real non-$\mathcal{L}^2$ function and n $\mathcal{L}^2$ functions are compared with those of the Schwinger method with only n $\mathcal{L}^2$ functions. We believe that this comparison is not fair. So in this chapter we always compare the results obtained with different methods for a fixed total number n of $\mathcal{L}^2$ and non-$\mathcal{L}^2$ expansion functions. In accordance with this, in Table 1 and in the following, n denotes the total number of $\mathcal{L}^2$ and non-$\mathcal{L}^2$ functions.

At zero energy ($k = 0$), the calculated quantities are the S-wave scattering lengths. At other energies they are the tangents of the S-wave phase shifts. The entries in Table 1 (and later in Tables 2 to 5) are the ratios of the quantities calculated for a fixed n to the converged results. So the closer these entries are to unity, the more converged are the results. From Table 1 we see that the Kohn methods produce better convergence than the present version of the Schwinger method at $k = 0$. At other energies ($k = 0.15, 0.55$ a.u.) the KVOF method produces equally good convergence.

In the early 1970s, basis-set methods based on off-shell Schwinger and Newton variational principles were advocated and studied in detail by Brady, Sloan, and the author [17,18]. The equivalence of the Schwinger basis-set method with a separable expansion scheme and of the Newton basis-set method with a particular form of the degenerate kernel method were emphasized in these works. Although separable expansion and degenerate kernel methods were in use for the numerical solution of scattering problems prior to these studies, their connection with the variational basis-set methods was not realized. Such numerical studies gained renewed importance and credibility after their connection with variational methods was established. Here we present an account of these later numerical calculations. Calculations were performed by Brady, Sloan, and the author in nuclear two- and three-body problems. Brady and Sloan [17] made numerical studies of the Newton variational method. Later, Sloan and the author [18] made extensive numerical studies of nucleon–nucleon scattering employing the Newton and the Schwinger basis-set methods using $\mathcal{L}^2$ basis functions and confirmed good convergence properties for both. Later, these variational methods were used extensively in many realistic scattering problems in atomic and molecular physics.

TABLE 2. Tangents of S-Wave Phase Shifts for the Exponential Potential for Different n[a]

k (a.u)	n	AFKV[b]	OAF[b]	MNK[b]	MNR[b]	OMN[b]	n	L2SV[b]	KVOF[c]
0.15	3	1.0005	1.0009	1.0004	1.0010	1.0009	2	−0.7321	1.004693451
	5	1.0006	1.0004	1.0002	1.0005	1.0004	4	1.2585	1.000007977
	7	1.0001	1.0000	1.0000	1.0000	1.0000	6	1.0163	1.000000046
	9	1.0000	1.0000	1.0000	1.0000	1.0000	8	1.0016	1.000000000
	11	1.0000	1.0000	1.0000	1.0000	1.0000	10	1.0013	1.000000000
0.55	3	0.9968	0.9940	0.9902	0.9969	0.9941	2	0.8996	1.067164125
	5	0.9999	0.9970	0.9910	0.9999	0.9970	4	0.9976	0.999958051
	7	1.0000	0.9999	0.9999	0.9999	0.9999	6	0.9999	0.999999935
	9	1.0000	1.0000	1.000	1.0000	1.0000	8	1.0000	0.999999996
	11	1.0000	1.0000	1.0000	1.0000	1.0000	10	1.0000	1.000000000

[a] Ratios of the calculated result for a fixed n to the converged result are shown in each case. The KVOF calculation is performed with variational parameters $\beta = 1.0$ a.u. and $\alpha = 1.5$ a.u. The converged result at $k = 0.15$ a.u. is $\tan\delta = -1.744939320716682$ and at $k = 0.55$ a.u. is $\tan\delta = 2.200382707305979$.
[b] From Thirumalai and Truhlar [19].
[c] From Adhikari [15].

Next we compare the results for S-wave phase shifts with different basis-set methods for the atomic exponential potential (6.11) in Table 2. In this comparison we employ the following methods: the anomaly-free Kohn variational method (AF) of Section 3.5.1, the optimized anomaly-free method (OAF) of Section 3.7.2, minimum-norm Kohn (MNK) and inverse Kohn methods (MNR) of Section 3.4.1, the optimized minimum norm method (OMN) of Section 3.7.1, the Schwinger variational method with $\mathcal{L}^2$ functions (L2SV) of Eq. (4.24), and the KVOF method for the K matrix of Eqs. (4.98) and (4.99) with basis functions (3.73) and (4.111). Most of these results are taken from Ref. 19. Again, n is the total number of $\mathcal{L}^2$ and non-$\mathcal{L}^2$ basis functions.

Of these methods, all are on-shell variational methods for the tangents of phase shifts or their inverse, except the L2SV and KVOF methods, which may also yield off-shell K matrices. In Table 2 the calculations were performed using trial function (4.39) with $\alpha = 2.5$ a.u. and $\beta = 1.0$ a.u. The KVOF calculation was performed with $\alpha = 1.5$ a.u. and $\beta = 1.0$ a.u. In the L2SV method only the $\mathcal{L}^2$ part of function (4.39) is considered with $\alpha = 2.5$ a.u. at $k = 0.15$ a.u. and with $\alpha = 0.5$ a.u. at $k = 0.55$ a.u. All the on-shell methods yield good and comparable rates of convergence with this set of parameters, except the L2SV method, which yields inferior convergence for $\alpha = 2.5$ a.u. The rate of convergence of the L2SV method improves with a smaller α ($= 1.5$ a.u. or 0.5 a.u.). For $\alpha = 0.5$ a.u. the convergence of the L2SV method is comparable to or better than the on-shell methods, as can be seen from the result at $k = 0.55$ a.u. This was the scenario until 1980.

There have been extensive studies of the Schwinger variational method (L2SV) (4.24) in realistic and model atomic scattering problems [7] using only $\mathcal{L}^2$ basis functions. However, inclusion of the appropriate asymptotic behavior of the scattering wave function in the basis functions should enhance the convergence rate. In the usual Schwinger variational principle (4.24) the basis function is multiplied by the potential. If the potential is of short range, the basis functions up to the range of the potential enters into the calculation and the necessity of including the correct asymptotic behavior is not severe. It has been demonstrated [20] numerically that for very short-range potentials, the L2SV method leads to good convergence. But as the range of the potential increases, the convergence rate of the L2SV method deteriorates. For potentials with longer range, if sine and cosine functions with the same asymptotic property as the scattering wave function are included in the basis set of the Schwinger method, excellent convergence is obtained [20].

Since 1980 there have been new tests on the rate of convergence of various variational methods, and alternative methods based on variational principles have been suggested. Among these studies three are worth mentioning. Takatsuka et al. performed numerical calculations using the Schwinger method (SV) of Eq. (4.24) with trial function (4.39) by including two non-$\mathcal{L}^2$ functions and the usual $\mathcal{L}^2$ functions in the basis set and found excellent convergence properties [20]. In the Newton variational method (NV) the unknown function is an $\mathcal{L}^2$ form factor, so one should use an $\mathcal{L}^2$ basis set. Staszewska and Truhlar [21] demonstrated the excellent convergence of the Newton variational method in electron scattering with atomic exponential potential using $\mathcal{L}^2$ basis functions. Finally, Miller and collaborators [22] demonstrated the very good convergence of the complex Kohn variational method (CK).

In Table 3 we exhibit the results for CK, L2SV, SV, NV, and KVOF methods for electron scattering with exponential potential (6.11). In

TABLE 3. Tangents of *S*-Wave Phase Shifts for the Exponential Potential for Different n[a]

k (a.u.)	n	CK[b]	L2SV[c]	SV[d]	NV[c]	KVOF[e]
0.15	2	—	1.5974	1.0124	1.0006	1.0046934513
	3	—	1.0565	—	1.0000	1.0006096943
	4	1.0006	1.0068	1.0000	1.0000	1.0000079769
	6	1.0003	1.0016	1.0000	1.0000	1.0000000461
	8	1.0000	1.0009	1.0000	1.0000	1.0000000003
	10	1.0000	1.0002	1.0000	1.0000	1.0000000001
0.55	2	—	0.9015	0.9972	0.9795	1.0671641249
	3	—	0.9112	—	1.0000	0.9989808081
	4	0.9795	0.9351	0.9999	1.0000	0.9999580511
	6	1.0000	0.9935	1.0000	1.0000	0.9999999349
	8	1.0000	0.9995	1.0000	1.0000	0.9999999957
	10	1.0000	0.9997	1.0000	1.0000	0.9999999999

[a] Same as in Table 2. The KVOF calculation is performed with variational parameters $\beta = 1.0$ a.u. and $\alpha = 1.5$ a.u.

[b] From Miller and Jansen op de Haar [22].

[c] From Staszewska and Truhlar [21].

[d] From Takatsuka et al. [20].

[e] From Adhikari [15].

all these calculations the $\mathcal{L}^2$ functions are taken from Eq. (4.39) in addition to appropriate non-$\mathcal{L}^2$ functions. In Table 3, n denotes the total number of basis functions, including both $\mathcal{L}^2$ and non-$\mathcal{L}^2$ functions. In the CK method, although only one non-$\mathcal{L}^2$ function is used, as it is complex with real and imaginary parts, the numerical complication is similar to the inclusion of two real non-$\mathcal{L}^2$ functions in the basis set. Hence, the complex non-$\mathcal{L}^2$ function of this method has been counted as two real functions in Table 3, and in this case $n = 4$, for example, indicates two $\mathcal{L}^2$ functions in addition to the single complex outgoing wave composed of both cosine and sine functions. In Table 3, when more than one calculation are available in the literature performed with different choices of the parameters, we exhibit the one with the best convergence. In a related work on the CK method, Zhang et al. [22] advocated using two complex functions and claimed convergence superior to that shown in Table 3 with one complex function. The inclusion of this new complex function in the basis set is equivalent to including two more real functions as commented before. Once the total number of functions n is increased accordingly, the convergence obtained by Zhang et al. is slightly inferior to that of Miller and Jansen op de Haar [22].

From Tables 1 to 3, and considering other calculations available in the literature, we conclude that in general the off-shell methods of Chapter 4 have better convergence properties than the on-shell methods of Chapter 3. Among the various on-shell methods, there is none that is systematically better than others for all problems. All on-shell methods have similar rates of convergence. Among the off-shell methods the best convergence is of the NV method, with the SV and KVOF methods maintaining a close second position. From Table 3 we find that the NV method converges to four significant figures for $n = 3$. The same convergence is obtained in the CK (L2SV, SV, KVOF) method for $n = 6$ $(10, 4, 4)$. The SV and the NV methods are more difficult in practice in involving more complicated integrals than the KVOF method. The KVOF method is numerically the simplest of the three off-shell methods of Table 3, produces very competitive convergence, and hence is attractive for the numerical solution of scattering problems.

There are certain numerical difficulties with the basis-set variational methods discussed in this section. For obtaining accurate numerical results, the number of basis functions could be large, and subsequently, one may need to invert a large matrix involving large round-off error. In an ideal situation the basis functions should be ortho-

TABLE 4. *S*-Wave Scattering Lengths and Tangents of Phase Shifts[a]

N	Yukawa ($k = 0$)	Exponential ($k = 0$)	Yukawa ($k = 0.15$)	Exponential ($k = 0.15$)	Exponential ($k = 0.55$)
0	0.4550407017	0.5889624075	2.4070888958	1.9537093230	0.6875213332
1	1.0628835422	1.1094969597	0.9508377182	0.8898028226	0.9543748521
2	1.0009906100	1.0013279806	0.9991722408	0.9972672993	1.0035508106
3	1.0000400237	0.9999635234	0.9999916475	1.0000133976	1.0001518592
4	1.0000019050	0.9999946001	1.0000023971	1.0000067319	1.0000019555
6	1.0000000013	0.9999999926	1.0000000709	1.0000000150	0.9999999776
8	1.0000000000	1.0000000000	1.0000000014	1.0000000001	0.9999999999
≥ 10	1.0000000000	1.0000000000	1.0000000001	1.0000000002	0.9999999999

[a] Results are obtained from iterative solution of the Γ-matrix equation (2.28) for the exponential and Yukawa potentials for $k = 0$, 0.15 a.u. and 0.55 a.u., with $\gamma(k, q)$ given by Eqs. (6.18) and (6.19) [24]. At $k = 0$ the converged scattering lengths are 8.693254331840347 a.u. (exponential) and 7.911380206 a.u. (Yukawa).

normal and linearly independent. But in practice they are neither orthogonal nor linearly independent, and for large n they tend to be similar. If two of the functions are identical, one encounters a singular matrix to be inverted, leading to a breakdown of the calculational scheme. In practice the use of linearly dependent basis functions for large n results in the inversion of a nearly singular matrix. Hence, as n increases this procedure leads to poor numerical results. If the integrals are evaluated numerically, this problem is more severe. One can use approximately orthogonalized basis functions as in Refs. 17 and 18. These problems can be avoided in the nonvariational iterative methods of Section 6.3.2, which do not require the inversion of a matrix.

6.3.2 Iterative Methods

Born–Neumann iterative solution of Fredholm integral equations leads to rapidly convergent results for weak kernels. The iterative solutions of Lippmann–Schwinger equations involve only matrix vector multiplications and converge for a sufficiently weak potential or at sufficiently high energies. Successive matrix vector multiplications required in iterative methods are numerically much more economical and accurate than a matrix inversion encountered in most other methods. However, for most scattering energies the iterative solution of a realistic scattering problem is either divergent or slowly convergent to be of practical use.

TABLE 5. Scattering Lengths and Tangents of Phase Shifts[a]

N	Exponential ($k = 0$)	Yukawa ($k = 0$)	Exponential ($k = 0.15$ a.u.)	Yukawa ($k = 0.15$ a.u.)
1	1.055908698645270	1.2126812508	1.079071594402372	1.2609085936
2	1.003410010969849	1.0204190349	1.004242112709321	1.0230669313
3	1.000168451686852	1.0017009398	1.000193507341894	1.0018485113
4	1.000008094837622	1.0001334233	1.000008718250539	1.0001410945
6	1.000000018488864	1.0000007919	1.000000017575058	1.0000007968
8	1.000000000042132	1.0000000047	1.000000000035371	1.0000000043
10	1.000000000000096	1.0000000000	1.000000000000071	1.0000000000
≥ 12	1.000000000000000	1.0000000000	1.000000000000000	1.0000000000

[a] Results are obtained from iterative solution of Eq. (4.79) for the exponential and the Yukawa potentials [25].

There are several iterative schemes for the solution of Lippmann–Schwinger equations, when the Born–Neumann series diverges. The best known of these methods is the Padé method [11], which has been used successfully in many problems of physics, including scattering theory. Details of the Padé method have appeared in numerous monographs and review articles [11,12]. Instead of presenting an account of the Padé and all other methods for the iterative solution of scattering problems, in this section we present two physically motivated iterative methods. The first is based on a convergent Born–Neumann series of the Γ-matrix equation (2.28) [10,23], and the second is based on the off-shell variational principles of Section 4.4. Both methods have been applied for calculation of the K matrix.

The kernel of the Γ-matrix equation is real and weak compared to that of the original equation and may permit a convergent Born–Neumann series solution. To increase the flexibility of the Γ-matrix equation, a function $\gamma(k, q)$ with an arbitrary parameter is usually introduced. By adjusting this free parameter, this kernel can be made sufficiently weak to yield a rapidly convergent iterative Born–Neumann series for a wide class of scattering problems. The K- or the t-matrix elements are then obtained by performing an integral over known elements of the Γ matrix. This method was first suggested [23] for nuclear scattering problems and has recently been demonstrated to produce excellent convergence [24] in electron scattering with exponential and Yukawa potentials.

In this method the flexibility in the choice of the function γ of Eq.

(2.27) can be turned to good advantage in producing a rapidly convergent iterative Born–Neumann series. In numerical application we considered $\bar{k} = k$ in Section 2.4 and [24]

$$\gamma(k, q) = \left(\frac{k^2 + \alpha^2}{q^2 + \alpha^2}\right)^2 \tag{6.18}$$

for the exponential potential, and

$$\gamma(k, q) = \frac{k^2 + \alpha^2}{q^2 + \alpha^2} \tag{6.19}$$

for the Yukawa potential. The arbitrary parameter α was varied to obtain rapid convergence. With these choices, the leading asymptotic behavior for $p, q \to \infty$ of the two terms in kernel A of Eq. (2.26) of the Γ-matrix equation cancels. In addition, numerical calculations were performed with the simplest choice, $\gamma(k, q) = 1$, which also produced very good convergence. In the case of electron scattering with exponential (Yukawa) potential, 128 (400) Gauss points were used for numerical integration, and this yielded results accurate to 1 in 10^{10} in both cases after some 10 iterations. The parameter α was taken to be 3.0 a.u. (3.5 a.u.) for the exponential (Yukawa) potential. The results of calculation are exhibited in Table 4 for the γ functions given by Eqs. (6.18) and (6.19) for different iteration N. The first entry, $N = 0$, corresponds to $\Gamma = V$. The convergence is good in all cases. The small deviation from unity of some of the converged results does not indicate a defect of the iterative scheme but means that the number or the distribution of mesh points is not enough to obtain the accuracy desired.

Next we present results for the variational iterative method of Section 4.4 for a different iteration N. Variational iterative methods for the K matrix based on Eqs. (4.79) and (4.80) were used in numerical calculation with 400 Gauss mesh points for electron scattering with exponential and Yukawa potentials [25]. In the case of the exponential (Yukawa) potential, results with an estimated error of less than 1 in 10^{15} (10^{10}) were obtained after some 13 (10) iterations. There seems to be very little round-off error in the method, and the precision is increased by augmenting the number of mesh points and order of iteration. The calculations are performed for the K matrix and the iteration is initiated with $K = 0$. This lowest-order calculation corresponds to $N = 0$. The results for the Schwinger iterative scheme

(4.79) are exhibited in Table 5. The iterative scheme (4.80), based on the Newton variational principle, leads to a similar convergence [25].

We find that the iterative methods based on the Neumann series solution of the Γ-matrix equation and on the iterative variational method of Section 4.4 produce excellent convergence and should be considered as good options for solving scattering problems. However, any such iterative method, including the Padé method [11], constructs the final solution from the (divergent) iterative Born–Neumann series solution of the original Lippmann–Schwinger equation. If this original Born–Neumann series diverges strongly, the present iterative methods may lead to unacceptable/poor convergence properties.

6.4 MULTICHANNEL MODEL STUDIES

There are several two- and three-channel models that were extensively used for numerous numerical tests of different variational basis-set methods. The iterative (nonvariational) methods have not undergone such rigorous and extensive tests in multichannel scattering problems.

Multichannel studies of the different on-shell variational methods were undertaken first by Nesbet, Harris, Michels, and collaborators [26]. Multichannel calculations employing various off-shell methods, such as the Schwinger and Newton methods, were performed later. After equivalence was established between the (Schwinger and Newton) variational principles and appropriate basis-set methods, two-channel scattering calculations were performed in nuclear two- and three-body problems by Brady, Sloan, and the author [17,18,27].

First, we consider tests of the basis-set methods for the model two-channel problem due to Huck [28]. This model is analytically soluble and couples two S-wave channels. The total Hamiltonian $H = H_0 + V$ for this model is given by

$$H_0 = |\chi_1\rangle\left(-\frac{\hbar^2}{2m}\frac{d^2}{dr^2}\right)\langle\chi_1| + |\chi_2\rangle\left(-\frac{\hbar^2}{2m}\frac{d^2}{dr^2} + \Delta E\right)\langle\chi_2| \tag{6.20}$$

$$V = \sum_{i\neq j=1}^{2} |\chi_i\rangle V_{ij}\langle\chi_j|, \tag{6.21}$$

with $V_{12} = V_{21} = C/2\ (0)$, for $r \leq (\geq)\ B$. In Eq. (6.20), $\hbar^2/m = 1$. The form factors are orthonormalized, so that $\langle\chi_i|\chi_j\rangle = \delta_{ij}$, and the parameter $B = 1$. In numerical tests the following values of the parameters were employed: energy $E = 0.5$, $\Delta E = 0.375$, and $C^2 = 10$. The $\mathcal{L}^2$ basis functions in both channels were taken to be given by Eq. (3.75), and the oscillating functions for these two channels were taken to be $C(r)$ and $S(r)$ of Eqs. (3.73) and (3.74), with $k = k_1$ and k_2, respectively.

In the comparison, anomaly-free (AF, Section 3.5.1), minimum-norm (MN, Section 3.4.1), and optimized anomaly-free (OAF, Section 3.7.2) methods and the Schwinger method (4.24) with $\mathcal{L}^2$ functions (L2SV) as well as an anomaly-free Kohn method with only $\mathcal{L}^2$ functions due to Takatsuka and McKoy (TMKV) were employed [26,29]. In the TMKV method [29] the asymptotic boundary condition is incorporated by dividing the configuration space in the interior and exterior regions and fixing the log derivative of the wave function at the boundary by choosing the $\mathcal{L}^2$ basis functions appropriately, as in the R-matrix approach. They also modified the denominator as in Eqs. (4.231) to avoid anomalies.

The convergence of the K-matrix elements is slow with all the methods, especially for small basis sets. Comparisons were made for the on-shell K matrix elements, and it was found that the L2SV and the TMKV methods have better convergence properties. Both these methods yielded results significantly better than AF, MN, and OAF methods. The results obtained with the L2SV and TMKV methods using six basis functions were essentially exact and more precise than those obtained with on-shell methods with 25 basis functions. Convergence of the K-matrix elements is particularly slow for on-shell methods. However, convergence of the cross section is more rapid. Convergence of the on-shell methods was similar, and in Table 6 we exhibit the deviation of cross sections from exact values for AF, MN, L2SV, and TMKV methods [29,30]. The closer the entries are to zero, the better is the convergence.

From Table 6 we find that the L2SV and TMKV methods exhibit convergence properties superior to those of on-shell methods. Of the two methods, TMKV is numerically simpler in not involving double integral over the Green's function. The TMKV method should be compared with the KVOF method of Section 4.6 discussed in Section 6.3. Both are based on the Kohn variational principle, use an equation of the form (4.98), do not explicitly use the Green's function, yield very precise numerical results, and can be used in the modified form (4.231) so that spurious poles can be avoided. The KVOF method uses a single non-$\mathcal{L}^2$ function, whereas in the TMKV method the real $\mathcal{L}^2$ basis functions are chosen as in the R-matrix method, to include the

TABLE 6. Convergence of the Cross Sections with Different Methods for the Huck Model [28,30][a]

	n	AF	MN	L2SV	TMKV
$\Delta\sigma_{11}$	1	−0.271	−0.264	−0.04145	−0.04333
	2	−0.014	−0.013	−0.00516	−0.00011
	3	−0.050	−0.026	0.00049	−0.00003
	4	−0.008	0.001	0.00000	0.00000
$\Delta\sigma_{12}$	1	0.086	0.081	0.01342	0.02254
	2	0.005	−0.013	0.00215	0.00014
	3	0.007	0.014	0.00003	0.00002
	4	0.005	0.002	0.00000	0.00000
$\Delta\sigma_{22}$	1	0.005	−0.123	−0.04418	−0.01809
	2	0.036	0.189	−0.00262	0.00035
	3	0.160	−0.025	−0.00005	−0.00003
	4	−0.002	0.003	−0.00001	0.00000

Deviations from the exact values are presented in each case. The exact results are $\sigma_{11} = 2.16791$, $\sigma_{12} = 0.76746$, and $\sigma_{22} = 2.55844$. The results for the time reversed channel 21 are related to that of channel 12 by Eq. (1.223).

[a] Results are from Takatsuka and McKoy [29]. Reprinted with permission of the American Physical Society. Copyright © 1981.

proper boundary condition. Both involve similar numerical complications and should be considered as useful alternatives for precise solution of scattering problems. As in the single-channel models, the present coupled-channel study indicates that the off-shell methods have better convergence properties than the on-shell methods.

Callaway [31] made an extensive comparative numerical study of different on-shell methods in a three-state close-coupling (CC) model of electron–hydrogen scattering. The three states are the $1s$, $2s$, and $2p$ states of the hydrogen atom. For $L = 0$ (> 0), the model leads to three (four) coupled equations. Accurate calculations for this model have been performed by various authors, which can be taken as essentially exact [32]. Callaway et al. solved this three-state CC model using different variational on-shell basis-set methods and compared the results with the accurate exact calculation. Calculations were carried out using Slater-type orbitals (6.17) together with the usual long-range sine and cosine functions, where L is the angular momentum of scattering electrons in a specific channel and different values of α were taken for different functions.

For the three-state CC electron–hydrogen model, calculations were performed with Kohn (Section 3.3.1), inverse Kohn (Section 3.3.2), optimized minimum-norm Kohn and inverse Kohn (Section 3.7.1), optimized anomaly-free (Section 3.7.2), and variational least squares (Section 3.8) methods, among others [31]. After calculations with a large number of basis sets (15 for each channel) at a specific energy, reasonably good agreement between various basis-set methods was obtained. The standard deviation of the K-matrix element was 0.1% (2%) in the spin-singlet (triplet) state. These methods yield good convergence for small n (6 to 8). However, the final convergence is usually slow and nonmonotonic in nature. Also, all these on-shell methods have comparable convergence properties.

In these simple model scattering problems the numerical result is usually known, at least approximately, beforehand. But when one starts a new (complicated) problem this is not the case, and special judgment is needed regarding the acceptability of the numerical result of a particular method. Care should be taken to conclude, especially, (1) how far a particular result is from the converged result, and (2) if the energy under consideration is near a spurious resonance and consequently, if a particular result is grossly in error. Usually, calculations using more than one method for several n should be performed before deciding on these issues.

6.5 NUCLEAR FEW-BODY PROBLEMS

After the initial applications of on-shell variational methods to various model scattering problems, the deficiency of these methods was detected while calculating high-precision results. Despite calculational simplicity, there were problems with slow convergence and spurious resonances. Calculation of scattering soon began with off-shell variational methods, and their convergence was studied in detail. The first applications of the off-shell variational methods were made to the problem of nucleon–nucleon scattering for potentials with soft core. The nucleon–nucleon system, although it involves at most only two coupled channels, is not numerically trivial because of a soft repulsive core in the interaction. Because of the presence of the soft core, this problem presents convergence difficulties in both momentum- and configuration-space treatments. The very large momentum components of the interaction contributes to low-energy results. Consequently, a large number of momentum mesh points extending over a wide range of momenta are needed for obtaining high-precision results. Hence, this is a welcome test problem. The off-shell variational methods passed this test with honors. Then there were applications of these methods to model three-nucleon scattering problems. The Γ-matrix approach was also used in solving these problems.

Different types of degenerate kernel methods and methods of moments were used in solving nucleon–nucleon scattering problems in the late 1960s and early 1970s [33]. In these studies the variational nature of the calculations was not realized. The Schwinger variational method is a special type of the method of moments, and the Newton variational method is a special type of the degenerate kernel method. This connection between off-shell variational methods and these earlier calculations was not known to a general audience at that time. Brady, Sloan, and the author emphasized this connection [17,18] and did numerous tests of these methods in both two- and three-nucleon scattering. After these initial applications these methods were applied to problems in atomic and molecular physics. Here we describe only those calculations that were performed after establishing the above-mentioned connection. Although these authors were aware of the variational nature of their calculations, the variational aspect of the calculations was not always emphasized.

Brady and Sloan [17] formulated the Newton variational principle as a variational basis-set method and applied this method to the solution of two- and three-nucleon scattering problems. In their application of the Newton variational method to nucleon–nucleon scattering via the Reid ${}^1S_0$ soft-core potential, Brady and Sloan established very good convergence for both on- and off-shell t-matrix elements. In their study of the three-nucleon scattering problem with nucleon–nucleon separable potential they also found good convergence for on- and off-shell t-matrix elements. The three-nucleon scattering problem for the spin-doublet system involves two physical channels, and hence this is the first application of the Newton basis-set method to a multichannel problem. Sloan and the author [18] applied the Newton variational basis-set method to the nucleon–nucleon scattering problem and established good convergence. At the three-particle level they applied this method to a three-boson decay problem as well as to the neutron–deuteron scattering problem and obtained excellent convergence in both cases [18]. The more recent studies of Rawitscher [34] and collaborators also use a special form of the Newton variational method in the study of nucleon–nucleon scattering.

In their application to nucleon–nucleon scattering with a Malfliet–Tjon soft-core potential using the Schwinger basis-set method with $\mathcal{L}^2$ functions, Sloan and the author established good convergence properties. However, the convergence obtained by Brady and Sloan using the Newton method for a similar potential was slightly better. They also applied the Schwinger basis-set method to nucleon–nucleon scattering in $^3S_1$–$^3D_1$ channels [18] employing the Reid soft-core potential. In this case orbital angular momentum is not conserved. This two-channel problem is particularly difficult because it couples S- and D-wave channels with total angular momentum $J = 1$ employing phenomenological potentials with a soft repulsive core, which usually leads to slow convergence. This is the first application of the Schwinger basis-set method to a nontrivial multichannel problem.

The author also applied the Schwinger $\mathcal{L}^2$ basis-set method to a model proton–proton scattering problem in the presence of a nuclear Malfliet–Tjon soft-core potential and a repulsive Coulomb potential [35]. Using a two-potential formulation, the Coulomb potential was treated as a distorted wave, as in Section 4.7. He obtained rapid convergence in this study.

There is one advantage of writing the t matrix in the form of separable expansion, as one obtains in the Schwinger basis-set method. Use of the two-body (three-body) t matrix in separable form in the three-body (four-body) problem simplifies the latter by reducing its dimensionality [4,5]. With this intention there have been many applications of the Schwinger basis-set method to solution of two- and three-nucleon problems. The three-nucleon problem is a difficult problem, as the singularity of the kernel depends on both momentum variables and is not easy to subtract out. The Schwinger basis-set method yields good convergence for three-nucleon scattering. In some cases, special energy-dependent basis functions were used [5].

Numerical activity in the construction of separable nucleon–nucleon t-matrix expansions increased after the theoretical work of Sloan and the author and of Ernst, Shakin, and Thaler (EST) [18,36]. In the EST approach the rank-n EST t matrix is made to be exact half-on-shell at n preselected energies. If these energies are chosen carefully, the method provides a good interpolation at other energies. Sloan and the author (AS) advocated the use of analytic expansion functions so that the final result for the separable t matrix has a simple structure, which facilitates the use of this t matrix in three- and four-body problems. Plessas and co-workers [4] combined the virtues of the

AS and EST methods to construct separable expansions for realistic Paris and Bonn two-nucleon t matrices, which they used successfully in solving the three-nucleon problem with realistic potentials. Baldo et al. [37] have constructed rapidly convergent separable expansions for Reid and Argonne V_{14} nucleon–nucleon potentials using Gamow functions in the AS method. Bund, Canton, Hartt, Oryu, and their collaborators [38] also obtained separable expansions for the t matrix. At the three-nucleon level, Sofianos et al. [38] using the AS method, constructed separable expansion of the neutron–deuteron t matrix to be used in the four-nucleon problem, and obtained good convergence. Such expansions have been used with great success by Fonseca [5] for solution of the four-nucleon problem.

All these studies of the Schwinger method in nuclear physics used only $\mathcal{L}^2$ basis functions. Consequently, in many applications final slow convergence has been observed. Large oscillations in the nucleon–nucleon phase shift obtained with this method in some numerical applications have later been identified with the occurrence of a spurious resonance [39].

There have been applications of the complex Kohn variational method to the solution of nucleon–nucleon scattering employing the Reid soft-core potential in $^1S_0$ and $^3S_1$–$^3D_1$ channels [40]. Rapid convergence was obtained in both cases for phase shifts and mixing parameters with a small number of basis functions. The convergence was faster than that obtained by the Schwinger method with the same potential. Recently, Kievsky [40] made an interesting application of the complex Kohn variational method to neutron–deuteron scattering below the breakup threshold using realistic local nucleon–nucleon potentials and three-nucleon forces. He obtained very precise and stable numerical results for scattering length, phase shifts, and mixing parameters with an error of less than 0.1%. Kievsky et al. also extended the method to energies above the deuteron breakup threshold [40].

The Γ-matrix method of Section 2.4 was also first applied to the solution of few-nucleon problems. Many of these applications use the iterative Neumann series solution of a manifestly nonsingular equation. In practical cases, the dimension of the discretized matrix equation could be large, so direct matrix inversion may lead to large round-off error. A convergent iterative Neumann series leads to low round-off error and involves much less computing time. The iterative Γ matrix method has been used for the numerical solution of both two- and three-nucleon scattering problems [23]. In the two-channel three-nucleon scattering problem, very satisfactory convergence was found from the iterative solution of the Γ-matrix equation. However, in general, the mathematical criteria for convergence of the Neumann series for the Γ-matrix equations are seldom clear. The best choice of the arbitrary function γ is achieved only by trial.

If the original Lippmann–Schwinger equation diverges very strongly or involves many coupled channels, convergence of the Neumann series solution of the Γ-matrix equation could be slow. When the iterative Neumann series does not converge satisfactorily, there are several alternatives. First, one can introduce further subtractions in the kernel to improve convergence. This has led to two alternative methods, known as the continued fraction method and the iteration subtraction methods [41]. These alternative methods have been found to accelerate the convergence rate of the Neumann series solution when the iterative solution of the original Γ-matrix equation is not attractive. Compared to the Padé method [11,12], these alternative methods have been found to lead to very competitive convergence behavior in solving few-nucleon scattering problems. Second, one can also directly solve the Γ-matrix equation by inversion or otherwise. This second possibility also offers advantages, as, unlike the K- and t-matrix Lippmann–Schwinger equations, the Γ-matrix equation is a real Fredholm equation. The K-matrix equation involves a principal-value prescription in the kernel, which makes its numerical application complicated. The Γ-matrix method has not seen widespread application in solving realistic collision problems and perhaps deserves more attention in the future.

6.6 SCATTERING BY HYDROGEN ATOM

There have been many numerical tests of the variational on-shell basis-set methods of Chapter 3 to physical scattering problems of interest in atomic and molecular physics, and we present a brief account of them in the following. In atomic physics we mostly consider the following simple problems: electron–hydrogen atom, positron–hydrogen atom, electron–helium atom, and positron–helium atom scattering. In these cases the idea is to perform a variational calculation based on essentially the exact Hamiltonian of the system. However, in most cases one needs the atomic wave functions. In the case of hydrogen, these atomic wave functions are known exactly. In other cases, model wave functions are used. The nuclear calculations of Section 6.5 are based on model Hamiltonians. So apart from numerical errors and lack of convergence due to the use of a finite basis set, numerical solutions of this problem can be taken to be essentially exact.

In the case of simpler atoms, for example alkali atoms, where the last valance electron can be considered to be the only active electron, the exact Hamiltonian is first approximated by a model Hamiltonian and the variational calculation is performed with this Hamiltonian. One can perform essentially exact calculations with these simplified models which can be compared with different approximation schemes. This gives an estimate of the usefulness of the models and approximations. Such a comparison cannot be made for atoms with more complex structure. In this section we deal with hydrogen atom scattering, and we consider helium scattering in the next section. We do not present an exhaustive review of the scattering processes but present mostly an account of the application of various numerical methods to these problems.

6.6.1 Electron–Hydrogen Scattering

There are two major studies of scattering by hydrogen atom at low and intermediate energies: electron–hydrogen and positron–hydrogen scattering. These are reasonably simple systems where one can hope to find reliable converged results by variational methods or otherwise so that relative accuracy of different methods and approximation schemes could be tested. An excellent account of these calculations can be found in the review article by Callaway and in the monograph by Nesbet [1,3]. Among these studies are those of Schwartz [13,14] and Armstead, Shimamura, Rudge, Register, Callaway, and collaborators [42–45]. As most of these calculations were based on the exact Hamiltonian of the problem and not on the few-state close-coupling (CC) model Hamiltonians, they can produce results superior to those of the few-state CC model.

In electron–hydrogen atom scattering there are two total spin states: $S = 0, 1$. The total differential cross section for elastic scattering is given by [1,3]

$$\frac{d\sigma}{d\Omega} = \frac{1}{4}\sum_{S=0}^{1}(2S+1)|f_S|^2, \tag{6.22}$$

where f_S is the scattering amplitude in spin state S. In this case one obtains a distinct partial-wave phase shift for spin singlet and triplet states. The need for higher partial waves in reproducing differential cross sections at small scattering angles and for energies as low as 8–10 eV was realized. In atomic physics, instead of calculating all the higher phase shifts, it is common to use the following partial-wave Born formula for the scattering phase shift in higher partial waves [1,3]:

$$\tan\delta_L = \frac{\pi\alpha_d k^2}{(2L+3)(2L+1)(2L-1)}\left[1 - \frac{3(\gamma-\alpha_q)k^2}{(2L+5)(2L+3)\alpha_d}\right], \tag{6.23}$$

where α_d (α_q) is the dipole (quadrupole) polarizability and γ is the leading term in the nonadiabatic correction to α_d. For hydrogen atom, $\alpha_d = 4.5a_0^3$ and $(\gamma - \alpha_q)/\alpha_d = 23/6$. Good agreement with experiment was obtained by calculating the lower (higher) partial-wave phase shifts variationally (from the Born formula above).

An adequate representation of the electric-dipole polarization potential is essential in electron–hydrogen scattering. However, the use of few exact atomic states in the CC scheme makes it impossible to describe the polarization of the target correctly. A correct description of atomic p-wave states is needed for this purpose. Castillejo et al. [46] gave the following estimates for generation of the polarization potential in the CC approximation. Inclusion of the $2p$ state (all bound p-wave states) produces 66% (81%) of the full polarization potential. The remaining 19% comes from the continuum p-wave states of hydrogen. The effect of all the p-wave states can be incorporated by including a p-wave pseudostate in both a variational and a CC treatment of electron–hydrogen scattering. A pseudostate is a mathematically constructed state which is orthogonal to other atomic states and simulates the effect of the continuum and the neglected atomic bound states in a CC calculation. After including the pseudostate(s), the CC model can reproduce the correct value of the polarizability, and this improves the result significantly.

In the pioneering study of electron–hydrogen scattering at low energies below the lowest inelastic threshold at 10.2 eV, Schwartz [13,14] employed the Kohn method for calculating the S-wave phase shifts. By varying the parameter in the basis set, he calculated accurate values of phase shifts. His trial functions incorporated the exact atomic bound states in the physical channel(s). A similar calculation was repeated by Shimamura [42], who used not only the Kohn method but also the Harris method. For both ^1S and ^3S scattering states, with about 50 basis functions, a converged result was obtained for both the Kohn and Harris methods. The Kohn method produced better convergence. By studying the convergence of the result with a finite basis set, it was possible to extrapolate the result to an infinite basis set. Oberoi and Callaway [47] obtained very similar results from a crude basis set suggested by the polarized orbital method [1]. Abdel-Raouf and Belschner [48] calculated *S*-wave phase shifts by the minimum-norm method. Callaway [44] calculated *D*- and *F*-wave phase shifts using different trial functions and pseudostate expansions. Results of Callaway from five variational basis-set methods (Kohn, inverse Kohn, optimized minimum-norm, inverse optimized minimum-norm, and optimized anomaly-free) were found to be in satisfactory agreement. Rudge [43] repeated the calculations for the *S*-wave phase shifts using the anomaly-free method of Section 3.5.1 with slightly more flexible functions, and his calculations are in agreement with those of Schwartz and Shimamura. Armstead extended the Kohn variational calculations of Schwartz to *P* waves, and the results were later confirmed by Das and Rudge, who used the anomaly-free method suggested by Rudge in Section 3.5.1 [43]. Register and Poe extended these variational studies to *D* waves using the Kohn and the inverse Kohn methods [43].

Independent variational calculations have been carried out and the results compared for each of the phase shifts up to *D* wave. Except where the result is in doubt because of the presence of a nearby anomaly, the convergence was monotonic in many cases. Hence it was possible to extrapolate the result of a finite basis calculation to find the converged result with an infinite basis [1,3].

As one approaches the lowest inelastic threshold, several resonances were found in numerical calculations. Shimamura [42] determined the positions (widths) of the lowest ^1S and ^3S resonances to be 0.702452 Ry and 0.745923 Ry (0.00347 Ry and 0.0000026 Ry), respectively. Narrow *P*- and *D*-wave resonances were also found in elastic electron–hydrogen scattering at energies close to the *S*-wave resonances near the lowest inelastic threshold. Das and Rudge [43] determined the position (width) of the lowest ^3P resonance to be 0.715735 Ry (0.00043 Ry). Callaway [1] determined the position (width) of the lowest ^1P resonance to be 0.747901 Ry (0.0000027 Ry). Register and Poe [43] determined the position (width) of the lowest ^1D resonance to be 0.743924 Ry (0.00066 Ry).

Callaway [1,44] used pseudostates in his basis for calculating phase shifts for $L = 2$ and 3. For large L, short-range correlations become less important and long-range correlations dominate. The dipole and quadrupole polarizabilities are represented accurately by his basis set. His calculated phase shifts with Kohn, inverse Kohn, optimized minimum-norm, inverse optimized minimum-norm, and optimized anomaly-free methods were in good agreement with each other. In Table 7 we present the low-energy phase shifts at various energies taken from different references. Although some of these phase shifts were given up to four significant figures after the decimal point, we have rounded the results in Table 7 up to three significant figures after the decimal. The ^1F and ^3F phase shifts are essentially identical within this precision and are shown in a single column.

Callaway and collaborators [1,45] performed electron–hydrogen atom scattering calculations at intermediate energies using on-shell variational basis-set methods above the excitation and ionization thresholds for incident energies from 10.2 to 54.4 eV using pseudostates. At these energies, because of the presence of an effective static dipole potential, the asymptotic form of the continuum basis function used in variational methods needs to be modified to obtain good convergence. Callaway et al. used energy-dependent $\mathcal{L}^2$ basis func-

TABLE 7. Phase Shifts (in radians) for Elastic Electron–Hydrogen Scattering at Different Energies in a.u.[a]

k (a.u.)	$^1S^b$	$^3S^b$	$^1P^c$	$^3P^c$	$^1D^d$	$^3D^d$	$^{1,3}F^e$
0	5.965	1.769					
0.1	2.553	2.939	0.007	0.011	0.001	0.001	0
0.2	2.067	2.717	0.015	0.045	0.005	0.005	0.002
0.3	1.696	2.500	0.017	0.106	0.011	0.011	0.004
0.4	1.415	2.294	0.010	0.187	0.018	0.020	0.007
0.5	1.202	2.105	−0.001	0.271	0.027	0.030	0.010
0.6	1.041	1.933	−0.009	0.341	0.038	0.042	0.015
0.7	0.930	1.780	−0.013	0.393	0.052	0.056	0.020
0.8	0.886	1.643	−0.004	0.427	0.075	0.070	0.026

[a] The entries at zero energy are the scattering lengths in a.u. The results at $k = 0$ are from Schwartz (1961a) [13].
[b] From Shimamura (1971a) [42].
[c] From Armstead [43].
[d] From Register and Poe [43].
[e] From Callaway [44].

tions for this purpose. An algebraic CC scheme was used for the variational calculation. The $1s$, $2s$, $2p$ channels were treated as open channels with appropriate non-$\mathcal{L}^2$ functions and the usual exponential $\mathcal{L}^2$ functions were used for both open and closed channels. Several types of on-shell methods were used in the calculation for partial waves $L \leq 3$ and complemented by the Born formula (6.23) for higher partial waves.

Callaway et al. [1,45] found reasonably well-converged results for elastic and excitation cross sections to the lowest excited state(s) of the hydrogen atom. They obtained 1 to 2% accuracy in elastic scattering and 5 to 10% accuracy in excitation. Compared to these variational calculations, convergence of the CC calculation could be extremely slow and exact calculations impossible to perform in many cases. In this problem, many simpler calculations, such as Born or few-state CC approximation, perform rather poorly, especially at low energies.

A series of resonances in different partial waves were confirmed as a new threshold was crossed. Just above the $n = 2$ threshold and below the $n = 3$ threshold several resonances were found. These resonances have been studied by different authors [1]. Above the $n = 3$ threshold, the pseudostate expansion used by Callaway et al. [1] cannot adequately represent the hydrogen atom, and it becomes increasingly difficult to locate the resonances.

McKoy and collaborators [7] used the off-shell multichannel Schwinger basis-set method for the calculation of $^{1,3}S$ elastic electron–hydrogen scattering phase shifts at low energies below the inelastic threshold employing the exact Hamiltonian. Their basis functions were antisymmetrized products of single-electron wave functions. In the basis they included the product of:

1. Two hydrogen $1s$ functions (one basis function)
2. A hydrogen $1s$ function and long-range sine and cosine functions (two basis functions)
3. A hydrogen $1s$ function and Slater scattering orbital functions $r^i \exp(-\alpha r) Y_{lm}, i = 1, 2, ..., M$ (M functions)
4. Hydrogen atom exact or pseudostates for closed channels and Slater scattering orbital functions with $i = 1, 2, ..., M_p$ (M_p functions)

First they studied the convergence of the variational calculation with respect to both M and M_p for a $1s$–$2p$ calculation. The total number of basis functions $n = 3 + M + M_p$. They found that the convergence is good with respect to the number of scattering orbitals for a given closed channel. Next they considered the problem using a $1s$–$2s$–$\overline{2p}$ expansion and also found good convergence properties. Compared with the converged variational results of Table 7, McKoy et al. results differ by up to 5% for the ^1S phase shifts. However, it is pleasing to find that these variational calculations produce very reliable results with comparatively little numerical effort, which agree well with experiment [49].

6.6.2 Positron–Hydrogen Scattering

At low energies, positron–hydrogen elastic scattering is apparently easier to handle than electron–hydrogen scattering because in the former case one does not need to antisymmetrize the model with respect to the incident positron and the target electron. At low energies the effect of antisymmetrization in electron–hydrogen scattering is important. At higher energies the effect of antisymmetrization in electron–atom scattering is not important and results for electron–atom and positron–atom scattering tend to be similar. But this advantage of the positron–hydrogen system is more than compensated due to the appearance of open rearrangement positronium formation channels above 0.5 Ry, which are difficult to handle. The appearance of these channels brings in most complications of a three-body problem. However, even at very low energies, the presence of the closed positronium channel provides additional attractive interaction, which causes convergence difficulties in variational calculations of positron–hydrogen scattering, as noted by Schwartz [13] in calculating of S-wave phase shifts using the Kohn method. Shimamura [42] repeated this calculation, and his results are in close agreement with that of Schwartz.

Bhatia et al. [50] suggested a different variational approach to positron–hydrogen scattering. They were able to construct an approximate optical potential for the system which is everywhere more repulsive than the true optical potential. Then the phase shift calculated with this potential gives a bound on the exact result. They improved this optical potential systematically so that it converges to the exact optical potential. The resultant phase shifts also converge to the exact result. By this procedure they calculated essentially converged phase shifts at low energies below the excitation and rearrangement thresholds for low partial waves.

Roy and Mandal [50] made a useful and practical application of the

Schwinger variational method to positron–hydrogen–atom collisions for both the direct and positronium-formation channels. They gave accurate results for S, P, D and higher-partial-wave phase shifts. They also calculated low-energy differential scattering cross sections. Kar and Mandal [51] studied the effect of correlated basis functions for studies on positron collisions using the Schwinger method.

First, we consider the S-wave phase shifts. The S-wave phase shifts of Schwartz [13], Shimamura [42], Roy and Mandal [50], and Bhatia et al. [50] are in good agreement with each other. Using a complex energy extrapolation method, Doolen et al. [51] also calculated S-wave phase shifts using 120 trial functions and obtained results in excellent agreement with Bhatia et al. Humberston, using the Kohn method, and Stein and Sternlicht, using the anomaly-free method, also found good results [52]. Houston and Drachman [53] performed variational calculations of S-wave phase shifts using the Kohn and Harris methods.

The most accurate P-wave phase shifts were calculated by Armstead [43] using the Kohn variational method and by Roy and Mandal [50] using the Schwinger method. Bhatia et al. [50] also calculated P-wave phase shifts later using their variational approach, and their results are in excellent agreement with those of Armstead. These results stand as a benchmark. Variational D-wave phase shifts have been calculated by Register and Poe [43] using the Harris method and by Roy and Mandal [50] using the Schwinger Method. We tabulate some of these results in Table 8. At low energies scattering occurs predominantly in the S wave, and from Table 8 we find a zero in the S-wave phase shift at $k^2 \approx 0.4$ a.u. Consequently, the total cross section has a Ramsauer–Townsend minimum at this energy.

Above the positronium formation and other inelastic thresholds, the conventional variational calculations become increasingly complicated and the convergence slow. In this energy region the CC method is appropriate for an approximate study of positron–hydrogen scattering. The CC method leads to slow convergence but is much simpler to implement than the variational methods and reveals interesting features of the scattering process, as in electron–hydrogen scattering. As in the case of electron–hydrogen scattering, CC calculations have produced a series of narrow resonances in positron–hydrogen scattering as one approaches the $n = 2$ threshold of the hydrogen atom. In their algebraic $(1s, 2s, 2p)$ CC calculation, Seiler et al. [32] established five resonances in the positron–hydrogen system. The positions (widths) of three S-wave resonances found by them are

TABLE 8. Phase Shifts (in radians) for Elastic Positron–Hydrogen Scattering at Different Energies in a.u.[a]

k (a.u.)	S^b	S^c	S^d	S^e	S^f	P^g	P^c	D^h	D^f
0	−2.10	−2.103							
0.1	0.151	0.148	0.149	0.151	0.147	0.009	0.009	0.0013	0.0013
0.2	0.188	0.188	0.189	0.187	0.187	0.033	0.034	0.0054	0.0054
0.3	0.168	0.168	0.169	0.167	0.167	0.065	0.067	0.0125	0.0126
0.4	0.120	0.120	0.123	0.120	0.120	0.102	0.102	0.0235	0.0235
0.5	0.062	0.062	0.065	0.062	0.063	0.132	0.131	0.0389	0.0382
0.6	0.007	0.004	0.008	0.002	0.004	0.152	0.155	0.0593	0.0593
0.7	−0.054	−0.051	−0.049	−0.049	−0.051	0.173	0.180	0.0863	0.0870

[a] The entries at zero energy are the scattering lengths in a.u.
[b] From Schwartz (1961a) [13].
[c] From Bhatia et al. [50].
[d] From Houston and Drachman [53].
[e] From Shimamura (1971a) [42].
[f] From Roy and Mandal [50].
[g] From Armstead [43].
[h] From Register and Poe [43].

0.7195 Ry, 0.7484 Ry, and 0.7502 Ry (0.0011 eV, 0.000065 eV, and 0.00000035 eV). The positions (widths) of two *P*-wave resonances found by them are 0.7346 Ry and 0.7502 Ry (0.00027 eV and 0.000016 eV). The resonances have been studied by a number of workers using different methods, including Mitroy and Stelbovics, Archer et al., Ho and Greene, Pelikan and Klar, and Ho [54,55]. Due to a small discrepancy in the positions of the resonances with the various groups, it is sometimes difficult to determine the exact number of resonances. From the CC study of Mitroy and Stelbovics, it is possible to identify several more distinct resonances. The positions (widths) of two more *S*-wave resonances are 0.8577 Ry and 0.8707 Ry (5.5^{-4} Ry and 1.76^{-4} Ry). The positions (widths) of two more *P*-wave resonances are 0.8598 Ry and 0.8714 Ry (4.1^{-3} Ry and 1.26^{-3} Ry). The positions (widths) of three *D*-wave resonances are 0.7499 Ry, 0.8639 Ry, and 0.8727 Ry (1.1^{-6} Ry, 1.73^{-4} Ry, and 4.5^{-5} Ry).

In addition to resonances, the CC calculations have yielded elastic and different transition cross sections. We refer the interested reader to the recent studies of Mitroy and Stelbovics and of Walters et al. [54] for a detailed description of these calculations.

6.7 SCATTERING BY HELIUM ATOM

After hydrogen, the next most complex atom is helium, which deserves special mention. The wave function of helium is not known analytically and one has to use a theoretical model. However, because of the relatively small number of electrons in helium, one can perform accurate variational calculation of helium scattering by electrons and positrons which can be compared with various approximation schemes and also with experiment. Experimentally, being an inert gas, a helium atom target is easily available, whereas nascent hydrogen (atom) easily forms a hydrogen molecule, which makes experiments with hydrogen atom more difficult to perform.

6.7.1 Electron–Helium Scattering

Variational calculations for electron–helium scattering have been reported by several groups. Most of these calculations considered only elastic scattering. Some considered low-energy inelastic scattering also. Usually, independent orbitals for the two electrons with no correlation has been used to describe the helium wave function. This provides an excellent description of excited states with large principal quantum number n, and a good description of low-lying states. Unlike in hydrogen, for a fixed small n, the different l orbitals are not degenerate.

The first variational basis-set calculation of electron–helium elastic phase shifts at low energies is due to Michels et al. [57] using the Harris method of Section 3.4. This calculation was improved by Sinfailam and Nesbet [1,3,57], who employed a practically complete basis set within the Bethe–Goldstone approximation of level [1s] of helium. Nesbet derived general Bethe–Goldstone equations for electron–atom scattering [3]. This method includes the important effect of polarization. Sinfailam and Nesbet used Hartree–Fock wave functions for the helium states and solved these equations by the optimized anomaly-free method of Section 3.7.2. They calculated phase shifts for partial waves $L \leq 3$. The phase shifts are in good agreement with those calculated by Michels et al. [57] by the Harris method and by Duxler et al. [58] using the polarized orbital method. In Table 9 we exhibit the low-energy phase shifts of Sinfailam and Nesbet and Duxler et al. and the scattering length of Michels et al. There have been other calculations of electron–helium scattering. Using a method

TABLE 9. Phase Shifts (in radians) for Elastic Electron–Helium Scattering at Different Energies in a.u.[a]

k (a.u.)	S^b	S^c	P^b	P^c	D^b	D^c	F^c
0	1.145						
0.1	3.016	3.020	0.0032	0.0033			
0.2	2.882		0.0129		0.0016		0.0003
0.3	2.704	2.755	0.0297	0.0327	0.0036	0.0040	0.0005
0.4	2.614	2.622	0.0539	0.0594	0.0064	0.0073	0.0009
0.5	2.484	2.494	0.0847	0.0926	0.0097	0.0115	0.0015
0.6	2.364	2.372	0.1200	0.1362	0.0141	0.0167	0.0023
0.7	2.252	2.256	0.1567	0.1696	0.0197	0.0228	0.0032
0.8	2.134	2.147	0.1947	0.2081	0.0258	0.0298	0.0043
0.9	2.043		0.2311		0.0322		0.0057
1.0	1.955	1.953	0.2646	0.2749	0.0393	0.0458	0.0076
1.1	1.877	1.8662	0.2927	0.3016	0.0473	0.0544	0.0098

[a] The zero-energy entry is the scattering length from Michels et al. [57].
[b] From Sinfailam and Nesbet [57].
[c] From Duxler et al. [58].

of Drachman, Houston calculated the scattering length to be 1.159 a_0 [59]. Using the R-matrix method, Berrington et al. calculated a scattering length of $1.189a_0$, in close agreement with experimental data of Crompton et al. [1,60]. A later analysis yielded a scattering length of $1.145a_0$, in agreement with Sinfailam and Nesbet and Michels et al. [1,57,60].

Wichmann and Heiss [61] performed electron–helium scattering calculation employing the five-state CC model. They considered the usual expansion in terms of the target states. By including some correlation functions in their expansion, they found the effect to be insignificant. They calculated the K matrix using both Kohn and inverse-Kohn methods and obtained good convergence within their model. For absolute convergence (with experimental results) the CC model should include some pseudostates, which are essential for representing the polarizability in a finite basis calculation. For helium, 52.4% of the polarizability comes from the scattering states [1,3], which can be taken into account only through pseudostates. For large principal quantum number n, levels of different l become nearly degenerate, and oscillatory energy-dependent functions need to be included in the basis for obtaining good convergence [1,3].

As in electron– and positron–hydrogen scattering, close-coupling methods predict the position of the resonances quite well. Close-coupling calculation also yields different transition cross sections at no extra cost at all energies [1]. An electron–helium system has a single S-wave resonance below the $1s2s$ threshold. Wichmann and Heiss found the S-wave elastic scattering resonance to be 19.2 eV. Previously, using the CC model, Burke et al. found the same resonance at 19.3 eV [56]. Temkin et al. put the resonance at 19.363 eV, with width 0.0144 eV while Sinfailam and Nesbet put it at 19.42 eV with width 0.015 eV [62]. Experimentally, this resonance has been observed by various groups. Brunt et al. [56] confirmed the position (width) of this resonance at 19.366 eV (0.009 eV). This resonance has also been observed experimentally by Gibson and Dolder and by Golden and Zecca [63].

O'Malley et al. performed an R-matrix calculation including both target atom correlation and polarization effects consistently [64]. They included pseudostates to represent both dipole and quadrupole polarizabilities. Andrick and Bitsch [64] showed that up to about 20 eV, the D- and higher-wave phase shifts could be estimated from the lowest-order Born formula (6.23) with experimental polarizability $\alpha_d = 1.384138a_0^3$ and $\gamma = \alpha_q$. Hence, O'Malley et al. calculated only S-

and P-wave phase shifts. Nesbet [3] refined these calculations by the matrix variational method using a correlated helium target wave function. S- and P-wave phase shifts were computed in this way below the first inelastic threshold. Nesbet obtained good convergence in his calculation. O'Malley et al. used a modified effective range expansion and found the scattering length to be $1.177a_0$, whereas Nesbet found $1.1835a_0$ [1,3].

Using the Born formula for $L > 2$, the scattering amplitude is given by [1,3]

$$f(\theta) = k^{-1} \sum_{L=0}^{2} (2L+1) \exp(i\delta_L) \sin \delta_L P_L(\cos\theta)$$

$$+ \pi\alpha_d k \left[\frac{1}{3} - \frac{1}{2} \sin\frac{\theta}{2} - \sum_{L=1}^{2} \frac{1}{(2L+3)(2L-1)} P_L(\cos\theta) \right], \quad (6.24)$$

and the total elastic and momentum transfer cross sections are given by [1,3]

$$\sigma_E = \frac{4\pi}{k^2} \sum_{L=0}^{2} (2L+1) \sin^2 \delta_L + \frac{1}{2450} \pi^3 \alpha_d^2 k^2, \quad (6.25)$$

$$\sigma_M = \frac{4\pi}{k^2} \sum_{L=0}^{2} (L+1) \sin^2(\delta_L - \delta_{L+1}) + \frac{2}{33075} \pi^3 \alpha_d^2 k^2, \quad (6.26)$$

respectively, both in units of a_0^2. These cross sections were calculated with phase shifts of Nesbet, O'Malley et al., Sinfailam and Nesbet [3,57,64], and compared with the experimental results of Kaupila et al., Stein et al., Crompton et al., and Kennerly and Bonham [65]. The agreement between theory and experiment is good.

Oberoi and Nesbet extended the Bethe–Goldstone variational calculations of Sinfailam and Nesbet above the $n = 2$ threshold [3,57,66]. For energies between $n = 2$ and $n = 3$ thresholds, multichannel variational equations were solved including all significant effects of virtual excitations of reference configuration $1s2s$. Scattering states with open-channel orbitals up to $l = 3$ were included in these calculations. The experimental $n = 2$ thresholds are 2^3S at 19.818 eV, 2^1S at 20.614 eV, 2^3P^0 at 20.964 eV, and 2^1P^0 at 21.218 eV. Several resonances have been detected in this energy region in the lowest partial waves [3,57,66]. Nesbet made a careful study of scattering between the $n = 2$ and $n = 3$ thresholds. The calculations of Burke et al. (using the CC approach), Berrington et al. (using the R-matrix approach), and Nesbet et al.

(using the Bethe-Goldstone method) are in good qualitative agreement [3,66,67]. Total cross sections for $1^1S \to 2^1S$ and $1^1S \to 2^3S$ transitions reported by Berrington et al. and by Nesbet and Oberoi are in good agreement with each other. Oberoi and Nesbet used the optimized anomaly-free method of Section 3.6.2. Thomas and Nesbet reported results in the ionization region [3]. Differential elastic and excitation cross sections have also been reported in this energy region in good agreement with experiment.

6.7.2 Positron–Helium Scattering

Because of both calculational and experimental difficulties, positron–helium scattering has been explored much less than has electron–helium scattering. However, at low energies there have been many applications of variational methods to positron–helium scattering [68]. In general, there have been convergence difficulties in this problem and the result seems to be very sensitive to the helium wave function employed. Convergence difficulties can be avoided in the variational methods by using a suitable model, as suggested by Drachman [69]. Houston and Drachman, and Campeanu and Humberston [70] used this approach to calculate low-energy phase shifts. Houston and Drachman used the Kohn method to calculate the scattering length and the Harris method to calculate the phase shifts. Campeanu and Humberston used Kohn method for both. Aulenkamp et al. [71] performed variational elastic scattering calculation using the Kohn method for partial waves $L \leq 3$ using distorted wave functions, taking into account virtual positronium formation. Amusia et al. [71] also calculated accurate phase shifts for partial waves $L \leq 3$. They used a simplified version of the random-phase approximation with exchange and with the virtual positronium formation effect taken into account. The phase shifts are shown in Table 10.

Of the phase shifts shown in Table 10, results for the H5 and H14 models of Humberston and Campeanu are probably the most accurate. For higher partial waves these phase shifts are not available in tabulated forms and we exhibit the results of Amusia et al., which are in good agreement with those of Humberston and Campeanu. Ho and Fraser [72] performed accurate calculation using both correlated and uncorrelated wave functions. They found that the helium wave function should have correct dipole polarizability for producing good phase shifts. Drachmann's variational S-, P-, and D-wave phase shifts are in good agreement with other results [72].

TABLE 10. Phase Shifts (in radians) for Elastic Positron–Helium Scattering at Different Energies in a.u.[a]

k (a.u.)	S^b	S^c	S^d	S^e	P^e	D^e	F^e
0	−0.472	−0.48	−0.524				
0.1	0.032	0.033	0.035	0.026	0.0026	0.0006	0.0002
0.2	0.041	0.042	0.049				
0.3	0.030	0.031	0.039	0.016	0.020	0.0051	0.0016
0.4	0.007	0.009	0.020				
0.5	−0.022	−0.020	−0.003	−0.032	0.039	0.0116	0.0043
0.6	−0.056	−0.053	−0.034				
0.7	−0.093	−0.090	−0.069	−0.091	0.053	0.0215	0.0085
0.8	−0.127	−0.124	−0.106				
0.9	−0.162	−0.159	−0.143	−0.152	0.057	0.032	0.014
1.0	−0.193	−0.190	−0.177				

[a] The zero-energy entry is the scattering length in a.u.
[b] From Humberston (1973) [70].
[c] From Campeanu and Humberston (1977) [70].
[d] From Houston and Drachman [70].
[e] From Amusia et al. [71].

At low energies the total cross section can be calculated by using the phase shifts above together with a Born estimate (6.23). The interesting feature of the low-energy total cross sections is the appearance of a Ramsauer–Townsend minimum at 2 eV. This is related to the change in sign of the S-wave elastic phase shift near this energy, as can be seen from Table 10. The variational calculations above have reproduced the Ramsauer–Townsend minimum in the cross section. However, the experimental value of the cross section at this minimum ($0.04\pi a_0^2$) is lower than the theoretical estimate ($0.066\pi a_0^2$). At slightly higher energies (> 6 eV) the experimental cross section is in close agreement with theory [68].

Above the lowest excitation threshold, the variational calculations of positron–helium scattering becomes very difficult to perform, due to convergence difficulties and the existence of an open positronium formation channel. In this region the CC calculation, including a few positron formation channels, describes well the experimental situation. Ghosh and the author [73] performed such a calculation with five helium states and three positronium states. As no pseudostates were included, the low-energy phase shifts of this model are not realistic. At low energies the cross sections are much too large compared with experiment; a correct description of polarization is needed for obtaining smaller cross sections. However, they calculated elastic, excitation, and positronium formation cross sections at medium and high energies, in good agreement with experiment. They also found a very narrow S-wave resonance just below the $n = 2$ excitation of the helium atom at 19.27 eV. It should be recalled that there was a similar resonance below the $n = 2$ threshold in electron–helium scattering [56]. The appearance of the resonance at essentially the same energies in both systems suggest that same long-range polarization potential is responsible for both.

6.8 ELECTRON SCATTERING BY HEAVIER ATOMS

There have been applications of variational methods to the study of electron scattering by heavier atoms. The simplest of these systems are the alkali atoms (Li, Na, K, Rb, etc). Alkali atoms have one electron outside the closed shell, and in approximate treatment this bears some resemblance to the hydrogen atom. Only the outermost electron is assumed to take active part in electron scattering. Electric-dipole polarizability is important in alkali atoms. Sinfailam and Nesbet [74] have performed variational elastic electron scattering calculations at low energies by lithium, sodium, and potassium below the first excitation threshold. They used the optimized anomaly free method employing the Bethe–Goldstone model [3], which they used successfully for the hydrogen atom, and calculated phase shifts for spin singlet and triplet states for partial waves $L < 3$. The results were in fair agreement with calculations of the close-coupling type by others [1,3]. Several resonances were observed in P and D waves, both by Sinfailam and Nesbet and by other workers.

Nesbet and collaborators [3] also applied variational methods for the calculation of electron scattering by carbon, nitrogen and oxygen atoms. These atoms have a much more complex structure than that of hydrogen and alkali atoms. Dipole polarizability of these atoms is small compared to that of alkali atoms, but polarization potential still dominates low-energy scattering. A low-energy scattering calculation for these atoms must incorporate enough configuration interaction to give a good description of the low-lying states, negative-ion energies, and polarizability. Nesbet et al. achieve this through their Bethe–Goldstone approach.

Thomas and Nesbet [75] reported several types of calculations for each of carbon, nitrogen, and oxygen atoms. They considered first an uncorrelated wave function equivalent to a three-state CC calculation, which does not specially build in the atomic polarizability. Then they included some atomic configurations that build in both correlation and polarizability and obtained good agreement with an experimental total cross section at low energies. From these calculations one finds that the total cross section becomes smaller as more contributions to polarizability are included in the wave function. This is also true in electron and positron scattering by helium, where the inclusion of atomic polarizability in the model reduces the total cross section at low energies.

In Thomas–Nesbet variational Bethe–Goldstone calculations, the effects of $2s \rightarrow np$ virtual excitations in electron–oxygen scattering beyond $n = 2$ were found to be small and were neglected. Basis orbitals and partial waves for $l \leq 3$ were included in the calculation. Important virtual polarization effects due to $2s \rightarrow 2p$, $2p \rightarrow ns$, and $2p \rightarrow nd$ were included in all the states of oxygen and short-range correlation between the incident electron and oxygen. This was necessary for obtaining agreement with experiment and with the polarized orbital calculation of Henry [75].

In electron–nitrogen scattering, variational Bethe–Goldstone calculation by Thomas and Nesbet [3,75] of an elastic differential cross section was found to be in good agreement with other calculations. They included the $2s^2 2p^3$, $2s2p^4$, and $2p^5$ configurations of nitrogen. Electron affinities of various atoms, including nitrogen, were computed by Moser and Nesbet using a variational method [3,76]. The energy calculated for C^- (O^-) relative to the ground states of the neutral atoms was -1.29 eV (-1.43 eV), in agreement with experiment. The energy computed for $N^-(^3p)$ was 0.12 eV, which indicates that this state should either be a very weakly bound state or a resonance. This state of nitrogen appears as a sharp peak in the total cross section at a fraction of 1 eV. It is difficult to produce this resonance at the correct position. Nesbet et al. [3,75] introduced a parameter in their model to produce this resonance.

Thomas and Nesbet also calculated the cross sections for the electron–carbon system and compared them with the polarized orbital calculations of Henry [3,75]. At low energies these two calculations are not in agreement. There were no reliable experimental results to resolve the discrepancy. Thomas–Nesbet calculation shows a peak in the cross section at a fraction of 1 eV, whereas the calculation of Henry does not exhibit this peak.

6.9 MOLECULAR PHOTOIONIZATION

The simplest of all scattering processes in molecular physics is molecular photoionization. This problem can be solved essentially using the Schwinger variational method with a good knowledge of the target wave function. The amplitude can be calculated as indicated in Section 4.9, with the molecular Hartree–Fock wave function constructed as in Section 5.5. McKoy et al. [7] made an extensive study of molecular photoionization using the iterative Schwinger methods [11] of Section 4.9 employing H_2, N_2, CO_2, C_2H_2, and other molecules. For H_2 [7,77], a small basis set gave rapid convergence in the iterative Schwinger method. With a basis of s- and z-type Cartesian Gaussian functions of exponents 0.3 and 1.0 on each nuclei and a z-type Cartesian Gaussian function of exponent 1.0 at the center of mass, the photoionization cross section in the $1\sigma_g \rightarrow k\sigma_u$ channel converged within one iteration. For calculating the cross section they used frozen-core Hartree–Fock approximation for the final ionized-state wave function, which presents a simplified picture of the actual state of affairs. For photon energies above 18 eV, their cross-section results agreed well with experiment.

The iterative Schwinger method has been applied to the case of N_2 [7,78]. In this case there is a shape resonance in the cross section in the $3\sigma_g \rightarrow k\sigma_u$ channel, leading to the $X^2\Sigma_g^+$ state of N_2^+. For the initial state they used the usual Hartree–Fock wave functions of Section 5.5 and also a configuration interaction wave function constructed from 386 spatial configurations and considered the effect of initial state correlation on the cross section. The initial basis set contained 18 Gaussian functions in the σ_u symmetry and eight Gaussian functions in the π_u symmetry. One iteration gave good convergence in this case. The shape resonance at about 30 eV was well reproduced by the theoretical calculation. The total photoionization cross section was obtained by averaging over all orientations of the molecule and summing over all directions of the emitted photoelectron. They also calculated the cross section in the $1\pi_u \rightarrow k\delta_g$ channel, which exhibits a broad peak in the total cross section. But an analysis of the eigenphase sum has revealed that there is no resonance in the latter channel. The shape of cross section in this channel was determined by the energy dependence of the dipole matrix element. They calculated the total cross section leading to the $A^2\Pi_u$, and $B^2\Sigma_u^+$ states of N_2^+ and compared their results with the Stieltjes–Tchebycheff moment theory approach and the continuum multiple scattering theory method.

Luchesse et al. [79] also studied the photoionization cross sections of CO_2 using the iterative Schwinger method. They used 32 spherical Gaussian basis functions in the $1\pi_u \rightarrow k\delta_g$ channel and 14 in the $1\pi_g \rightarrow k\delta_u$ channel. Again, the cross section converged after one iteration. As in the case of N_2, they also studied the effects of initial-state correlation on the photoionization cross sections. The interesting feature in this case is the appearance of a shape resonance in the cross section in the $1\pi_g \rightarrow k\sigma_u$, $4\sigma_g \rightarrow k\sigma_u$, $1\sigma_g \rightarrow k\sigma_u$ photoionization channels at a photoelectron kinetic energy of 15 to 20 eV. Their results for cross section leading to the $C^2\Sigma_g^+$ state of CO_2 were in reasonable agreement with experiment.

There has also been a study of photoionization of C_2H_2 by the Schwinger method [80]. In this case there is a double-peaked structure in the cross section leading to the $X^2\Pi_u$ state of $C_2H_2^+$. Basis sets with 12 to 22 spherical Gaussian functions were used. Satisfactory convergence was achieved after two iterations of the Schwinger method. However, the calculation exhibits a single broad peak in place of the double-peaked structure mentioned above.

In the case of CO, Luchesse et al. [81] compared the results for cross section and eigenphase sums obtained by the iterative Schwinger and Padé methods of Section 4.9. They used 28 and 20 spherical Gaussian functions for the σ and π continuum photoionization channels, respectively. Comparing the two methods, they find that the convergence of the Padé method is slightly better than that of the iterative Schwinger method in general.

6.10 ELECTRON–MOLECULE SCATTERING

There have been many numerical studies of electron–molecule scattering using complex Kohn, Schwinger, and R-matrix methods [9]. Most of these studies involve calculation of only elastic scattering by the static exchange approximation. Static exchange approximation makes the calculation possible, but usually leads to unrealistic results at low energies where polarization effect is important. The static exchange cross sections predict shape resonances at higher energies, with larger widths compared to experiment. The typical discrepancy in the position of resonances is on the order of 2 to 4 eV. Moreover, many of these systems exhibit Ramsauer–Townsend minima in cross section at very low energies. The static-exchange calculation cannot predict these minima. Using the Schwinger multichannel approach,

McKoy et al. [9] studied elastic electron scattering with alkanes from CH_4 to n-C_4H_{10}, the C_3H_6 isomer cyclopropane, the hydride series CH_4, SiH_4, GeH_4, NH_3, PH_3, and the fluorides CF_4, and SiF_4. Although these large-molecule calculations are encouraging, we do not discuss them here and refer the interested readers to a review article [9]. For simpler molecules, such as H_2, N_2, CO_2 and so on, there have been extensive studies of inelastic scattering with electrons, which we describe briefly in the following. In all these studies, the multichannel Schwinger variational method yielded the most realistic results.

As hydrogen is the simplest diatomic molecule, there have been many theoretical calculations and experimental investigations of inelastic e^-–H_2 scattering. McKoy et al. [9] applied the iterative Schwinger method of Section 4.9 to several electron–molecule scattering processes, including H_2 and CO_2. In electron–hydrogen–molecule scattering, the convergence properties of this method was studied carefully. In calculations of the excitation cross section for the $b^3\Sigma_u^+$ state, convergence of the computed K-matrix elements by the iterative Schwinger method was found to be good. This problem was also studied by linear algebraic (Baluja et al.), R-matrix (Schneider and Collins), Schwinger multichannel (Lima et al.), and complex Kohn (Rescigno and Schneider) methods [82]. The results of these different computations are in agreement with each other and with experiment. The complex Kohn method has been used to calculate cross sections for the $b^3\Sigma_u^+$, $a^3\Sigma_g^+$, and $c^3\Pi_u$ states in a five-channel coupling scheme. The R-matrix method has been used to calculate cross sections for the $b^3\Sigma_u^+$, $a^3\Sigma_g^+$, $c^3\Pi_u$, $B^1\Sigma_u^+$, $C^1\Pi_u^+$, and $E, F^1\Sigma_g^+$ states, employing coupling between all channels. McKoy et al. [9,82] made a Schwinger multichannel calculation for the a, b, and c states in a five-channel coupling scheme.

For the $b^3\Sigma_u^+$ $(1\sigma_g \rightarrow 1\sigma_u)$ channel of hydrogen, both for total and differential cross sections, calculations employing the complex Kohn and Schwinger methods are in reasonable agreement with each other and with experiment. R-matrix calculation deviates to some extent from variational calculations and experiment. For the $a^3\Sigma_g^+$ $(1\sigma_g \rightarrow 2\sigma_g)$ channel, the two-channel Schwinger calculation is in qualitative agreement with experiment and complex Kohn calculation. Theoretical calculations of differential cross sections agree with experimental results at medium angles where measurements are available. Hence, the discrepancy between theory and experiment in the integrated cross section may imply that the differential cross sections calculated at very small and large angles are not correct. For the $c^3\Pi_u$ channel, both variational and *R*-matrix calculations are in rough agreement with each other but in disagreement with experiment. The reason for this discrepancy is not clear.

Because of its technological importance, the electron–nitrogen system is the most studied following the electron–hydrogen system [9]. Theoretically, there have been studies by the Schwinger (Mu-Tao and McKoy) and *R*-matrix (Gillan et al.) methods [83]. These studies have been confined to the lowest-order approximation. There have been a two-state CC study of the $X^1\Sigma_g^+ \rightarrow a^1\Pi_g$ $(3\sigma_g - 1\pi_g)$ excitation and a *R*-matrix study of excitations $A^3\Sigma_u^+(1\pi_u \rightarrow 1\pi_g)$, $B^3\Pi_g(3\sigma_g \rightarrow 1\pi_g)$, and $W^3\Delta_u(1\pi_u \rightarrow 1\pi_g)$ [9]. The convergence obtained in Schwinger multichannel calculations with $A^3\Sigma_u^+$, $B^3\Pi_g$, $W^3\Delta_u$, $w^1\Delta_u$, $B'^3\Sigma_u^-$, and $a'^1\Sigma_u^-$ excitations was satisfactory. There have been theoretical calculations of electronic excitation to the $A^3\Sigma_u^+$ state of nitrogen using the Schwinger method [9]. Above 15 eV, experiments and theories are in agreement. At lower energies there are discrepancies between different theoretical predictions of total cross sections and experimental findings. The reason for these discrepancies is not clear. There have been experimental studies for excitation to the $^1\Pi_g$ state of nitrogen. Both the Schwinger multichannel and two-state calculations are in reasonable agreement with experiment. There have also been encouraging theoretical studies of $a^1\Pi_g$ and $B^3\Pi_g$ excitation cross sections. In the former case, agreement between Schwinger differential cross section and experiment is generally good, although there are quantitative differences in the forward direction, where the various experiments also tend to disagree with each other. In the latter case, the integrated and differential Schwinger multichannel excitation cross sections are in agreement with experiment.

The CO molecule has been studied less than H_2 or N_2. Satisfactory results have been obtained by the Schwinger multichannel method, employing a five-channel coupling scheme for the $a^3\Pi$ and $A^1\Pi$ states that arise from the $5\sigma \rightarrow 2\pi$ excitation and a nine-channel scheme for the $a'^3\Sigma^+$, $e^3\Sigma^-$, $d^3\Delta$, $I^1\Sigma^-$, and $D^1\Delta$ states that arise from the $1\pi \rightarrow 2\pi$ excitation [9,84]. Of these states, $a^3\Pi$ is the lowest excited state with the largest excitation cross section. For an integrated cross section, the agreement between theory and experiment is good. For a differential cross section, theory and experiment have similar trends with somewhat different magnitudes. There are discrepancies between theory and experiment for the $a^3\Pi$ cross section. The total excitation cross section for all states agrees with measurements and empirical estimates below 13 eV. At higher energies, experimental values are much higher, indicating a large contribution from the excitation of Rydberg states, which are not considered in Schwinger multichannel formulation.

The iterative Schwinger method of Section 4.9 has also been successfully used for the calculation of differential and total scattering cross sections for electron–CO_2 scattering [9]. In these studies, spherical Gaussian basis sets with some 20 to 30 basis functions for various scattering symmetries were used. The total elastic scattering cross section of e^-–CO_2 scattering exhibits a shape resonance experimentally at about 3.8 eV. The calculated static exchange elastic cross section using the iterative Schwinger method predicts a resonance at 5.39 eV. Considering the simplicity of the model, the static exchange elastic differential cross section for e^-–CO_2 scattering at 10 eV is in good agreement with experiment.

Inelastic scattering of electrons with complex molecules such as formaldehyde, ethylene, and propellane has also been treated by the Schwinger method [9]. Although these preliminary studies clearly demonstrate the usefulness of the multichannel Schwinger method in dealing with these complex scattering processes, much remains to be done is settling the discrepancy between theory and experiment. Assuming that the discrepancy is due to the deficiency of the multichannel model used and not due to convergence difficulties, it remains to be seen if the model can be enriched within present computational limitations to obtain agreement with experiment.

6.11 REACTIVE ATOM–DIATOM COLLISIONS

Reactive atom–diatom collision is a challenging problem theoretically. The possibility of rearrangement makes this problem extremely difficult from both a theoretical and a computational point of view. The simplest system in this category is $H + H_2$. The accurate solution of $H + H_2$ scattering was obtained about 20 years ago at low energies and for total angular momentum zero. In the beginning, model one- and two-dimensional problems [85] were solved before the actual three-dimensional problem was solved [86]. Later, accurate results have been obtained for heavier atoms, at higher energies, and for nonzero total angular momentum [4].

Recently, using the generalized Newton variational method (GNVM) of Section 5.3, Truhlar et al. performed [87] full three-dimensional quantum mechanical solution for such systems as $D+H_2$, $O+H_2$, $H+OH$, $H+HBr$, $F+H_2$, and $O+HD$. They employed both real and complex boundary conditions. The use of complex boundary conditions in GNVM reduces it essentially to the complex Kohn method. They extended their calculation of (full) cross sections to higher energies and nonzero total angular momentum. They calculated state-to-state reaction cross sections for an atomic beam colliding with a diatomic molecular target. Miller et al. [88] also studied many of these reactive processes using the complex Kohn variational method. Below we give an account of these variational calculations, with reference to other works whenever necessary.

To take advantage of symmetries to simplify the calculation, Truhler et al. used an angular momentum representation where the problem decouples into blocks for fixed total angular momentum J and parity P. For $J = 0$ there is only one parity block. At low energies the small J components dominate. Hence, the $J = 0$ solution is especially interesting. For calculating cross section and related observables, one has to sum over all J, P components.

Sun et al. [8] expanded the radial scattering wave function in terms of the sum of basis functions in more than one arrangement, thus reducing the Schrödinger equation to coupled integrodifferential equations with a nonlocal kernel, which were then converted to a set of coupled integral equations. Then they applied the GNVM in solving these equations. By expanding the scattering amplitudes in a set of $\mathcal{L}^2$ basis functions, the Newton variational principle was implemented in terms of the solution of a set of algebraic equations. They used the direct product of translational basis functions in the form of distributed Gaussians and internal basis functions in the form of diatom vibrational functions and analytic angular functions. In test calculations of the GNVM in realistic scattering problems, they found better convergence properties than with some methods of moments and other variational methods. Also, with the GNVM they could obtain stable converged results to more significant figures. The cost of these advantages is evaluation of multiple integrals over Green's functions. They reached the conclusion that the advantages obtained using GNVM are well worth the price.

Sun et al. [8] used a model Hamiltonian with a nonreactive zero-order part and reactive coupling employing a coupled-channel Green's function. As rotational orbital coupling is stronger than vibrational and reactive coupling, quantum states with different vibrational and reactive quantum states are decoupled in the zeroth order. The basis function expansion was not used to solve the full problem, only the problem of coupling to the zeroth-order problem. The zeroth-order potential differs from the full potential only in a fairly small region of space, and the basis functions have to work less in finding an accurate solution to this problem. Solution of the zeroth-order problem is the equivalent of a distorted-wave Born approximation. When this approximation is good, very few basis functions are required to obtain the converged solution.

The Green's functions involve products of a function of the position of one of the atoms, which is regular at the origin, times a function that goes to infinity there. To avoid numerical difficulties with the irregular part, Truhlar et al. reformulated the variational principle in terms of what they call half-integrated Green's functions, which can be calculated without knowledge of the irregular solution. Then they computed the regular solution of the zero-order problem and the half-integrated Green's functions and evaluated integrals numerically over these functions and the coupling potentials. Finally, a set of linear equations was solved to obtain the scattering matrix. These steps were repeated for each total energy and *JP* block. For obtaining observables, the sum over *JP* was calculated.

These reactions had been treated successfully in the past with the use of semiclassical theories [87]. It is interesting to see the limitations of these theories in comparison to an exact quantum description. Application of the approach to $O + H_2 \rightarrow OH + H$ illustrates the motivation behind the use of a quantum theory. This reaction proceeds on two similar potential energy surfaces. By using analytic potential energy surfaces, the reaction rates for thermal and vibrationally excited molecules have been calculated by variational transition-state theory with semiclassical transmission coefficients, and good agreement with experiment has been obtained. The calculations were performed with real and complex boundary conditions using the GNVM. Satisfactory convergence was obtained with both boundary conditions at total energy 0.65 eV and $JP = 0^+$ and even exchange symmetry.

The scenario corresponding to the semiclassical approach is worth mentioning. At the threshold for a chemical reaction, the energies of motions orthogonal to the reaction coordinate, s, can be computed to a good approximation assuming that the vibrations are adiabatic with respect to s. Then an effective potential barrier is calculated semiclassically by adding a zero-point vibrational energy to the energy along the minimum-energy path at each point s along this path. At energies above this barrier, the reaction proceeds naturally. At energies below this, for the reaction to take place the system must tunnel through a multidimensional barrier. When one considers the reaction of a molecule with a high-frequency excited vibration, one can calculate a new effective potential assuming that this vibration is adiabatic as well. The semiclassical calculation also provides a picture of tunneling at low energies in agreement with experiment. However, it is not clear whether these models are realistic or whether the agreement with experiment is accidental. Very careful comparison with quantum models may answer these questions.

Truhlar et al. [87] performed semiclassical and accurate quantum calculations for the reaction probability for the same potential energy surface of the reaction $O + H_2 \rightarrow OH + H$. They find good agreement between the quantum and semiclassical calculations. For the $O + HD$ system, semiclassical theories are in good agreement with experiment for kinetic isotope effects. Accurate quantum calculations are also in agreement with semiclassical theory. In this case they considered reactive processes $O + HD \rightarrow H + OD$ and $\rightarrow D + OH$ at a total energy of 15 kcal/mol for $JP = 2^-$. They obtained good convergence with both real and complex boundary conditions of the GNVM.

Truhlar et al. also studied the $D + H_2 \rightarrow HD + H$ reaction. This reaction also has a high effective barrier for vibrationally adiabatic reaction when the initial vibration is excited. Again the semiclassical result is in good agreement with accurate quantum calculation. At that time (1986), there was a large discrepancy between the calculated threshold reaction probability and experiment. However, the experiments were later found to be in error. There is good agreement between new experimental results and both semiclassical and quantum calculations. Later they extended the calculations to very high total energy, 1.8 eV for the state-to-state cross sections $D + H_2$ $(v = 1, j = 1) \rightarrow H + HD$ $(v' = 1, j' = 0$ to $13)$ and compared the results with semiclassical calculation as well as experiment. Both agreements are satisfactory. Zhang and Miller [89] applied the complex Kohn variational method for calculating the differential and integral cross sections for this process. For a wide range of energy (0.4 to 1.35 eV), they used 32 partial waves for the total angular momentum to obtain convergence.

The reaction $H + H_2$ (*para*) $\rightarrow H + H_2$ (*ortho*) is possibly the most studied atom–diatom reaction [90]. For $H + H_2$, Sun et al. [8,84,90] used the double many-body expansion potential energy surface. They found that calculations with complex and real boundary conditions using the GNVM converged to the same result. They performed two sets of calculations. The first was at a total energy of 0.6 eV and a total angular momentum $J = 0$, and the second was at a total energy of 1.369 eV and $J = 3$. In both cases they obtained satisfactory convergence. This process was also studied by Zhao et al. [91], who performed state-to-state $H + p\text{-}H_2$ $(v = 0, j = 0, 2, J = 0$ to $4) \rightarrow H + o\text{-}H_2$ $(v' = 0, 1)$ calculations and compared the results with semiclassical theory. The error in the semiclassical calculation was by a factor of 1.5 to 2, which calls for full quantum treatment for this system. Zhang and Miller [89] reported state-to-state fully converged integral cross sections for this last process for a wide range of energy using the complex Kohn method. They included up to 19 partial waves, and the converged calculations for the energy range 0.9 to 1.4 eV were in disagreement with the experimental results. It was then found that these experimental results were wrong. Later, Zhang and Miller [88] calculated differential cross sections for state-to-state $H + H_2$ $(v = j = 0) \rightarrow H + H_2$ $(v', \text{odd}\, j')$ for a total energy of 0.9 to 1.35 eV. Twenty-five partial waves were used in this case to obtain convergence.

Next we consider exothermic reactions. One interesting aspect of these reactions is their tendency to produce inverted population distributions of the vibrational energy levels of the newly formed bond. Full quantum calculations provide a useful way to study the extent of population inversion. Truhlar et al. [87] studied the reaction $H + HBr \rightarrow H_2 + Br$ quantum mechanically. They calculated the population ratio to different vibrational states of H_2 and found that reactions to higher-energy excited states were suppressed, whereas higher-momentum states were enhanced. It would be interesting to see if semiclassical calculation could reproduce this result. Both Truhlar et al. and Zhang and Miller also studied the $F + H_2 \rightarrow HF + H$ reaction carefully for both zero and nonzero total angular momentum.

From these calculations it becomes increasingly clear that a full quantum-mechanical description of atom–diatom reactions is called for. Low-energy light–molecular reactions are really governed by quantum mechanics, and both the complex Kohn method and the GNVM provide useful mechanisms for studying atom–diatom reactions.

6.12 NOTES AND REFERENCES

[1] For a review, see, for example, Callaway (1978a) and Nesbet (1978).
[2] Truhlar et al. (1974), Nesbet (1969b).

[3] Nesbet (1980).

[4] For a review, see, for example, Adhikari and Tomio (1987), Adhikari and Kowalski (1991), Plessas (1986), Plessas and Haidenbauer (1987), and Ferreira (1986).

[5] Fonseca (1986a,b), Oryu (1986).

[6] Miller (1990, 1992, 1994).

[7] Lucchese et al. (1986).

[8] Sun et al. (1990), Kouri et al. (1988), Tawa et al. (1994).

[9] For a review of multichannel Schwinger calculation, see Winstead and McKoy (1994).

[10] Adhikari and Tomio (1986, 1987), Adhikari and Kowalski (1991), Adhikari (1996), Tomio and Adhikari (1980a,b, 1981, 1995).

[11] See, for example, Baker, and Gammel, (1970), Glöckle (1983).

[12] Adhikari and Kowalski (1991).

[13] Schwartz (1961a,b).

[14] Schwartz (1966).

[15] Adhikari, unpublished.

[16] Smith and Truhlar (1972), Callaway (1980), Thirumalai and Truhlar (1980), Takatsuka et al. (1981), Staszewska and Truhlar (1986, 1987), Rescigno and Schneider (1987), Miller and Jansen op de Haar (1987).

[17] Sloan and Brady (1972), Brady and Sloan (1972, 1974), Sloan and Adhikari (1974).

[18] Adhikari and Sloan (1975a–d).

[19] Thirumalai and Truhlar (1980).

[20] Takatsuka et al. (1981).

[21] Staszewska and Truhlar (1986).

[22] Miller (1990), Miller and Janson op de Haar (1987), Rescigno and Schneider (1987), Zhang et al. (1988).

[23] Tomio and Adhikari (1980a,b, 1981), Adhikari, (1991), Kowalski and Feldman (1961, 1963a,b), Kowalski (1965, 1972).

[24] Tomio and Adhikari (1995).

[25] Adhikari (1996).

[26] Nesbet (1969a), Nesbet and Oberoi (1972), Harris (1967), Harris and Michels (1969a,b, 1971).

[27] The first applications of the multichannel Newton and Schwinger variational principles seems to be in Brady and Sloan (1974) and Adhikari and Sloan(1975b,c).

[28] Huck (1957).

[29] Takatsuka and McKoy (1981b).

[30] Nesbet (1969a), Nesbet and Oberoi (1972), Harris and Michels (1969a).

[31] Callaway (1978a).

[32] Burke and Schey (1962), Burke et al. (1963), Seiler et al. (1971), Smith and Truhlar (1972), Smith and Henry (1973).

[33] Levinger (1974).

[34] Rawitscher (1987, 1989).

[35] Adhikari (1976), This seems to be the first derivation and application of the Schwinger basis-set method using distorted waves.

[36] Ernst et al. (1973a,b).

[37] Baldo (1985, 1987).

[38] Oryu (1986), Hartt and Yidana (1987), Bund (1985), Canton et al. (1988), Sofianos et al. (1979).

[39] Adhikari (1990).

[40] de Araujo et al. (1995), Kievsky (1997), Kievsky et al. (1998).

[41] Tomio and Adhikari (1981), Hořacek and Sasakawa (1983, 1984).

[42] Shimamura (1971a–c).

[43] Rudge (1973, 1975), Register and Poe (1975), Armstead (1968), Das and Rudge (1976).

[44] Callaway (1978b).

[45] Callaway and Wooten (1974, 1975), Callaway and Williams (1975), Callaway et al. (1976), Morgan et al. (1977).

[46] Castillejo et al. (1960).

[47] Oberoi and Callaway (1969).

[48] Abdel-Raouf and Belschner (1978).

[49] Williams and Willis (1974).

[50] Bhatia et al. (1971), Roy and Mandal (1993a,b).

[51] Doolen et al. (1971), Kar and Mandal (1997).

[52] Humberston (1984), Stein and Sternlicht (1972).

[53] Houston and Drachman (1971).

[54] Mitroy and Stelbovics (1994a), Kernoghan et al. (1995).

[55] Mitroy and Stelbovics (1994b), Archer et al. (1990), Ho and Greene (1987), Ho (1990), Pelikan and Klar (1983).

[56] Brunt et al. (1977), Burke et al. (1969).

[57] Michels et al. (1969), Sinfailam and Nesbet (1972, 1973).

[58] Duxler et al. (1971).

[59] Drachman (1972), Houston (1973).

[60] Crompton et al. (1970), Berrington et al. (1975).

[61] Wichmann and Heiss (1974).

[62] Temkin et al. (1972), Nesbet (1978).

[63] Gibson and Dolder (1969), Golden and Zecca (1971).
[64] O'Malley et al. (1979), Andrick and Bitsch (1975).
[65] Crompton et al. (1970), Kauppila et al. (1977), Stein et al. (1978), Kennerly and Bonham (1978).
[66] Oberoi and Nesbet (1973a,b).
[67] Burke et al. (1969), Berrington et al. (1975).
[68] For a review, see Ghosh et al. (1982).
[69] Drachman (1966, 1968).
[70] Houston and Drachman (1971), Humberston (1973, 1974), Campeanu and Humberstron (1975, 1977).
[71] Aulenkamp et al. (1974), Amusia et al. (1976).
[72] Drachman (1968), Ho and Fraser (1976).
[73] Adhikari and Ghosh (1996).
[74] Sinfailam and Nesbet (1973), Nesbet (1967).
[75] Thomas et al. (1974), Thomas and Nesbet (1975a–d), Henry (1967, 1968).
[76] Moser and Nesbet (1971).
[77] Lucchese and McKoy (1981a).
[78] Lucchese and McKoy (1981b), Lucchese et al. (1982).
[79] Lucchese and McKoy (1981c, 1982).
[80] Lynch et al. (1984), Lucchese et al. (1986).
[81] Lucchese and McKoy (1983).
[82] Baluja et al. (1985), Schneider and Collins (1985), Lima et al. (1985), Reseigno and Schneider (1988).
[83] Mu-Tao and McKoy (1983), Gillan et al. (1990).
[84] Sun et al. (1992).
[85] Mortensen and Gucwa (1969), Truhlar and Kuppermann (1970, 1972), Kuppermann et al. (1974).
[86] Schatz and Kuppermann (1975), Elkowitz and Wyatt (1975).
[87] For a review, see Tawa et al. (1994) and Truhlar et al. (1990).
[88] For a review, see Miller (1990, 1992, 1994).
[89] Miller and Zhang (1991), Zhang and Miller (1990).
[90] Sun et al. (1989).
[91] Zhao et al. (1989).

BIBLIOGRAPHY

Abdallah, J., and D. G. Truhlar (1974). *J. Chem. Phys. 60*, 4670.

Abdel-Raouf, M. A., and D. Belschner (1978). *J. Phys. B 11*, 3677.

Abramowitz, M., and I. A. Stegun, Eds. (1965). *Handbook of Mathematical Functions*, Dover, New York.

Adhikari, S. K. (1976). *Phys. Rev. C 14*, 782.

Adhikari, S. K. (1979). *Phys. Rev. C 19*, 1729.

Adhikari, S. K. (1985). *Phys. Lett. A 113*, 1.

Adhikari, S. K. (1990). *Phys. Rev. A 42*, 6.

Adhikari, S. K. (1991). *Chem. Phys. Lett. 181*, 435.

Adhikari, S. K. (1992a). *Chem. Phys. Lett. 189*, 340.

Adhikari, S. K. (1992b). *J. Comput. Phys. 103, 415*

Adhikari, S. K. (1996). *Chem. Phys. Lett. 258*, 595.

Adhikari, S. K. and A. S. Ghosh (1996). *Chem. Phys. Lett. 262*, 460.

Adhikari, S. K., and K. L. Kowalski (1991). *Dynamical Collision Theory and Its Applications*, Academic Press, San Diego, Calif.

Adhikari, S. K., and I. H. Sloan (1975a). *Nucl. Phys. A 241*, 429.

Adhikari, S. K., and I. H. Sloan (1975b). *Nucl. Phys. A 251*, 297.

Adhikari, S. K., and I. H. Sloan (1975c). *Phys. Rev. C 11*, 1133.

Adhikari, S. K., and I. H. Sloan (1975d). *Phys. Rev. C 12*, 1152.

Adhikari, S. K., and L. Tomio (1981). *Phys. Rev. C 24*, 1186.

Adhikari, S. K., and L. Tomio (1982). *Phys. Rev. C 26*, 83.

Adhikari, S. K., and L. Tomio (1986). *Phys. Rev. C 33*, 467.

Adhikari, S. K., and L. Tomio (1987). *Phys. Rev. C 36*, 1275.

Adhikari, S. K., A. C. Fonseca, and L. Tomio (1982). *Phys. Rev. C 26*, 77.

Amado, R. D. (1963). *Phys. Rev. 132*, 485.

Amado, R. D. (1969). *Annu. Rev. Nucl. Sci. 19*, 61.

Amusia, M. Y., N. A. Cherepkov, L. V. Chernysheva, and S. G. Shapiro (1976). *J. Phys. B 9*, L531.

Andrick, D., and A. Bitsch (1975). *J. Phys. B 8*, 393.

Apagyi, B., P. Lévai, and K. Ladányi (1988). *Phys. Rev. A 37*, 4577.

Archer, B. J., G. A. Parker, and R. T. Pack (1990). *Phys. Rev. A 41*, 1303.

Armstead, R. L. (1968). *Phys. Rev. 171*, 91.

Arthurs, A. M. (1970). *Complementary Variational Principles*, Clarendon Press, Oxford.

Aulenkamp, H., P. Heiss, and E. Wichmann (1974). *Z. Phys. 268*, 213.

Bagchi, B., T. O. Krause, and B. Mulligan (1977). *Phys. Rev. C 15*, 1623.

Bagchi, B., and B. Mulligan (1979). *Phys. Rev. C 20*, 1973.

Baker, G. A., Jr., and J. L. Gammel, Eds. (1970). *The Padé Approximants in Theoretical Physics*, Academic Press, San Diego, Calif.

Baldo, M., L. S. Ferreira, and L. Streit (1985). *Phys. Rev. C 32*, 685.

Baldo, M., L. S. Ferreira, and L. Streit (1987). *Phys. Rev. C 36*, 1783.

Baluja, K. L., C. J. Noble, and J. Tennyson (1985). *J. Phys. B 18*, L851.

Bardsley, J. N., E. Gerjuoy, and C. V. Sukumar (1972). *Phys. Rev. A 6*, 1813.

Berrington, K. A., P. G. Burke, and A. L. Sinfailam (1975). *J. Phys. B 8*, 1459.

Bhatia, A. K., A. Temkin, R. J. Drachman, and H. Eiserike (1971). *Phys. Rev. A 3*, 1328.

Blasczak, D., and M. G. Fuda (1973). *Phys. Rev. C 8*, 1665.

Blatt, J. M., and L. C. Biedenharn (1952). *Rev. Mod. Phys. 24*, 258.

Brady, T. J., and I. H. Sloan (1972). *Phys. Lett. B 40*, 55.

Brady, T. J., and I. H. Sloan (1974). *Phys. Rev. C 9*, 4.

Bransden, B. H. (1983). *Atomic Collision Theory*, 2nd ed., W.A. Benjamin, New York.

Brown, G. E., and A. D. Jackson (1976). *The Nucleon–Nucleon Interaction*, North-Holland, Amsterdam.

Brunt, J. N. H., G. C. King, and F. H. Read (1977). *J. Phys. B 10*, 1289.

Bund, G. W. (1985). *Phys. Rev. C 31*, 2022.

Burke, P. G. (1965). *Adv. Phys. 14*, 521.

Burke, P. G. (1968). *Adv. At. Mol. Phys. 4*, 173.

Burke, P. G. (1977). *Potential Scattering in Atomic Physics*, Plenum Press, New York.

Burke, P. G., and W. D. Robb (1975). *Adv. At. Mol. Phys. 11*, 143.

Burke, P. G., and H. M. Schey (1962). *Phys. Rev. 126*, 147.

Burke, P. G., and M. J. Seaton (1971). *Methods Comput. Phys. 10*, 1.

Burke, P. G., H. M. Schey, and K. Smith (1963). *Phys. Rev. 129 A*, 1258.

Burke, P. G., J. W. Cooper, and S. Ormonde (1969). *Phys. Rev. 183*, 245.

Buttle, P. J. A. (1967). *Phys. Rev. 160*, 719.

Callaway, J. (1978a). *Phys. Rep. 45*, 89.

Callaway, J. (1978b). *Phys. Lett. 65A*, 199.

Callaway, J. (1980). *Phys. Lett. 77A*, 137.

Callaway, J., and J. W. Wooten (1974). *Phys. Rev. A 9*, 1924.

Callaway, J., and J. W. Wooten (1975). *Phys. Rev. A 11*, 1118.

Callaway, J., and J. F. Williams (1975). *Phys. Rev. A 12*, 2312.

Callaway, J., M. R. C. McDowell, and L. A. Morgan (1976). *J. Phys. B 9*, 2181.

Campeanu, R. I., and J. W. Humberston (1975). *J. Phys. B 8*, L244.

Campeanu, R. I., and J. W. Humberston (1977). *J. Phys. B 10*, L153.

Canton, L., G. Cattapan, and G. Pisent (1988). *Nucl. Phys. A 487*, 333.

Castillejo, L., I. C. Percival, and M. J. Seaton (1960). *Proc. R. Soc. London Ser. A 254*, 259.

Coester, F. (1971). *Phys. Rev. C 3*, 525.

Crompton, R. W., M. T. Elford, and A. G. Robertson (1970). *Austr. J. Phys. 23*, 667.

Das, J. N., and M. R. H. Rudge (1976). *J. Phys. B 9*, L131.

de Araujo, C. F., S. K. Adhikari, and L. Tomio (1995). *J. Comput. Phys. 118*, 200.

Demkov, Yu. N. (1963). *Variational Principles in the Theory of Collisions*, Macmillan, New York.

Doolen, G., G. McCartor, F. A. McDonald, and J. Nuttal (1971). *Phys. Rev. A 4*, 108.

Drachman, R. J. (1966). *Phys. Rev. 144*, 25.

Drachman, R. J. (1968). *Phys. Rev. 173*, 191.

Drachman, R. J. (1972). *J. Phys. B 5*, L30.

Duxler, W. M., R. T. Poe, and R. W. La Bahn (1971). *Phys. Rev. A 4*, 1935.

Elkowitz, A. B., and R. E. Wyatt (1975). *J. Chem. Phys. 62*, 2504.

Epstein, S. T. (1957). *Phys. Rev. 106*, 598.

Ernst, D. J., C. M. Shakin, and R. M. Thaler (1973a). *Phys. Rev. C 8*, 46.

Ernst, D. J., C. M. Shakin, R. M. Thaler, and D. Weiss (1973b). *Phys. Rev. C 8*, 2056.

Faddeev, L. D. (1965). *Mathematical Aspects of the Three-Body Problem in Quantum Scattering Theory*, Davey, New York.

Fano, U., and C. M. Lee (1973). *Phys. Rev. Lett. 31*, 1573.

Ferreira, L. S. (1986). *Lecture Notes in Physics*, Vol. 273, Springer-Verlag, Berlin, p. 100.

Feshbach, H. (1958). *Ann. Phys. (N.Y.) 5*, 357.

Feshbach, H. (1962). *Ann. Phys. (N.Y.) 19*, 287.

Foldy, L. L., and W. Tobocman (1957). *Phys. Rev. 105*, 1099.

Fonseca, A. C., and T. K. Lim (1985). *Phys. Rev. Lett. 55*, 1285.

Fonseca, A. C. (1986a). *Lecture Notes in Physics*, Vol. 273, Springer-Verlag, Berlin, p. 161.

Fonseca A. C. (1986b). In T. K. Lim, C.-G. Bao, D. -P. Hou, and S. Huber, *Few-Body Methods: Principles and Applications*, Eds., World Scientific, Singapore, p. 111.

Gell-Mann, M., and M. L. Goldberger (1953). *Phys. Rev. 91*, 398.

Geltman, S. (1969). *Topics in Atomic Collision Theory*, Academic Press, San Diego, Calif.

Gerjuoy, E. (1958). *Phys. Rev. 109*, 1806.

Gerjuoy, E. (1971). *Philos. Trans. R. Soc. London 270*, 197.

Gerjuoy, E., A. R. P. Rau, and L. Spruch (1983). *Rev. Mod. Phys. 55*, 725.

Ghosh, A. S., N. C. Sil, and P. Mandal (1982). *Phys. Rep. 87*, 313.

Ghosh, A. S. (1992). Private communication.

Gibson, R. J., and K. T. Dolder (1969). *J. Phys. B 2*, 741.

Gillan, C. J., C. J. Noble, and P. G. Burke (1990). *J. Phys. B 23*, L407.

Glöckle, W. (1970a). *Nucl. Phys. A 141*, 620.

Glöckle, W. (1970b). *Nucl. Phys. A 158*, 257.

Glöckle, W. (1983). *The Quantum Mechanical Few-Body Problem*, Springer-Verlag, New York.

Goldberger, M. L., and K. M. Watson (1964). *Collision Theory*, Wiley, New York.

Golden, D. E., and A. Zecca (1971). *Rev. Sci. Instrum. 42*, 210.

Haftel, M. I., and F. Tabakin (1970). *Nucl. Phys. A 158*, 1.

Haidenbauer, J., and W. Plessas (1983). *Phys. Rev. C 27*, 63.

Harms, E. (1970). *Phys. Rev. C 1*, 1667.

Harris, F. E. (1967). *Phys. Rev. Lett. 19*, 173.

Harris, F. E., and H. H. Michels (1969a). *Phys. Rev. Lett. 22*, 1036.

Harris, F. E., and H. H. Michels (1969b). *J. Comput. Phys. 6*, 237.

Harris, F. E., and H. H. Michels (1971). *Methods Comput. Phys. 10*, 143.

Hartt, K., and P. V. A. Yidana (1987). *Phys. Rev. C 36*, 475.

Henry, R. J. W. (1967). *Phys. Rev. 162*, 56.

Henry, R. J. W. (1968). *Phys. Rev. 172*, 99.

Ho, Y. K. (1990). *J. Phys. B 23 L419.*

Ho, Y. K., and P. A. Fraser (1976). *J. Phys. B 9*, 3213.

Ho, Y. K., and C. H. Greene (1987). *Phys. Rev. A 35 3169.*

Hochstadt, H. (1973). *Integral Equations*, Wiley, New York.

Hořacek, J., and T. Sasakawa (1983). *Phys. Rev. A 28*, 2151.

Hořacek, J., and T. Sasakawa (1984). *Phys. Rev. A 30*, 2274.

Houston, S. K. (1973). *J. Phys. B 6*, 131.

Houston, S. K., and R. J. Drachman (1971). *Phys. Rev. A 3*, 1335.

Huck, R. J. (1957). *Proc. Phys. Soc. London Sec. A 70*, 369.

Hulthén, L. (1948). *Ark. Mat. Astron. Fys. A 35*, No. 25.

Hulthén, L. (1944). *K. Fysiogr. Saellsk. Lund. Foerh. 14*, 257.

Humberston, J. W. (1973). *J. Phys. B 6*, L305.

Humberston, J. W. (1974). *J. Phys. B 7*, L286.

Humberston, J. W. (1984). *J. Phys. B 17*, 2353.

Ishikawa, S. (1987). *Nucl. Phys. A 463*, 145c.

Jackson, J. (1951). *Phys. Rev. 83*, 301.

Joachain, C. K. (1975). *Quantum Collision Theory*, North-Holland, Amsterdam.

Kapur, P. L., and R. E. Peierls (1938). *Proc. R. Soc. London Ser. A 166*, 277.

Kar, S., and P. Mandal (1997). *J. Phys. B 30*, L627.

Kato, T. (1950). *Phys. Rev. 80*, 475.

Kauppila, W. E., et al. (1977). *Rev. Sci. Instrum. 48*, 322.

Kennerly, R. E., and R. A. Bonham (1978). *Phys. Rev. A 17*, 1844.

Kernoghan, A. A., M. T. McAlinden, and H. R. J. Walters (1995). *J. Phys. B 28*, 1079.

Kievsky, A. (1997). Nucl. Phys. Axx, xxx.

Kievsky, A., M. Viviani, and S. Rosati (1998). *Phys. Rev. C*, submitted.

Kohn, W. (1948). *Phys. Rev. 74*, 1763.

Kouri, D. J., and D. G. Truhlar (1989). *J. Chem. Phys. 91*, 6919.

Kouri, D. J., Y. Sun, R. C. Mowrey, J. Z. H. Zhang, D. G. Truhlar, K. Haug, and D. W. Schwenke (1988). In *Mathematical Frontiers in Computational Chemical Physics*, D. G. Truhlar, Ed., Springer-Verlag, New York, p. 207.

Kowalski, K. L. (1965). *Phys. Rev. Lett. 15*, 798.

Kowalski, K. L. (1972). *Nucl. Phys. A190*, 645.

Kowalski, K. L., and D. Feldman (1961). *J. Math. Phys. 2*, 499.

Kowalski, K. L., and D. Feldman (1963a). *Phys. Rev. 130*, 276.

Kowalski, K. L., and D. Feldman (1963b). *J. Math. Phys. 4*, 507.

Kuppermann, A., G. C. Schatz, and M. J. Baer (1974). *J. Chem. Phys. 61*, 4362.

Lane, A. M., and D. Robson (1966). *Phys. Rev. 151*, 774.

Lane, A. M., and D. Robson (1969). *Phys. Rev. 178*, 1715.

Lane, A. M., and R. G. Thomas (1958). *Rev. Mod. Phys. 30*, 257.

Levinger, J. S. (1974). *Springer Tracts in Modern Physics*, Vol. 71, Springer-Verlag, Berlin, 88.

Lima, M. A. P., and V. McKoy (1988). *Phys. Rev. A 38*, 501.

Lima, M. A. P., T. L. Gibson, W. M. Huo, and V. McKoy (1985). *J. Phys. B 18*, L865.

Lippmann, B. A. (1956). *Phys. Rev. 102*, 264.

Lippmann, B. A., and J. Schwinger (1950). *Phys. Rev. 79*, 469.

Lovelace, C. (1964a). *Phys. Rev. 135, B1225.*

Lovelace, C. (1964b). In R. G. Moorhouse, Ed., *Strong Interaction and High Energy Physics*, Oliver & Boyd, London, p. 437.

Lucchese, R. R. (1989). *Phys. Rev. A 40*, 6879.

Lucchese, R. R., and V. McKoy (1980). *Phys. Rev. A 21*, 112.

Lucchese, R. R., and V. McKoy (1981a). *Phys. Rev. A 24*, 770.

Lucchese, R. R., and V. McKoy (1981b). *J. Phys. B 14*, L629.

Lucchese, R. R., and V. McKoy (1981c). *J. Chem. Phys. 85*, 2166.

Lucchese, R. R., and V. McKoy (1982). *Phys. Rev. A 26, 1406 and 1992.*

Lucchese, R. R., and V. McKoy (1983). *Phys. Rev. A 28*, 1382.

Lucchese, R. R., K. Takatsuka, and V. McKoy (1986). *Phys. Rep. 131*, 147.

Lucchese, R. R., G. Raseev, and V. McKoy (1982). *Phys. Rev. A 25*, 2572.

Lynch, D., M. T. Lee, R. R. Lucchese, and V. McKoy (1984). *J. Chem. Phys. 80*, 1907.

Malfliet, R. A., and J. A. Tjon (1968). *Nucl. Phys. A 127*, 161.

Malfliet, R. A., and J. A. Tjon (1970). *Ann. Phys. (N.Y.) 61*, 425.

McCurdy, C. W., T. N. Rescigno, and B. I. Schneider (1987). *Phys. Rev. A 36*, 2061.

Mercier, A. (1963). *Variational Principles of Physics*, Dover, New York.

Michels, H. H., F. E. Harris, and R. N. Scolsky (1969). *Phys. Lett. A 28*, 467.

Michlin, S. G. (1964). *Variational Methods in Mathematical Physics*, Interscience, New York.

Miller, W. H. (1990). *Annu. Rev. Phys. Chem. 41*, 245.

Miller, W. H. (1992). In Eds., S. Wilson and G. H. F. Diercksen, *Methods in Computational Molecular Physics*, Plenum Press, New York, p. 519.

Miller, W. H. (1994). *Advances in Molecular Vibration and Collision Dynamics*, Vol. 2A, 1, JAI Press, Greenwich, Conn., p. 1.

Miller, W. H., and B. M. D. D. Jansen op de Haar (1987). *J. Chem. Phys. 86*, 6213.

Miller, W. H., and J. Z. H. Zhang (1991). *J. Phys. Chem. 95*, 12.

Mito, Y., and M. Kamimura (1976). *Prog. Theor. Phys. 56*, 583.

Mitroy, J., and A. T. Stelbovics (1994a). *J. Phys. B 27*, 3257.

Mitroy, J., and A. T. Stelbovics (1994b). *J. Phys. B 27*, L55.

Morgan, L. A., M. R. C. McDowell, and J. Callaway (1977). *J. Phys. B 10*, 3297.

Mortensen, E. M., and L. D. Gucwa (1969). *J. Chem. Phys. 51*, 5695.

Moser, C. M., and R. K. Nesbet (1971). *Phys. Rev. A 4*, 1366.

Mu-Tao, L., and V. McKoy (1983). *Phys. Rev. A 28*, 697.

Nesbet, R. K. (1967). *Phys. Rev. 156*, 99.

Nesbet, R. K. (1968). *Phys. Rev. 175*, 134.

Nesbet, R. K. (1969a). *Phys. Rev. 179*, 60.

Nesbet, R. K. (1969b). *Adv. Chem. Phys. 14*, 1.

Nesbet, R. K. (1975). *Advance in Quantum Chemistry*, Vol. 9, Academic Press, San Diego, Calif., p. 215.

Nesbet, R. K. (1978). *Advance in Atomic and Molecular Physics*, Vol. 13, Academic Press, Calif., p. 315.

Nesbet, R. K. (1980). *Variational Methods in Electron–Atom Scattering Theory*, Plenum Press, New York.

Nesbet, R. K., and R. S. Oberoi (1972). *Phys. Rev. A 6*, 1855.

Newton, R. G. (1982). *Scattering Theory of Particles and Waves*, 2nd ed., Springer-Verlag, Berlin.

Nuttall, J. (1969). *Ann. Phys. (N.Y.) 52*, 428.

Oberoi, R. S., and J. Callaway (1969). *Phys. Lett. A 30*, 419.

Oberoi, R. S., and R. K. Nesbet (1973a). *Phys. Rev. A 8*, 215.

Oberoi, R. S., and R. K. Nesbet (1973b). *Phys. Rev. A 8*, 2969.

O'Malley, T. F., P. G. Burke, and K. A. Berrington (1979). *J. Phys. B 12*, 953.

Oryu, S. (1986). *Lecture Notes in Physics*, Vol. 273, Springer-Verlag, Berlin, p. 123.

Pearce, B. C. (1987). *Phys. Rev. C 36*, 471.

Pelikan, E., and H. Klar (1983). *Z. Phys. A 310*, 153.

Plessas, W. (1986). *Lecture Notes in Physics*, Vol. 273, Springer-Verlag, Berlin, p. 137.

Plessas, W., and J. Haidenbauer (1987). *Few-Body Syst. Suppl. 2*, 185.

Rawitscher, G. H. (1987). *Nucl. Phys. A 475*, 519.

Rawitscher, G. H. (1989). *Phys. Rev. C 39*, 440.

Redish, E. F., and K. Stricker-Bauer (1987). *Phys. Rev. C 36*, 513.

Register, D., and R. T. Poe (1975). *Phys. Lett. A 51*, 431.

Rescigno, T. N., C. W. McCurdy, and V. McKoy (1976). *Phys. Rev. A 11*, 2240.

Rescigno, T. N., and B. I. Schneider (1987). *Phys. Rev. A 36*, 2061.

Rescigno, T. N., and B. I. Schneider (1988). *J. Phys. B 21*, L691.

Rosenberg, L., L. Spruch, and T. F. O'Malley (1960). *Phys. Rev. 118*, 184.

Rountree, S. P., and G. Parnell (1977). *Phys. Rev. Lett. 39*, 853.

Roy, U., and P. Mandal (1993a). *Phys. Rev. A 48*, 233.

Roy, U., and P. Mandal (1993b). *Phys. Rev. A 48*, 2952.

Rubinow, S. I. (1955). *Phys. Rev. 98*, 183.

Rudge, M. R. H. (1973). *J. Phys. B 6*, 1788.

Rudge, M. R. H. (1975). *J. Phys. B 8*, 940.

Schatz, G. C., and A. Kuppermann (1975). *Phys. Rev. Lett. 35*, 1266.

Schneider, B. I., and L. A. Collins (1985). *J. Phys. B 18*, L857.

Schwartz, C. (1961a). *Phys. Rev. 124*, 1468.

Schwartz, C. (1961b). *Ann. Phys. (N.Y.) 16*, 36.

Schwartz, C. (1966). *Phys. Rev. 141*, 1468.

Schwinger, J. (1947a). *Phys. Rev. 72*, 742.

Schwinger, J. (1947b). *Lecture Notes, Harvard University, unpublished.*

Seaton, M. J. (1953). *Philos. Trans. R. Soc. London Ser. A 245*, 469.

Seaton, M. J. (1973). *Comput. Phys. Commun. 6*, 247.

Seiler, G. J., R. S. Oberoi, and J. Callaway (1971). *Phys. Rev. A 3*, 2006.

Shimamura, I. (1971a). *J. Phys. Soc. Japan 30*, 1702.

Shimamura, I. (1971b). *J. Phys. Soc. Japan 31*, 217.

Shimamura, I. (1971c). *J. Phys. Soc. Japan 31*, 852.

Sinfailam, A. L., and R. K. Nesbet (1972). *Phys. Rev. A 6*, 2118.

Sinfailam, A. L., and R. K. Nesbet (1973). *Phys. Rev. A 7*, 1987.

Sloan, I. H. (1968). *J. Comput. Phys. 3*, 332.

Sloan, I. H., and T. J. Brady (1972). *Phys. Rev. C 6*, 701.

Sloan, I. H., and S. K. Adhikari (1974). *Nucl. Phys. A 235, 352*

Smith, E. R., and R. J. W. Henry (1973). *Phys. Rev. A 8*, 572.

Smith, R. L., and D. G. Truhlar (1972). *Phys. Lett. A 39*, 35.

Sofianos, S., N. J. McGurk, and H. Fiedeldey (1979). *Nucl. Phys. A 318*, 295.

Staszewska, G., and D. G. Truhlar (1986). *Chem. Phys. Lett. 130*, 341.

Staszewska, G., and D. G. Truhlar (1987). *J. Chem. Phys. 86*, 2793.

Stein, J., and R. Sternlicht (1972). *Phys. Rev. A 6*, 2162.

Stein, T. S., et al. (1978). *Phys. Rev. A 17*, 1600.

Sun, Y., C. H. Yu, D. J. Kouri, D. W. Schwenke, P. Halvick, M. Mladenovic, and D. G. Truhlar (1989). *J. Chem. Phys. 91*, 1643.

Sun, Y., D. J. Kouri and D. G. Truhlar (1990). *Nucl. Phys. A 508*, 41c.

Sun, Q., C. Winstead, and V. McKoy (1992). *Phys. Rev. A 46*, 6987.

Takatsuka, K., and V. McKoy (1981a). *Phys. Rev. A 23*, 2352.

Takatsuka, K., and V. McKoy (1981b). *Phys. Rev. A 23*, 2358.

Takatsuka, K., R. R. Lucchese, and V. McKoy (1981). *Phys. Rev. A 24*, 1812.

Tang, Y. C., M. LeMere, and D. R. Thompson (1978). *Phys. Rep. 47*, 167.

Tawa, G. J., S. L. Mielke, D. G. Truhlar, and D. W. Schwenke (1994). *Advances in Molecular Vibration and Collision Dynamics,* Vol. 2B, JAI Press, Greenwich Conn., p. 45.

Temkin, A., A. K. Bhatia, and J. N. Bardsley (1972). *Phys. Rev. A 5*, 1663.

Thirumalai, D., and D. G. Truhlar (1980). *Chem. Phys. Lett. 70*, 330.

Thomas, L. D., and R. K. Nesbet (1975a). *Phys. Rev. A 11*, 170.

Thomas, L. D., and R. K. Nesbet (1975b). *Phys. Rev. A 12*, 1729.

Thomas, L. D., and R. K. Nesbet (1975c). *Phys. Rev. A 12*, 2369.

Thomas, L. D., and R. K. Nesbet (1975d). *Phys. Rev. A 12*, 2378.

Thomas, L. D., R. S. Oberoi, and R. K. Nesbet (1974). *Phys. Rev. A 10*, 1605.

Tomio, L., and S. K. Adhikari (1980a). *Phys. Rev. C 22*, 28.

Tomio, L., and S. K. Adhikari (1980b). *Phys. Rev. C 22*, 2359.

Tomio, L., and S. K. Adhikari (1981). *Phys. Rev. C 24, 43*

Tomio, L., and S. K. Adhikari (1995). *Chem. Phys. Lett. 241*, 477.

Truhlar, D. G., and A. Kuppermann (1970). *J. Chem. Phys. 52*, 3841.

Truhlar, D. G., and A. Kuppermann (1972). *J. Chem. Phys. 56*, 2232.

Truhlar, D. G., J. Abdallah, Jr., and R. L. Smith (1974). *Advances in Chemical Physics*, Vol. 25, Wiley, New York, p. 211.

Truhlar, D. G., D. W. Schwenke, and D. J. Kouri (1990). *J. Phys. Chem. 94*, 7346.

Vorobyev, Yu. V. (1965). *Method of Moments in Applied Mathematics*, Gordon and Breach, New York.

Watson, D., and V. McKoy (1979). *Phys. Rev. A 20*, 1474.

Weinberg, S. (1963). *Phys. Rev. 131*, 440.

Weinberg, S. (1964). *Phys. Rev. 133*, B232.

Whiting, J. S., and M. G. Fuda (1976). *Phys. Rev. C 14*, 18.

Wichmann, E., and P. Heiss (1974). *J. Phys. B 7*, 1042.

Wigner, E. P., and L. Eisenbud (1947). *Phys. Rev. 72*, 29.

Wildermuth, K., and Y. C. Tang (1977). *A Unified Theory of the Nucleus*, Vieweg, Wiesbaden, Germany.

Williams, J. F., and B. A. Willis (1974). *J. Phys. B 7*, L61.

Winstead, C., and V. McKoy (1994). In Ed. D. Yarkony, *Modern Electronic Structure Theory*, World Scientific, Singapore.

Wladawsky, I. (1973). *J. Chem. Phys. 58*, 1826.

Yakubovskii, O. A. (1967). *Yad. Fiz. 5*, 1312 [*Sov. J. Nucl. Phys. 5*, 937 (1967)].

Zhang, J. Z. H., and W. H. Miller (1990). *J. Phys. Chem. 94*, 7785.

Zhang, J. Z. H., S.-I. Chu, and W. H. Miller (1988). *J. Chem. Phys. 88*, 6233.

Zhao, M., et al. (1989). *J. Am. Chem. Soc. 111*, 852.

INDEX